Springer Series on Atomic, Optical, and Plasma Physics

Volume 99

The Springer Series on Atomic, Optical, and Plasma Physics covers in a comprehensive manner theory and experiment in the entire field of atoms and molecules and their interaction with electromagnetic radiation. Books in the series provide a rich source of new ideas and techniques with wide applications in fields such as chemistry, materials science, astrophysics, surface science, plasma technology, advanced optics, aeronomy, and engineering. Laser physics is a particular connecting theme that has provided much of the continuing impetus for new developments in the field, such as quantum computation and Bose-Einstein condensation. The purpose of the series is to cover the gap between standard undergraduate textbooks and the research literature with emphasis on the fundamental ideas, methods, techniques, and results in the field.

More information about this series at http://www.springer.com/series/411

Thomas Wriedt · Yuri Eremin
Editors

The Generalized Multipole Technique for Light Scattering

Recent Developments

Springer

Editors
Thomas Wriedt
Leibniz-Institut für Werkstofforientierte Technologien—IWT
Bremen
Germany

Yuri Eremin
Faculty of Computational Mathematics and Cybernetics
Lomonosov Moscow State University
Moscow
Russia

ISSN 1615-5653 ISSN 2197-6791 (electronic)
Springer Series on Atomic, Optical, and Plasma Physics
ISBN 978-3-030-09098-2 ISBN 978-3-319-74890-0 (eBook)
https://doi.org/10.1007/978-3-319-74890-0

Printed on acid-free paper

This Springer imprint is published by Springer Nature
The registered company is Springer International Publishing AG
The registered company address is: Gewerbestrasse 11, 6330 Cham, Switzerland

Preface

Computational electromagnetics is a very rapidly developing field that developed many theoretical approaches and computational tools. Over the years, it has extended its range of application from microwave, light scattering to nanophotonics and even electron energy loss spectroscopy. The Generalized Multipole Technique (GMT) is a surface-based theory which is not that well known but there are a couple of researchers continuously developing and extending the method such that it has reached some kind of maturity over the years.

In 1998, we arranged a workshop [1] in Bremen, Germany supported by the Volkswagen Foundation which had a focus on the Generalized Multipole Technique to document the state of the method at that time and to especially initiate discussion between the different research groups. Following the workshop, an edited volume [2] was published with contributions by major researchers in the field. Out of this workshop, a number of international collaborations arose which continued to work on different variants of the Generalized Multipole method.

The name Generalized Multipole Technique (GMT) was coined by Art Ludwig [3] for a number of related methods to solve the electromagnetic boundary value problem, which were developed independently by a number of research groups distributed all over the world. The common feature of these methods consists in field expansion by a number of multipoles positioned away from the boundary surface. Commonly, some kind of generalized point matching scheme is applied to find the expansion coefficient of the multipoles.

Over the years, research in the GMT continued and many new advances in theory, programming, and application have been achieved such that after 20 years, we think it is the right time to have another close look at the current state of the method. The edited book compiles a couple of chapters on various concepts related to the General Multipole Technique to demonstrate the progress achieved over the last two decades and show the new ideas developed during the last 10 years.

In Chap. 1, it is shown that the theory of principal modes can be derived for any smooth particle starting from a set of distributed electric and magnetic multipoles.

An important field of development is hybrid methods. That the invariant imbedding approach can be combined with the null-field method is demonstrated in Chap. 2.

The Null-field Method with Discrete Sources (NFM-DS) makes use of field expansion using multiple multipoles. Recent progress to compute light scattering by large axisymmetric particles using NFM-DS is presented in Chap. 3.

Some practical applications in nanotechnology require light scattering simulation by a particle partly embedded in an infinite stratified medium. How this problem can be handled using the Discrete Sources Method is investigated in Chap. 4.

Chapter 5 is an overview of the recent works in the Method of Auxiliary Sources.

Chapter 6 presents a novel numerical approach to investigate the resonance behavior of plasmonic particles on a substrate under electron beam illumination based on the Multiple Multipole Program.

Low-Loss Electron Energy Loss Spectroscopy is currently a hot research topic. How the Generalized Multipole Technique (GMT) can be used in this field is addressed in Chap. 7.

Yasuuras Method of Modal Expansion has been developed in Japan. In Chap. 8, this method is applied to investigate scattering by gratings.

An important aspect of the Generalized Multipole Technique is the suitable choice of locations for the sources. This topic is treated in Chap. 9 by James E. Richie.

We hope that these chapters give a fresh look at the evolution and development of the Generalized Multiple Technique. Of course in such a book, the fundaments needed cannot be fully covered. For this, the interested reader is referred to the book by Doicu et al. [4].

As no book can be published without some assistance, we have to thank all contributors who send their text on time. We especially like to thank Prabhan Vishwanath who helped with latex compilation.

Bremen, Germany Thomas Wriedt
Moscow, Russia Yuri Eremin

References

1. T. Wriedt, Y. Eremin (eds.), Electromagnetic and Light Scattering—Theory and Applications III, Proc. 3rd Workshop on Electromagnetic and Light Scattering—Theory and Applications, March 16–17, 1989 (Bremen, Universitt Bremen, 1998)
2. T. Wriedt (ed.), *Generalized Multipole Technique for Electromagnetic and Light Scattering* (Amsterdam, Elsevier, 1999)
3. A. C. Ludwig, IEEE Antennas and Propagation Newsletter **31**, 40 (1989)
4. A. Doicu, Y. Eremin, T. Wriedt, *Acoustic and Electromagnetic Scattering Analysis Using Discrete Sources* (San Diego, Academic Press, 2000)

Contents

Editors and Contributors

About the Editors

Thomas Wriedt studied Electrical Engineering in Kiel and Bremen, Germany. He received his Dr.-Ing. degree in Microwave Engineering from the University of Bremen in 1986. From 1986 to 1989 he was a postdoc researcher at the University of Bremen's Department of Process Engineering, working on optical particle characterization. He has been the head of the Powder and Particle Measurement research group at the Stiftung Institut für Werkstofftechnik (IWT) (from January 2018: Leibniz-Institut für Werkstofforientierte Technologien—IWT), Bremen since 1989. His current research focuses on the Null-field method with Discrete Sources (NFM-DS) for simulation in optical particle characterization and generally light scattering by particles. He has authored and co-authored four books and 170 peer-reviewed papers.

Yuri Eremin graduated from the Physics Faculty of Lomonosov Moscow State University (MSU), Russia in 1972. He received his Ph.D. in Mathematical and Theoretical Physics from Lomonosov Moscow State University (MSU) in 1976. From 1975 to 1982 he was a research scientist at the Research Computing Center (RCC) of MSU. From 1982 to 1992 he was a senior researcher at the Computational Mathematics and Cybernetics Faculty of MSU. He obtained his Dr. Sci. degree in Mathematical Modeling from MSU in 1989. Since 1993 he has been the head of a research group at the Computational Mathematics and Cybernetics Faculty of MSU. His current interests are direct and inverse light scattering problems. He is the author of the Discrete Sources Method that enables the construction of effective computational models to simulate light scattering. He has authored more than 215 papers in scientific journals and four books.

Contributors

Adrian Doicu Remote Sensing Technology Institute, German Aerospace Centre (DLR), Oberpfaffenhofen, Germany

Dmitry S. Efremenko Remote Sensing Technology Institute, German Aerospace Centre (DLR), Oberpfaffenhofen, Germany

Yuri Eremin Lomonosov Moscow State University, Lenin's Hills, Moscow, Russia

George Fikioris School of Electrical and Computer Engineering, National Technical University of Athens, Athens, Greece

Christian Hafner Department of Information Technology and Electrical Engineering, Institute of Electromagnetic Fields, Zurich, Switzerland

Ben Hourahine SUPA, Department of Physics, University of Strathclyde, Glasgow, UK

Mirza Karamehmedović Department of Applied Mathematics and Computer Science and Department of Physics, Technical University of Denmark, Kongens Lyngby, Denmark

Lars Kiewidt Center for Environmental Research and Sustainable Technology, University of Bremen, Bremen, Germany; Biobased Chemistry and Technology (BCT), Wageningen University, Wageningen, The Netherlands

Ueli Koch Department of Information Technology and Electrical Engineering, Institute of Electromagnetic Fields, Zurich, Switzerland

Juerg Leuthold Department of Information Technology and Electrical Engineering, Institute of Electromagnetic Fields, Zurich, Switzerland

Toyonori Matsuda National Institute of Technology, Kumamoto College, Kumamoto, Japan

Akira Matsushima Kumamoto University, Kumamoto, Japan

Duncan McArthur SUPA, Department of Physics, University of Strathclyde, Glasgow, UK

Jens Niegemann Lumerical Computational Solutions Inc., Vancouver, BC, Canada

Yoichi Okuno South China Normal University, Guangzhou, China

Francesco Papoff SUPA, Department of Physics, University of Strathclyde, Glasgow, UK

James E. Richie Marquette University, Milwaukee, WI, USA

Thomas Trautmann Remote Sensing Technology Institute, German Aerospace Centre (DLR), Oberpfaffenhofen, Germany

Nikolaos L. Tsitsas Department of Informatics, Aristotle University of Thessaloniki, Thessaloniki, Greece

Thomas Wriedt Leibniz-Institut für Werkstofforientierte Technologien—IWT, Bremen, Germany

Acronyms

Al	Aluminum
Al_2O_3	Alumina
AOI	Angle of Incidence
AS	Auxiliary Surface
Au	Gold
BEM	Boundary Element Method
CHIEF	Combined Helmholtz Integral Equation Formulation
CMASIE	Continuous Method of Auxiliary Integral Equation
CYM	Conventional Yasuuras Method
DDA	Discrete Dipole Approximation
DFT	Discrete Fourier Transform
DGTD	Discontinuous Galerkin Time Domain
DSCS	Differential Scattering Cross Section
DSIEM	Dual-Surface Integral Equations Method
EBW	Effective Spatial Bandwidth
EBC	Extended Boundary Condition
EBCM	Extended Boundary Condition Method
EELS	Electron Energy Loss Spectroscopy
EIE	Extended Integral Equation
FCM	Filamentary Current Method
FDTD	Finite Difference Time Domain
FEM	Finite Element Method
GLMT	Generalized Lorenz–Mie Theory
GMT	Generalized Multipole Technique
GPMT	Generalized Point Matching Technique
GPU	Graphics Processing Unit
GT	Green Tensor
LHS	Left-Hand Side
MAS	Method of Auxiliary Sources
MFS	Method of Fictitious Sources

MMP	Multiple Multipole Program
MPI	Message Passing Interface
NFM	Null-Field Method
NFM-DS	Null-Field Method with Discrete Sources
PEC	Perfect Electric Conductor
SFS	Scattered Field Singularity
SIE	Surface Integral Equation
SNOM	Scanning Near-Field Optical Microscopy
STEM	Scanning Transmission Electron Microscope
SVD	Singular Value Decomposition
SVWF	Spherical Vector Wave Function
TBB	Threading Building Blocks
TE	Transverse Electric
TM	Transverse Magnetic
VIE	Volume Integral Equations

Chapter 1
Principal Modes of Maxwell's Equations

Ben Hourahine, Duncan McArthur and Francesco Papoff

Abstract This chapter reviews the use of principal modes—states which are maximally correlated between two subspaces and hence form pairs unique up to phase factors—in solving Maxwell's equations and analysing these solutions for nanoparticles and structures. The mathematical structure of this method allows a computationally efficient generalisation of Mie's analytical approach for the sphere to obtain semi-analytical solutions for general geometries with smooth interfaces. We apply this method to investigate a range of single and multiple particle metallic structures in the linear, non-linear and non-local response regimes outside of the quasi-static limit.

1.1 Introduction

The interaction of light with particles with similar or smaller dimensions than the incident wavelength has been intensely investigated for more than a century. New and interesting results continue to be discovered in this regime. This coupling is crucial in the spectroscopy of single molecules and for single and multi-photon processes [1, 2], while interferences between different scattering channels give classical analogues of quantum processes [3, 4], and carefully designed particles are the building blocks of several classes of metamaterial [5–7]. All these effects are due to particle-light interaction, which depends on both the particle's shape and composition and also the properties of the incident light. This interaction can become very strong, particularly around resonances. Understanding and controlling fundamental elements

B. Hourahine (✉) · D. McArthur · F. Papoff
SUPA, Department of Physics, University of Strathclyde, Glasgow G4 0NG, UK
e-mail: benjamin.hourahine@strath.ac.uk

D. McArthur
e-mail: duncan.mcarthur@strath.ac.uk

F. Papoff
e-mail: f.papoff@strath.ac.uk

T. Wriedt and Y. Eremin (eds.), *The Generalized Multipole Technique for Light Scattering*, Springer Series on Atomic, Optical, and Plasma Physics 99,
https://doi.org/10.1007/978-3-319-74890-0_1

of this interaction are essential to design nanostructures with desirable near as well as far field optical responses.

In this chapter we review the theory of principal modes that we have developed over the last 10 years to describe single and multiple particle systems with smooth interfaces. We use the mathematical theory developed by Jordan [8] for the angles between subspaces of functions [9] to highlight the profound relation between resonances and the geometry of function spaces in Maxwell's equations. With this theory we can: predict resonances of particles, independently of the particular exciting field; calculate the near and far field properties of optical modes of a system since it gives efficient access to the electromagnetic field at arbitrary spatial locations; and optimise the excitation of the system with appropriately chosen incident fields. We can also find optimal excitation conditions for the coherent control of near or far field emission for elastic (single photon) or for inelastic (multi-photon) processes. In the latter case, we assume that the non-linear processes are sufficiently weak that the pump light at the driving frequency is not depleted. More recently, we have applied this method to modally decompose the local optical density of states, giving insight into Purcell and Lamb factors in the near ultraviolet [10].

1.2 Principal Modes of Single Particles

The theory of principal modes can be applied to general particles and structures without sharp edges, this includes both metallic and dielectric particles. When all the characteristic dimensions of the particle (and the skin depth in the case of metallic particles) are larger than the free propagation length of charges, then in this regime local macroscopic permittivities and susceptibilities can describe the interaction between light and the matter inside the particle.

The wave energy scattered by a particle propagates outward towards infinity [11], and the tangential components of electric and magnetic fields E, H are continuous at the bounding surface of the particle. Hence, for any incident field, the internal and scattered fields are determined by finding the solutions of Maxwell's equations in the internal and the external media that satisfy the boundary conditions and meet at the interface of the particle. We use vectors $F = [E^T, H^T]^T$ with six components for the full electromagnetic fields [12]; in the case of a local macroscopic permittivity and susceptibility, the tangential components, together with the radiation condition then fully determine the boundary conditions for valid solutions to Maxwell's equations. Hence only the projections, f, of the six components of the field onto the boundary of the particle are required, since by construction the internal and external fields are regular and radiating Maxwell solutions. These surface fields have four components, two electric and two magnetic, and form a space, $\mathscr{H}$, with defined scalar products in terms of the overlap integrals on the surface of the particle,

$$f \cdot g = \int_S f_j^* g_j \mathrm{d}s, \tag{1.1}$$

where f^* is the complex conjugate of field f, j labels the individual components and we sum over repeated indexes. In this notation the boundary conditions become

$$f^0 = f^i - f^s, \tag{1.2}$$

which can be interpreted geometrically in the Hilbert space $\mathscr{H}$: the projection, f^0, of the incident field, $F^0(x)$ at point x, onto the surface of the particle is equal to the difference between the projections of the internal and scattered fields, f^i and f^s. Therefore, an incident field having small values of the tangential components can then excite large amplitude internal and scattered surface fields, provided that both of these two fields closely match (i.e. they are nearly cancelling each other at the surface). This happens when the difference between these two fields, and therefore the "angle" between them, is small. These angles are rigorously defined as being between solutions of Maxwell's equations for the internal and external media; where these solutions are themselves standing and outgoing waves, respectively, and form two subspaces of $\mathscr{H}$. The electromagnetic response of each particle is completely characterised by these angles and their associated waves, which can be calculated with arbitrary precision from *any* complete set of solutions of Maxwell's equations for the internal and external media.

Several sets of exact solutions of the Maxwell equations are linearly independent and complete [11, 13] on Lyapunov surfaces. Such surfaces are mathematically characterised by the following three conditions: (1) the surface normal is well defined at every point; (2) the angle between the normals at any two points on the surface is bounded from above by a function of the distance between these points; (3) all the lines parallel to a normal at an arbitrary point on the surface intercept only once the patches of surface contained in balls centred at that point and which are smaller than a critical value [14]. These surfaces, along with C^2 continuous interfaces define *normal* domains [15] for which Gauss' theorem applies.

We use two sets of electric and magnetic multipoles, $\{\tilde{i}_n\}_{n=1}^{\infty}$ and $\{\tilde{s}_n\}_{n=1}^{\infty}$ for the internal and scattered fields respectively. These are centred at points within the particle [16, 17]. It is known that any function in $\mathscr{H}$ can be approximated to arbitrary precision by a finite but sufficiently large number of such multipoles [18]. In other words, $\{\tilde{i}_n\}_{n=1}^{\infty} \cup \{\tilde{s}_n\}_{n=1}^{\infty}$ is complete and no function in this set is the closure of the linear combinations of all the remaining functions. We note that both the internal and scattered fields exist at the surface of real metallic and dielectric particles, i.e. the union of internal and scattered fields provides completeness in the space $\mathscr{H}$ of surface fields of the type to which f belongs. Neither the scattered or the internal fields separately are complete. An example of this property is given by spherical particles, where both internal and scattered modes are necessary to form the complete basis of functions that underpins Mie theory.

For numerical applications, we could use any set of solutions for the Maxwell equations which satisfy the Silver-Müller radiation conditions [19, 20] in the external region and form standing waves in the interior. The numerical results shown here are derived specifically by using distributed multipoles [11].

One can show [18] that the coefficients of the internal and scattered fields, $\{\tilde{a}_n^i, \tilde{a}_n^s\}$, that minimise the discrepancy between an incident field and the expansion of internal and scattered fields,

$$\left| f^0 + \sum_{n=1}^{N} \tilde{a}_n^s \tilde{s}_n - \tilde{a}_n^i \tilde{\imath}_n \right|, \tag{1.3}$$

are the solutions to

$$\begin{bmatrix} \tilde{\Upsilon}^\dagger \tilde{\Upsilon} & \tilde{\Upsilon}^\dagger \tilde{\Xi} \\ \tilde{\Xi}^\dagger \tilde{\Upsilon} & \tilde{\Xi}^\dagger \tilde{\Xi} \end{bmatrix} \begin{bmatrix} \tilde{\mathbf{a}}^i \\ -\tilde{\mathbf{a}}^s \end{bmatrix} = \begin{bmatrix} \tilde{\Upsilon}^\dagger f^0 \\ \tilde{\Xi}^\dagger f^0 \end{bmatrix}, \tag{1.4}$$

where $\tilde{\mathbf{a}}^{i/s} = [\tilde{a}_1^{i/s}, \ldots, \tilde{a}_N^{i/s}]^T$ and $\tilde{\Upsilon}$, $\tilde{\Xi}$ are the matrices whose columns are the functions $\{\tilde{\imath}_n\}_{n=1}^N$ and $\{\tilde{s}_n\}_{n=1}^N$. The Gram matrix on the left of (1.4), i.e. the matrix of all possible inner products, can always be inverted because the functions used are linearly independent and complete. However, the inversion is numerically challenging [17] and for this reason this approach has received little attention. Instead, by using principal modes we can exploit the block structure of the matrix in (1.4) to give a complete characterisation of the geometry of the internal and scattered fields and analytical expressions for the expansion coefficients of the fields.

First, we find orthogonal modes for the scattering and internal basis fields: for any number, N, of multipoles, we obtain surface orthogonal modes through the matrix decompositions

$$\tilde{\Upsilon} = U^i Q^i, \tag{1.5}$$
$$\tilde{\Xi} = U^s Q^s, \tag{1.6}$$

where Q^i and Q^s are invertible matrices determined through singular value decomposition (SVD) or QR factorisation [21, 22] and U^i, U^s are unitary matrices whose columns are the orthogonal internal and scattering modes respectively.

The scalar products between internal and scattering modes then form a matrix which has the decomposition

$$U^{i\dagger} U^s = V^i C V^{s\dagger}, \tag{1.7}$$

where C is a diagonal matrix with positive elements, and V^i, V^s are unitary matrices acting respectively on the internal and scattered fields. We simplify the Gram matrix using these identities and the unitary transformation

$$\begin{bmatrix} V^{i\dagger} Q^{i\dagger -1} & 0 \\ 0 & V^{s\dagger} Q^{s\dagger -1} \end{bmatrix}, \tag{1.8}$$

which leads to the matrix equation (1.4) becoming

$$\begin{bmatrix} 1 & C \\ C^\dagger & 1 \end{bmatrix} \begin{bmatrix} \mathbf{a}^i \\ -\mathbf{a}^s \end{bmatrix} = \begin{bmatrix} \Upsilon^\dagger f^0 \\ \Xi^\dagger f^0 \end{bmatrix}. \tag{1.9}$$

On the left hand side of (1.9), 1 is the identity matrix and $\Upsilon = U^i V^i$ and $\Xi = U^s V^s$ are matrices whose columns are formed by the so called principal internal and scattering modes, $\{i_n\}$ and $\{s_n\}$, which are one of the main tools in this theory. The coefficients of the principal modes in the fields' expansions in the orthogonal modes $U^{i/s}$of (1.5) (1.6) are $\mathbf{a}^i$ and $\mathbf{a}^s$. The essential feature of our theory is that, because C is a diagonal matrix, the principal modes are coupled pairwise, i.e. each mode is orthogonal to all but at most one mode in the other space. This is also the main property of the Mie modes of spheres. The principal angles, ξ_n, between s_n and i_n are defined in terms of the positive diagonal elements of C as

$$i_n \cdot s_n = \cos(\xi_n). \tag{1.10}$$

The terms on the right-hand side of (1.10) are the principal cosines [8]: $\cos(\xi_n)$ and $\sin(\xi_n)$ are the statistical correlation [23] and the orthogonal distance between s_n and i_n respectively. Note that the spatial correlation of two modes is defined by an integral over the particle surface for the scalar product of their projections. In other words, each scattering mode, s_n, is spatially correlated to only *one* internal mode, i_n, and vice versa at the particle boundary (being orthogonal to all other internal and scattered modes).

The set of angles, $\{\xi_n\}$, are invariant under unitary transformation [9] and completely characterise the geometry of the subspaces consisting of the internal and scattered solutions in $\mathscr{H}$. This geometry is generated by the specific scattering particle, through the surface integrals of the scalar products between the functions in these two sub-spaces. Principal cosines are important for two reasons: they provide analytic equations for the coefficients of the internal and scattered principal modes, generalising the Mie formulae and clarifying the nature of all scattering channels of a particle; they allow us to reduce large matrices to their sub-blocks and eliminate the need for numerical inversion in the determination of the mode coefficients. We give more details for the numerical process of calculating the principal modes in Sect. 1.7, comparing the stability and performance of the discrete source method solved by QR factorisation against two numerical schemes to form the unitary transformations required to obtain the scattering problem in the form of (1.9).

For spherical particles, each pair of modes can be divided into two categories, the electric multipoles (transverse magnetic modes) having magnetic fields with null radial components, and the magnetic multipoles (transverse electric modes) having electric fields with null radial components. As a consequence of the spherical symmetry, modes for each of these sets can be labelled with two discrete indices, l and m, which are related to the eigenvalues of the total angular momentum and of the angular momentum along z, respectively. Electric and magnetic multipoles are divergence free, i.e. the condition $\nabla \cdot E = 0$ applies. From Maxwell's equations [24], one can find that this condition has two physical consequences: these modes cannot produce a charge density inside the medium in which they are supported; furthermore, at each point in space the electric fields of these modes are orthogonal to the local wave vector. When the response of the system is local, for example in the case that the current density at any point inside a metallic particle depends only on the electric field at that point, the interaction of the particle with light is fully described by these electric and magnetic multipoles. More specifically, internal and scattering electromagnetic fields are expanded in terms of electric and magnetic multipoles: in spherical coordinates the internal and external electric (magnetic) multipoles of indexes l, m have the same angular dependence, while the radial dependence is a function of the medium in which the multipole is defined. As a result of the spherical symmetry, the surface integral of the scalar product of a mode with the complex conjugate of another mode, i.e. the spatial correlation at the surface, is non null only if the two modes are both transverse electric or transverse magnetic and also have the same values of l and m. This property can be used to find the amplitudes of the modes by imposing the continuity of the tangent components of the electric and magnetic fields at the surface of the sphere and identifying the principal modes for the internal and scattered fields that are correlated at the surface. For non-spherical particles, instead, principal modes are linear superpositions of multipoles (although in some cases one multipole can be dominant) and again form unique internal and scattered mode pairings.

We can now interpret the interaction of particles with light in terms of eigenvalues and orthogonal eigenvectors, $w_n^{\pm} = (i_n \pm s_n)/\sqrt{2}$, of the Hermitian operator in (1.9). This provides useful analogies between the electromagnetic response of classical particles and the quantum-mechanical response of atoms or molecules. In experiments however, one typically observes only either the internal or scattered field in some way, so transforming the eigenfunctions, $\{w_n^{\pm}\}$, we find the principal modes' coefficients:

$$a_n^i = \frac{i_n - \cos(\xi_n) s_n}{\sin^2(\xi_n)} \cdot f^0 = \frac{i_n'}{i_n' \cdot i_n} \cdot f^0, \tag{1.11}$$

$$a_n^s = -\frac{s_n - \cos(\xi_n) i_n}{\sin^2(\xi_n)} \cdot f^0 = -\frac{s_n'}{s_n' \cdot s_n} \cdot f^0. \tag{1.12}$$

Here $i_n' = i_n - \cos(\xi_n) s_n$, $s_n' = s_n - \cos(\xi_n) i_n$ are biorthogonal to i_n, s_n ($i_n' \cdot s_n = s_n' \cdot i_n = 0$) with $i_n' \cdot i_n = s_n' \cdot s_n = \sin^2(\xi_n)$. Either the principal or the biorthogonal modes fully specify the response of the particle at any point inside, on the surface and outside the particle. A graphical representation of the principal and the biorthogonal modes is shown in Fig. 1.1. This is shown by recasting the expansions of internal and scattered field as

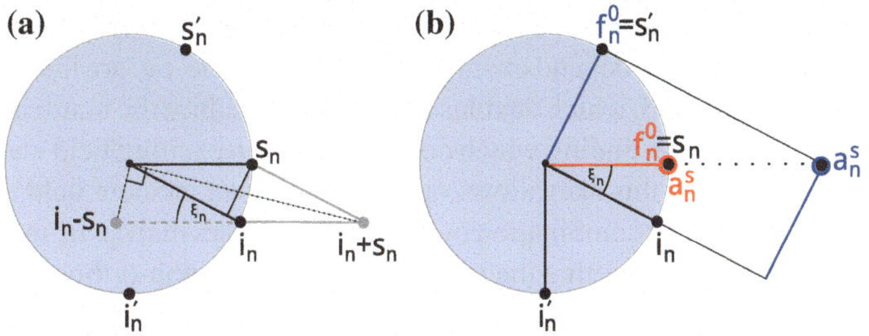

Fig. 1.1 A schematic for a pair of principal modes showing their relative geometry and interaction with incident fields. All of the modes and incident fields have been normalised to unit magnitude (the marked circle). Figure **a** shows the orientation of the internal and scattered (s_n, i_n) modes and the directions of the orthogonal eigenvectors $w_n^{\pm} \propto (i_n \pm s_n)$. Figure **b** demonstrates the orientation of the biorthogonal modes (s'_n, i'_n) and the induced mode amplitudes caused by different incident fields. We can see that a field of magnitude 1 with f_n parallel to s_n induces amplitudes $a_n^s = 1, a_n^i = 0$—shown in red, corresponding to (1.18)—while a unit field with f_n parallel to s'_n (and orthogonal to i_n)—shown in blue, induces the largest amplitude possible from a unit magnitude incident field together with a non-vanishing internal amplitude a_n^i (corresponding to (1.19)). Equivalently [25], the amplitudes are proportional to the scalar products of the incident field with the biorthogonal modes i'_n and s'_n

$$\begin{aligned} F^{s/i}(x) &= G_S(x,s)\cdot f^0(s), \\ &= \left(\frac{\mathscr{T}^i(x)I_n(x)i'_n(s)}{i'_n\cdot i_n} - \frac{\mathscr{T}^s(x)S_n(x)s'_n(s)}{s'_n\cdot s_n}\right)\cdot f^0(s), \end{aligned} \tag{1.13}$$

where $G_S(x, s)$ is the surface Green's function [12, 18] of the particle and the indicator function $\mathscr{T}^i(x)$ ($\mathscr{T}^s(x)$) is 1 inside (outside) the particle and null elsewhere. In numerical calculations, the propagation of fields away from the surface ($I(x)$ and $S(x)$) is performed as per (1.13) at a very low computational cost by evaluation of known solutions of Maxwell's equations, i.e. of Bessel or Hankel functions and vector spherical harmonics. From (1.13) one can see that the convergence of principal modes and principal angles as $N \to \infty$ is a consequence of the convergence of the surface Green's function [18] for any complete set of solutions of the Maxwell equations. We can monitor this convergence with the surface residual

$$\left| f^0 + \sum_{n=1}^{N} a_n^s s_n - a_n^i i_n \right|, \tag{1.14}$$

which provides an upper bound for the maximum error in the scattered and internal fields that decreases with the distance from the surface [11, 26]. Furthermore, the form of (1.11), (1.12) remains unchanged as $N \to \infty$, even if ξ_n, i_n, s_n themselves change. The angles relevant to this work are the point angles $0 < \xi < \pi/2$ of the infinite dimensional theory [9], together with the corresponding subspaces (the principal modes) and their orthogonal complements (the biorthogonal modes).

The modal decomposition in (1.13) has several unique advantages. The left-hand terms in the scalar products depend exclusively on the particle, i.e. are independent of the particular incident field, which enables us to strongly reduce the number of modes used in the calculations by finding which couple to a given exciting field and discarding the others. Similarly, this also allows optimisation of an incident field to excite a specific principal mode, or combination of modes. We remark that a_n^i, a_n^s in (1.11) and (1.12) are calculated by projecting the incident field f^0 onto non-orthogonal vectors, i_n and s_n, while sin (ξ_n) is defined as the Petermann factor in unstable optical cavities [27], which gives the order of magnitude of transient gain and excess noise for these systems. Therefore the existence of strongly aligned vectors with sin $(\xi_n) \ll 1$ determines large surface fields in nanoparticles as well as large transient gain and excess noise [28] in macroscopic unstable cavities and dissipative systems governed by non-Hermitian operators [29], demonstrating an analogy between optical cavities and nanoparticles. Macroscopic cavities and scattering particles are both open systems in which the internal and external modes are together necessary to provide a full description of the interaction with the environment: reducing the theory to one set of modes implies a loss of information, and this is the origin of the similarities of these systems. Mathematically, as a consequence of oneness and losses, internal and scattered principal modes are not themselves modes of a Hermitian operator and their optimal excitation is given by the corresponding biorthogonal modes. Physically, the biorthogonal modes are surface fields that are either totally reflected or totally absorbed. The experimental realisations of these surface fields may be challenging, but they can be used theoretically to find and optimise incident fields able to couple to the principal modes. For well aligned mode pairs (having a small value of ξ_n), the coefficient a_n^i, a_n^s are of the same order, but this is not the case for weakly aligned pairs ($\xi_n \rightarrow \pi/2$), which can have qualitatively different absorption and scattering cross sections. Furthermore, modes can have null amplitudes for *specific* incident fields but couple well to other incident fields, see (1.11) and (1.13).

To completely determine the response of particles to electromagnetic fields, we need also to assess the ability of the principal modes to transport energy. This is determined by the integral of the Poynting vector of each mode over the surface of the particle,

$$\Phi_n^{s/i} = \int_S Re\left(\hat{n}^{s/i} \cdot E_n^{s/i} \times H_n^{s/i^*}\right) \mathrm{d}s, \tag{1.15}$$

where $\hat{n}^{s/i}$ is the outward (inward) pointing normal of the surface for scattered (internal) modes and $E_n^{s/i}$, $H_n^{s/i}$ are the electric and magnetic components of the principal mode with $|a_n^{i/s}| = 1$. Radiative modes that are very effective at transporting energy from the particle surface to infinity are identified by large values of this "intrinsic" mode flux, Φ_n^s, while small values correspond to modes that can only transport energy effectively in close vicinity to the surface. Similarly, strongly (weakly) absorbing modes are identified by large (small) values of Φ_n^i.

This theory generalises the definition of Mie resonances to any particle without sharp edges and it is computationally efficient. Principal modes and their scalar products depend on geometric properties such as the size parameters of the particle and also the frequency dependent permittivity and susceptibility of the particle; as a consequence, the principal angles, $\{\xi_n\}$, vary with the frequency of incident light. The mode coefficients diverge when the denominators of (1.11) and (1.12) vanish. This happens when a pair of normalised internal and scattering modes are perfectly parallel. For a sphere it can be shown by factoring out the angular dependence of the modes that this is an example of the usual Mie resonance condition, which can be interpreted geometrically in terms of alignment between internal and scattered modes [30]. For spheres, the resonance $i_n = s_n$ occurs at complex wavelengths; for real wavelengths, resonances correspond to minima of the principal angles, and hence maxima in the excitability of the modes. This holds true for any smooth non-spherical particle, because perfect alignment is impossible with the linear independence and completeness of the principal modes. Therefore, as with spherical particles [31], physical resonances correspond to minimum angles ($\xi_n \neq 0$) of pairs in $\mathscr{H}$, which are also minima of the eigenvalues of the Hermitian operator in (1.9).

We also need to consider the energy transported by resonances. A resonant pair of modes can have very different values of Φ_n^s and Φ_n^i, which means that the effect of the resonance can be strong or weak depending on whether the detection is made in the near or far field, or if scattering or absorption are measured. Note that the total flux of energy scattered or absorbed by non-spherical particles (integrals of the Poynting vector over all directions) is given by the sum of principal mode contributions plus interference terms between modes, which are absent for spherical particles by symmetry. The interference terms are absent in spheres because both surface fields and power fluxes of Mie modes are orthogonal. Hence for non-spherical particles, interference effects can lead to strong efficiency peaks or to sharp asymmetric features resulting, for example, from Fano-like interference between broad and narrow resonances. Examples of these phenomena are given in the following.

In terms of numerical calculations this approach is a surface method that relies on the convergence of

$$\left\{\sum_{n=1}^{N} a_n^i I_n, \sum_{n=1}^{N} a_n^s S_n\right\} \tag{1.16}$$

to the exact electromagnetic fields at any point inside and outside the particle [12, 18] as $N \to \infty$. Equation (1.14) gives the numerical error in the evaluation of the surface fields; in practice the scattered power is calculated both at infinity and on the surface of the particle to estimate the error in the field propagation, i.e. in the evaluation of the special functions used for expanding the Maxwell solutions. The Stratton-Chu relations [32] can also be used to test both the propagation of the scattered light and also the reliability of the internal [22] field.

We have compared our calculations with Mie theory [33] for metallic spheres [34] and demonstrated that this method is numerically very accurate for particles with radii between 10^{-2} and 4×10^0 times the wavelength of light.

1.3 Optical Resonances of Single Particles

As the size of the particle becomes comparable to, or smaller than, the wavelength of light, the geometric description of resonances as closed orbits of light rays [35] becomes less and less effective, eventually requiring the solution of Maxwell's equations. Analytical solutions [36, 37], based on symmetry and separation of coordinates, provide exact results for a few specific shapes of particle. For spheres, electric and magnetic multipoles are used to expand internal and scattered fields and each multipole in one field is correlated only to the corresponding multipole in the other field. Resonances depend on the size of the particle and on the dielectric permittivity and magnetic permeability of the internal and external media and occur when the correlation of one pair of modes reaches a maximum. However, it is important to realise that a resonance can be observed only if the incident light has the appropriate frequency *and spatial structure* on the surface of the particle, otherwise the incident field does not couple to the resonant modes of the internal and scattered fields.

For particles of arbitrary shape, resonances are often defined empirically by the appearance of maxima in properties such as the far field extinction or scattering efficiency spectra. While there are several methods that can find spectra and their maxima [38], this approach to resonances is unsatisfying because it fails to disentangle the intrinsic properties of the resonance from those of the incident field. It also depends on an arbitrary choice of the property which is monitored to determine the resonance. For instance, a resonance of a surface mode that does not efficiently transport energy to infinity, would not be distinguishable in any far field measure; these resonances, however, can be extremely important in near field applications or through interference with radiative modes able to transport energy into the far field. We use the principal modes of Sect. 1.2 to define field expansions and resonances for any smooth particle, where Mie's treatment of the sphere is a special case of this more general theory.

For all particles presented in this section we use a fitted dielectric function [39, 40] for gold.

1.3.1 Gold Nanorods

We examine here light interaction with a type of particle that is extensively used in experiments, gold nanorods. For simplicity, we consider these particles in vacuum without considering any supporting substrates.

As a consequence of the axial symmetry of these particles, all fields in the problem can be decomposed into components demonstrating a well defined azimuthal angular dependence, $\exp i m\phi$, where m is an integer. In the case of axially incident plane wave light, functions transforming as $m = \pm 1$ fully describe the incident light and the optical response of the particle (other channels being 'dark' to this excitation).

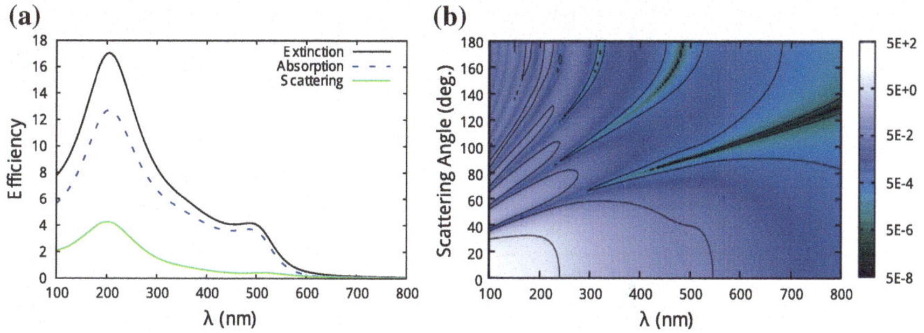

Fig. 1.2 **a** The scattering, absorption and extinction efficiencies of a small gold rod, showing a 205 nm mode and a weaker absorption peak at 486 nm. Other resonances can be excited for different angles of incidence (since the incident light should couple with the structure of the mode to excite it). **b** The DSCS, showing that this particle has strong forward scattering for plane wave exciting light

Figure 1.2a, b show the calculated optical efficiencies and differential scattering cross section (DSCS) of a 480 nm long rod with diameter 40 nm, which is illuminated with plane-wave light incident along its axis. There is a strong resonance at a wavelength of 205 nm with a weaker absorption peak at ~515 nm. The DSCS shows that the incident light is strongly scattered forward, particularly at short wavelengths.

We see from Fig. 1.3a, d that most mode pairs are either strongly radiative or absorbing, with the exception of one weakly aligned pair that is able to be both absorbing and radiating. The peak at 205 nm results from the constructive interference of several weakly aligned principal mode pairs as shown by Fig. 1.3c. Most of these mode pairs are weakly radiating into the far field and collectively they can very efficiently extract energy from the incident field. Interference between scattering modes leads to enhanced sensitivity to perturbations near to the scattering surface, therefore these multi-mode resonances can be very advantageous in sensing applications. Additionally, the effects of entropy associated with these multiple mode features [41] can be significant.

A few modes cause the absorption feature at ~515 nm. Two of these modes are weakly aligned and absorb comparable amounts of energy; they become more strongly aligned at around this resonance. The scattering mode of the pair with the weakest alignment is instead dominant, as shown in Fig. 1.3c, where the resonance at ~515 nm is enhanced, but in absolute terms this resonance is barely observable in the far field scattering efficiency of Fig. 1.2a at that wavelength.

Figure 1.4 shows the same particle, now illuminated equatorially with an incident light polarisation of 45° with respect to its long axis, where now modes with $m = 0$ character dominate the response of the system. The broad feature at around 200–450 nm, containing structures similar to the composite modes of Fig. 1.2, is insensitive to the particle length, as shown by examining particles with varying length and the same diameter. No clear hot or cold spots can be observed on the particle surface, see Fig. 1.5a. The sharp resonance at 676 nm shifts with rod length, and its surface

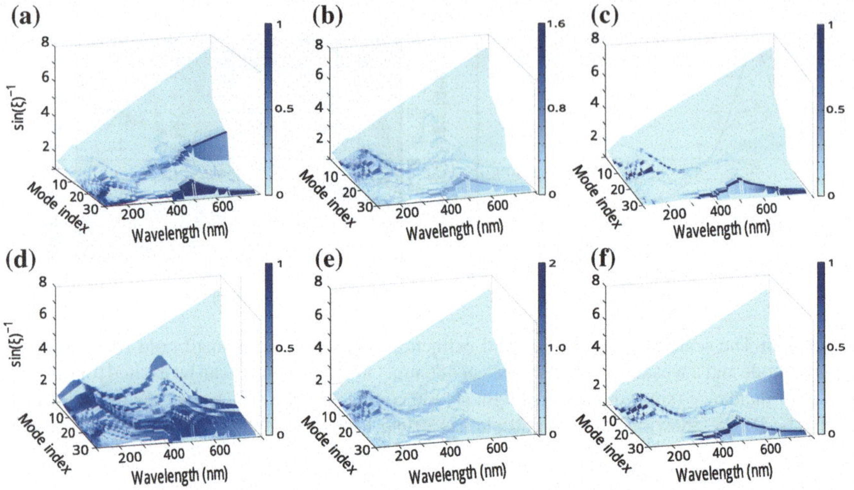

Fig. 1.3 The principal angle landscapes of a rounded nanorod: mode pairs are plotted according to their principal cosines (mode index), the wavelength of the incident light, with the height of the landscape as $\sin^{-1}(\xi)$, the largest possible values of $|a_n^{i/s}|$ for $|f^0| = 1$. Landscapes are overlaid with traces that are colour coded according to: **a** the intrinsic mode fluxes, Φ_n^s, normalised at each wavelength to the range [0, 1], **b** the amplitudes $|a_n^s|$ of the scattered field, **c** the mode fluxes $|a_n^s|^2\Phi_n^s$ again normalised to [0, 1]. **d**–**f** as **a**–**c** but for internal modes. Discontinuities in the traces are due to crossings between different modes which occur when the values of their principal cosines become degenerate, leading to a change in ordering of the mode indices. **b** and **c** show that a multi-mode resonance produces the peak at 205 nm, with several principal modes contributing similarly to the scattered power. A group of internal modes, which become more aligned at 515 nm, determine the weak absorption peak around this wavelength. In these pairs internal and scattering modes couple differently to the incident field and the radiative mode with the weakest alignment dominates the far field, as shown in **c**

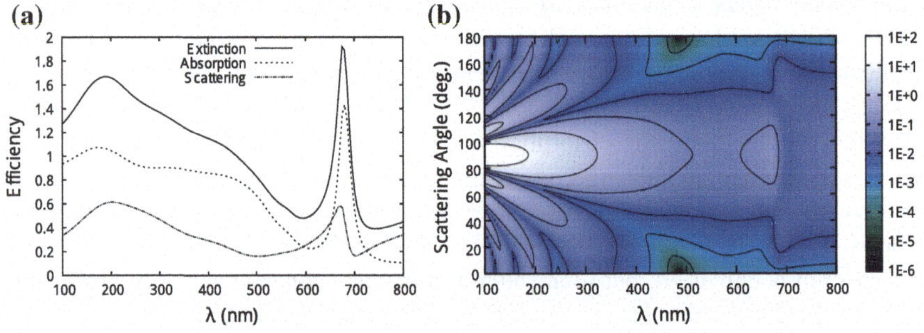

Fig. 1.4 The particle from Fig. 1.2, but now illuminated from the side with an incident light polarisation of 45° with respect to its long axis. **a** The spectra show the presence of both a broad feature similar to Fig. 1.2 at around 200–450 nm and also a sharp resonance at 676 nm. Note the strongly asymmetric Fano-like resonance (sharper than the resonance in the extinction spectrum) in the scattering efficiency at this wavelength. **b** The DSCS for equatorial illumination

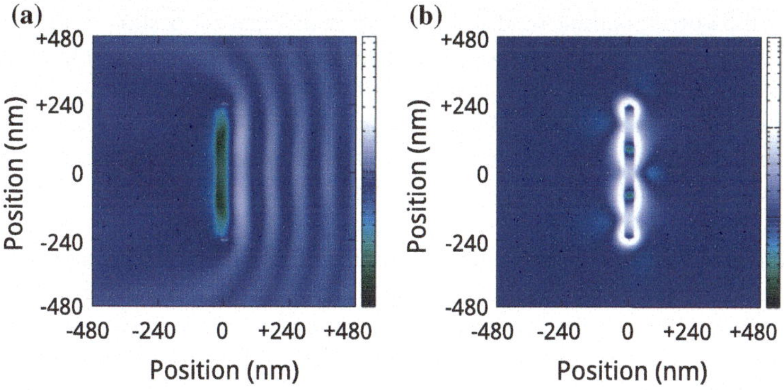

Fig. 1.5 Near field intensity for the 480 nm length rod, shown at the **a** broad feature (207 nm) and at the **b** "waveguide" Mie-like mode at 676 nm of Fig. 1.4a

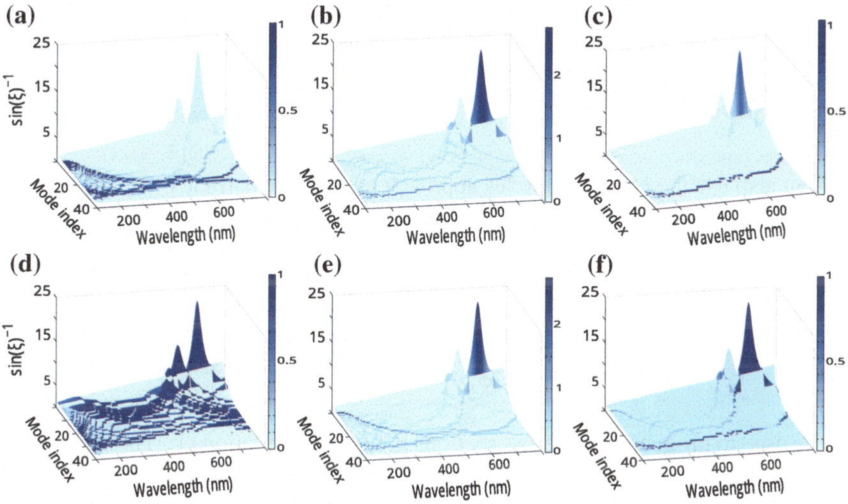

Fig. 1.6 Principal angle landscapes for the rounded nanorod for modes able to couple to illumination from the side of the rod, with resulting spectral properties shown in Fig. 1.4. The feature at 676 nm is due to a single well aligned pair of modes. The landscapes are colour coded and have the same axes as Fig. 1.3. **c** Shows that the energy is transported into the far field by the resonant mode and by a weakly aligned, non-resonant, scattering mode. These modes also interfere to give the total scattering cross section and produce the characteristic asymmetric and sharpened Fano-like resonance shown in the scattering efficiencies of Fig. 1.4a

field as shown in Fig. 1.5b, has the strong nodal local structure of a "waveguide" mode on the long axis, remarkably similar to the experimental results of [42].

Figure 1.6 shows only mode pairs that cannot be excited by axially incident light, primarily being in the $m = 0$ channels; multi-mode resonances similar to the one discussed for axial incidence originate the broad features at short wavelength;

most of the corresponding modes are not shown. Figure 1.6 show that for most principal mode pairs of this particle, excitation occurs only for either the absorbing (internal) or radiating (scattering) member of the pair, and all the three resonances at around 550–700 nm are absorbing. The excitation paths in Fig. 1.4b, e show that only the best aligned of the three resonant pairs is excited for this particular field. This resonance is undetectable for incident light propogating along the axis, but it is observable as the angle of incidence is rotated towards the equator at 90°, Fig. 1.4; at ~600 nm the more weakly aligned mode pair becomes excited for incident angles of ~50°. Near the sharp resonance at 676 nm, the resonant mode and a weakly aligned, non-resonant, scattering mode transport the energy into the far field, see Fig. 1.4d. Interference between these two modes gives a Fano-like asymmetric feature in the total scattering cross-section that is sharper and more asymmetric than the single pair resonance. On the contrary, only the resonant mode determines the absorption, producing a symmetric feature in the absorption cross-section, with its maximum coinciding with only the largest alignment of the mode pair.

1.3.2 Supershapes

We can also use the principal modes approach to examine particles with more complex shapes. For example "peanut" shaped gold particles with a length of 80 nm. These are an extreme case of a family of geometries from a spheroidal particle, via cylinder to composites of two spheroids connected together. To describe this range of geometries a "superformula" [43],

$$r(\phi) = \left[\left| a^{-1} \cos \mu\phi/4 \right|^{\nu_2} + \left| b^{-1} \cos \mu\phi/4 \right|^{\nu_3} \right]^{-\nu_1}, \tag{1.17}$$

can be used to produce all the cross sections in Fig. 1.7. Equation (1.3) is used to give an error measure for the internal and scattered field solutions. Normalised optical cross-sections are shown in Fig. 1.8 for different incident angles of plane wave light

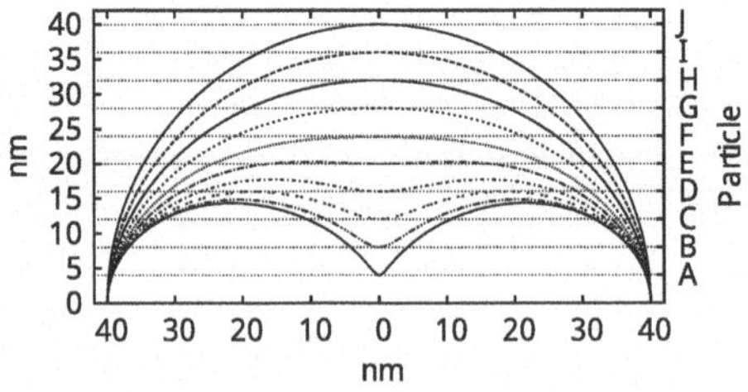

Fig. 1.7 Generatrices of 10 axially symmetric particles, labelled A–J, ordered by depth of the indentation and being between spherical (J), through a rounded cylinder (F) to two merged spheres (A). The superformula with parameters $\mu = 2$, $a = b = 1$ and $\nu_1 = 0.1$, $\nu_2 = 0.1\nu_3$ [43] is used to describe all of these shapes

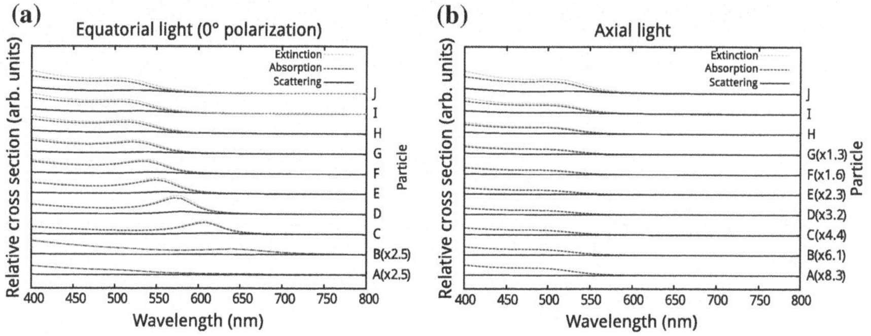

Fig. 1.8 The calculated cross sections for particles A–J. The light is polarised parallel to the long axis and its incident directions are chosen along (**a**) and across (**b**) the symmetry axis. Particularly in the case of axial light, there is a rapid decrease in extinction and absorption by the surface plasmon on distorting away from a spherical symmetry. The plasmon feature also shows a strong red-shift for equatorial incident light and is lost at the "peanut" shaped limit

(at 0 and $\pi/2$ rad. with respect to the particle axis). In all cases absorption is stronger than optical scattering for these particles and is seen to dominate their extinction; for both angles of incidence that have been considered, the smoother shaped ellipsoid show pronounced resonances, but these become almost indistinguishable when moving to particles with a deeper indentation. For axial incident light, a broad resonance near 510 nm is apparent: this resonance becomes weaker if the indentation is increased, but its wavelength location remains the same. For light with an equatorial incidence direction another resonance becomes apparent, that for the ellipsoidal particles, is present at ≈530 nm. Here, on indenting the shape, this feature moves to lower energy, again becoming weaker and finally is lost for the deepest indentations. A simple explanation for this behaviour would be that shape change perturbs the surface field, moving the resonance to longer wavelengths where the geometrical distortion is less significant and is averaged out, but finally the indentation is sufficiently strong as a perturbation and the feature disappears from the investigated spectral range.

To confirm this simple picture, the principal modes [44] of the system demonstrate that the 510 nm feature is due to a weak resonance of one mode pair, while the 530 nm feature is due to another more strongly aligned mode pair. In the case of the most indented particles, the mode landscapes show that while there are several modes of the particles which can strongly couple with the incident light, these are not efficient at absorbing or scattering being generally poor at transporting energy (low mode fluxes). However these modes may be apparent, for example to methods such as surface enhanced spectroscopies, since they are strongly excited but could easily be overlooked by measurements in the far field.

1.3.3 Scanning Near-Field Optical Microscopy of Gold Nanodiscs

Principal modes have also been applied to explain Scanning Near-Field Optical Microscopy (SNOM) measurements performed on gold nanodiscs [45]. Here the experiments were performed in collection mode, scanning the tip of an optical fibre in close proximity to the particle, and collecting light via a large numerical aperture in the far field.

Measured and simulated images of an 800 nm diameter, 30 nm thick disc, are shown in Fig. 1.9 at two wavelengths. The theoretical model neglects back-coupling from the disc onto the tip, but yields good qualitative agreement with the measured images. The change in the images of Fig. 1.9 is mainly due to which principal modes of the disc are dominant at the two wavelengths. At 705 nm, an $m = 1$ character mode is near to its resonance, while at 765 nm instead it is an $m = 0$ mode responsible for the image. The excitability of these two modes is shown in Fig. 1.10, these are the two relevant modes from the the full landscape, for example in Fig. 1.11 we show the rest of the $m = 1$ character modes.

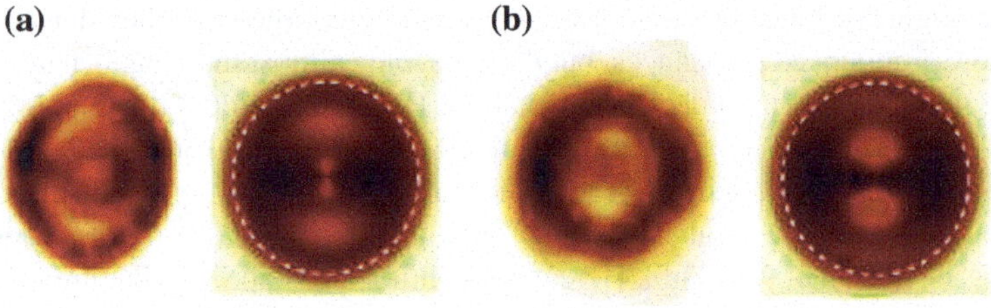

Fig. 1.9 **a** Experimental (left) and theoretical (right, with the particle shape marked with a fine white dotted line) images of near-field transmission with incident light at 705 nm. **b** The same disc illuminated instead at 765 nm. The exciting beam is unpolarised, while the far field collection is linearly polarised in the vertical page direction

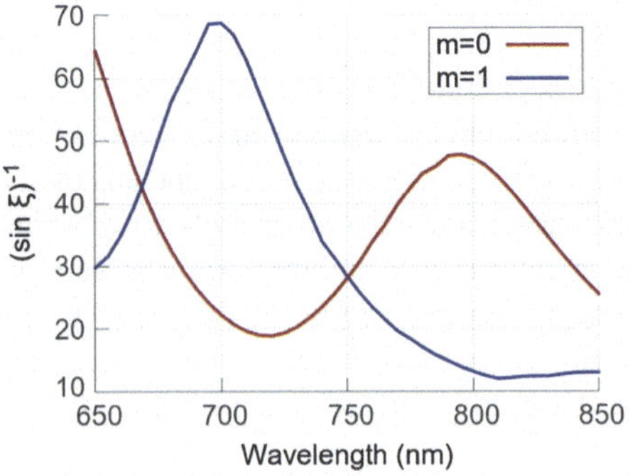

Fig. 1.10 The $m = 1$ and $m = 0$ resonant principal modes that dominate the experimental SNOM images shown in Fig. 1.9

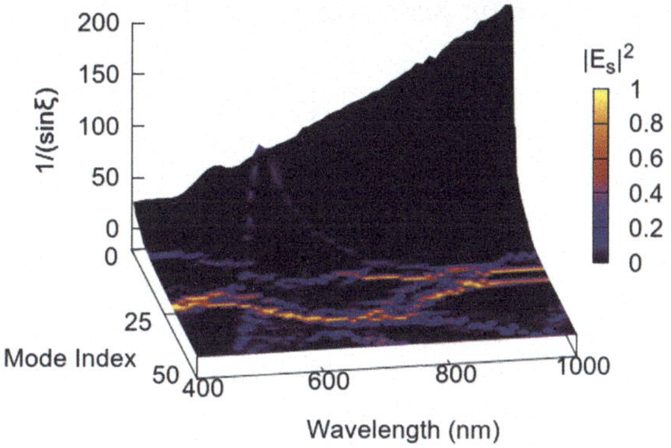

Fig. 1.11 Landscape of the $m = 1$ modes, including the mode shown in Fig. 1.10 which is responsible for the observed SNOM image at 705 nm. The colour of the landscape corresponds to energy transported to the collection optics from each mode when excited by an unpolarised light source located above a point at 75% of the disc radius

We note that the effect of multiple scattering of light between the particle and the tip may have to be included for more complex particles. In that case, we should consider the principal supermodes of the whole system, as discussed in Sect. 1.5, consisting of both the SNOM tip and the particle.

1.4 Coherent Control

Several approaches for nonlinear [46] and linear control of light have been suggested based on pulse shaping [47, 48], application of learning algorithms [49], coherent absorption mechanisms [50] and time reversal of an outgoing field [51]. In quantum optics the interference between fields has been proposed as a way to remove beam splitter losses [52] and for control of light by light in graphene [53] or inside plasmonic metamaterials in the linear regime [54].

Using the principal modes decomposition [30], general analytical control of both internal and scattered fields can be achieved, either enhancing or suppressing the internal and/or the scattered fields for general structures at arbitrary wavelengths. It is also, for example, possible to induce narrow features in the spectral responses of systems by, for example, converting phase control into amplitude modulation. In order to control N modes of the system, this approach requires control of the relative phases and amplitudes of $N + 1$ independent exciting light fields. Depending on the modes being controlled, the flow of energy out from a particle (or structure) can be adjusted in the near and/or in the far field, similarly absorption can be changed by controlling the internal fields of the particle in a similar manner.

1.4.1 Elastic Scattering

A constraint on the control approach should be that the fields required are of types that are commonly available. Fortunately it is possible to apply both monochromatic or broad band light sources, provided there is a reasonable degree of coherence. This should be on a length scale of at least the order of the particle size, so can even be from conventional lamps in some case, as long as the differences in optical paths are within the coherence length.

To be practical, simple control of the relative phase of the incident optical sources would be preferred since this then requires only one source of light and can be achieved with standard optical bench technology (splitters, phase modulators, etc.). Alternatively, more complex approaches including mixtures of internal and external sources, such as embedded impurities or quantum dots can be used (see Sect. 1.4.2 for how to apply non-linear effects in a similar role). Figure 1.12 shows some idealised conditions to achieve this type of source control. We have previously demonstrated [30] that simple scanning of parameters can be used to locate conditions that control scattering responses. This can be achieved without detailed information on the optical modes of the system, making it practical to be experimentally applied.

The correlation between the internal and scattered modes in the nth pair (i.e. the non-orthogonal basis of vectors [25]) gives a simple geometry which can be used to determine exciting surface fields which project onto s_n but not i_n (or i_n and not s_n). Similarly, fields that produce a maximum amplitude for modes s_n or i_n, see Fig. 1.1.

To construct a field which couples in these specific ways, we note that incident fields can generally be decomposed as $f = f_n + f_n^{\vee}$, i.e. the components co-planar with the nth mode pair (f_n) and that lies outside of the plane ($f_n^{\vee}$).

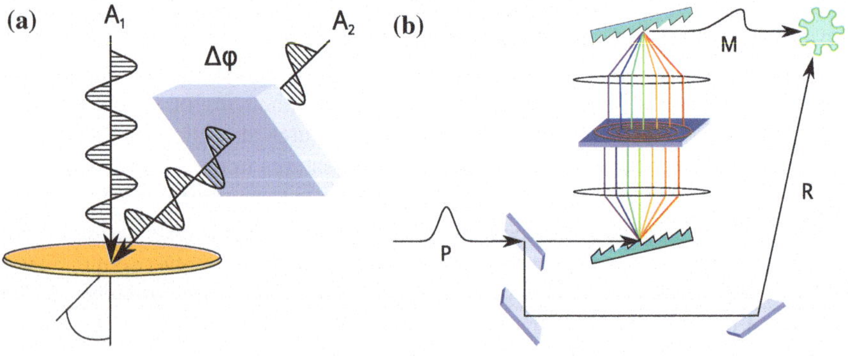

Fig. 1.12 Some potential set ups to control optical channels of nanostructures. **a** Using monochromatic light fields, the relative amplitudes A_1 and A_2 and phase difference $\Delta\Phi$ specified. **b** More generally, a spatial-light modulator is deployed to modify both phase and/or the amplitude over a range of frequencies from a coherent broad-band or pulsed source of light. By splitting the light pulse (P) along two paths, reference (R) and modified (M) pulses are recombined at the nanostructure

$$f_n = s_n \rightarrow a_n^s = 1, a_n^i = 0, \tag{1.18}$$

$$f_n = s_n' \rightarrow a_n^s = \frac{1}{\sin(\xi_n)}, a_n^i = -\frac{\cos(\xi_n)}{\sin(\xi_n)}. \tag{1.19}$$

Equation (1.18) then gives conditions for fields that excite only the scattering mode of the pair (independent of the $f_n^\vee$ component), i.e. does not couple to the paired internal mode. Equivalently, the largest conversion factor into scattered intensity, $|a_n^s/f_n|$, occurs for the field given in (1.19). Both types of incident field and the resulting induced scattered mode amplitude are shown in Fig. 1.1. The corresponding fields to induce equivalent amplitudes for the internal mode i_n is obtained by exchanging s and i in (1.18), (1.19).

These are exact conditions, valid for any frequency. Possible applications are to the control of one or more modes over a range of frequencies, or to introduction narrow band features in the spectral response. It is noteworthy that the largest excitation of a mode is not caused by a single mode, parallel in the sense of Fig. 1.1, excitation. Instead optimal amplitude is obtained by producing amplitude in the two principal modes of the pair.

Physical fields with tangent components according to (1.18), or (1.19) can be produced by several methods. The mode parallel $f = s_n$ case could be produced via a time reversal for a lasing mode from a similar shaped particle having gain opposite to the target particles' loss [50] but only at the resonant frequency. This can be challenging to achieve experimentally, requiring convergence from all directions for the radiation. Instead interference can be used to construct such surface fields. A field $F(x)$ having tangent components of $f = i_n$ or $f = s_n$ cannot in general be realised by common external radiation sources or by purely internal sources (for example fluorescence). The reason being that sources purely in one region do not simultaneously contain components that are both outgoing radiating waves (as required to couple to the scattered modes) and standing internal waves (required to couple to the internal mode). Instead, linear combinations of two or more sources having appropriate amplitudes and phases provide a simple way to construct exciting fields for (a few) dominant interaction channels of general particles. The necessary fields which realise (1.18) or (1.19) can be constructed from two linearly independent fields $A_1 f^1$, and $A_2 f^2$,

$$\frac{s_n \cdot f^1}{s_n \cdot f^2} \neq \frac{i_n \cdot f^1}{i_n \cdot f^2}, \tag{1.20}$$

both with a component projecting into the nth channel. To suppress i_n,

$$A_2 = -A_1 \frac{i_n' \cdot f^1}{i_n' \cdot f^2}, \tag{1.21}$$

while to maximise the amplitude of s_n

$$A_2 = -A_1 \frac{i_n \cdot f^1}{i_n \cdot f^2}, \tag{1.22}$$

where A_1 and A_2 are complex amplitudes. Figure 1.12a presents a simple setup for constructing these fields. Generalisation of (1.22) to larger numbers of modes and for control of internal modes is discussed in [30].

Using Stratton-Chu representations [32] it can be shown [30] that for particles such as spheres and infinite cylinders, (1.20) is *always* violated for exciting sources all located on the same side of the particle boundary. This is due to the common dependence of the components of i_n, s_n and f_n with respect to surface components. The ratios given in (1.20) then depend on the flux of incident energy into the particle. For the example of spherical particles, any external source can be expanded as a set of regular multipoles, while any internal source is representable by radiating multipoles [30]. Hence internal and scattering modes cannot be independently addressed by purely internal—or external—sources and (1.22), (1.21) become equivalent. For mixed internal and external sources, maximal excitation (1.22) or mode suppression (1.21) *can* be fulfilled.

For non-spherical particles, for example a rounded gold disc, sharp spectral features can be induced, as shown in Fig. 1.13, using simple interference. A setup similar to Fig. 1.12b might be the basis of producing the necessary phase changes over a narrow wavelength range.

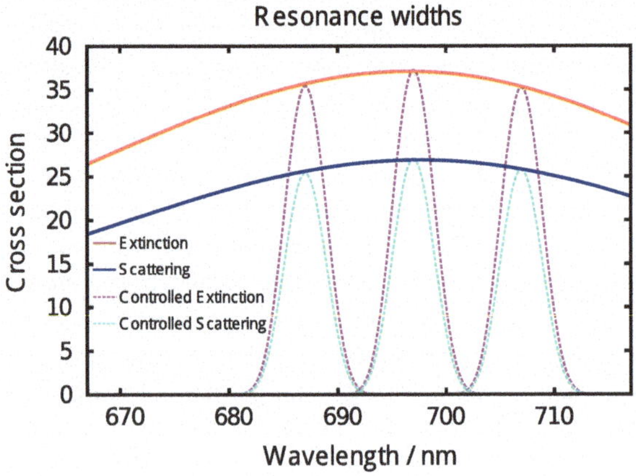

Fig. 1.13 A rounded gold disc (20 nm thick and r = 60 nm) and axial plane wave light, with a broad feature in extinction (red) and scattering (blue) cross sections. Mixing independent plane waves from three directions (axial, 45 and 90°) with specified relative phases (but not amplitudes) causes constructive interference only at a chosen wavelength. By rotating the phase to become destructive between the sources, specified line width features within the original feature envelope can be induced

1.4.2 *Inelastic Scattering: Multiphoton Processes*

Numerical examples of control of coherent second-harmonic generation have recently been described using an extra pump beam in cylindrical symmetry [55]. In the case of spherical particles the emission direction when combining two pump beams is controlled by selection rules depending on particles size and the specific order of the process [56]. Instead, by using the same pump beam to drive two copies of the same type of non-linear process, we can use the ideas of Sect. 1.4.1 to control both internal and external multipolar waves at the driven frequency. This is possible provided the strength of the non-linearity is insufficient to deplete the pump beam [57]. Since one non-linear process is inside the particle while the other is outside, this can be applied to nanospheres since (1.20) holds. Schematically this process is depicted in Fig. 1.14.

This control is also extremely sensitive to phase variations, since these reduce absorption and/or change the scattered energy by several orders of magnitude. This can enable detection of strongly subwavelength scale changes in the position of the particle or sensitivity to changes in surrounding electric permittivity, ε, or magnetic permeability [57].

The nonlinear response can be described by introducing bulk and surface tensors [58, 59], or alternatively the polarisation modelled using a hydrodynamical description [60–65]. In the latter case, transverse and longitudinal waves in a metal are described as occurring in a fluid with a pressure term of quantum origin and proportional to the Fermi velocity. The hydrodynamical model displays non-local response, hence a third type of mode is present—irrotational solutions which fulfil the condition $\nabla \times E = 0$. These purely electric modes are non-radiative but have associated charge density fluctuations which penetrate into the metal. This leads to

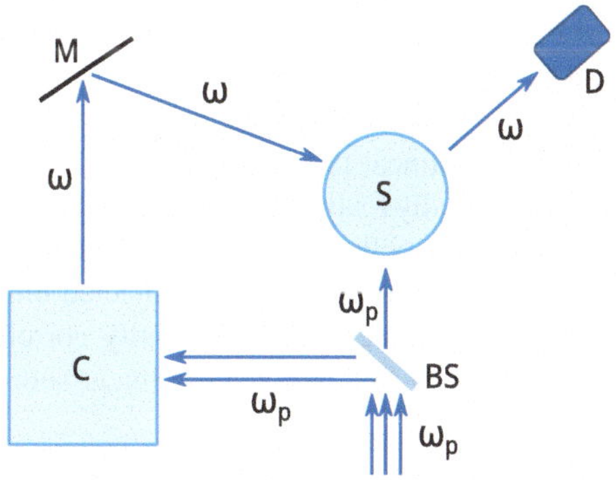

Fig. 1.14 Schematic, not to scale, that shows an implementations of the control scheme of Sect. 1.4.2. The pump beam is divided (BS) to drive equivalent nonlinear processes, converting light at frequency ω_p to ω. This is then collected (C) and directed to the particle (S) along the direction where the controlled multipole is at a maximum. The output is then received at detector D

revised boundary conditions where the normal components of the electric fields are considered, leading to new types of responses to light [60–65].

Internal media Maxwell solutions which include a nonlinear bulk polarisation, P^B, then can written in the form

$$E^B(x) = \int_V G_E(\omega; x, x') \cdot P^B(x')\mathrm{d}x', \tag{1.23}$$

where G_E is the electric dyadic Green's function and for the hydrodynamical model include longitudinal terms [66].

For many common particles, non-locality is relevant only in the surface region [67], hence $E^B(x)$ can be approximated by a Green's function lacking the longitudinal terms. Such a function can also be relevant when P^B is due to the products of tensors and the electric field (and its derivatives) such that linear response is local [68]. In these cases, surface nonlinearities are confined to a thin layer, and P^S is well represented by an infinitesimal polarisation sheet outside the particle (if the external medium supports interface charges). Surface [69, 70] and also volume nonlinearities then are contained in the boundary conditions:

$$\varepsilon^{in} E^i_\perp - \varepsilon^{ex} E^s_\perp = -\varepsilon^{in} E^B_\perp + \varepsilon^{ex} E^c_\perp - \nabla_\parallel \cdot P^S, \tag{1.24}$$

$$E^i_\parallel - E^s_\parallel = -E^B_\parallel + E^c_\parallel - (\varepsilon^{ex})^{-1} \nabla_\parallel P^S{}_\perp, \tag{1.25}$$

$$H^i_\parallel - H^s_\parallel = -H^B_\parallel + H^c_\parallel + i\omega(\hat{n} \times P^S), \tag{1.26}$$

with i, s, c being labels for internal, scattered and external control fields. The internal and external regions are labelled as the subscripts in and ex. The electric fields at the boundary are then defined as $E_\perp = \hat{n}(\hat{n} \cdot E)$, $E_\parallel = -\hat{n} \times (\hat{n} \times E)$ and similarly for other fields. Unlike in Sect. 1.2, the additional boundary condition for $E_\perp$ becomes necessary due to the presence of longitudinal electric waves. Amplitudes of the modes depend upon the left of (1.24) and (1.26). This enables us, for any E^B, H^B and P^S fields, to obtain E^c and H^c to control the interaction of light with the particle via amplitudes of the internal and scattering modes, and this control is independent of the nature of the underlining nonlinear processes.

Again, to focus on experimentally realisable situations, we now describe control of two modes of a sphere. These are labelled by the index of angular momentum, l (total angular momentum) and m (angular momentum projected onto z). By symmetry, only modes with the same l, m values can be spatially correlated on a spherical surface. In our notation the control field and the nonlinear sources in (1.24)–(1.26) are:

$$f^c = (\varepsilon^{ex} E^c_\perp, E^c_\parallel, H^c_\parallel), \tag{1.27}$$

$$f^{NL} = -(\varepsilon^{in} E^B_\perp + \nabla_\parallel \cdot P^S, E^B_\parallel + (\varepsilon^{ex})^{-1} \nabla_\parallel P^S{}_\perp, H^B_\parallel - i\omega\hat{n} \times P^S). \tag{1.28}$$

The component $\varepsilon E_\perp$ in the surface field is included due to the required longitudinal wave boundary condition. The amplitude and phase of f^c are given by the complex

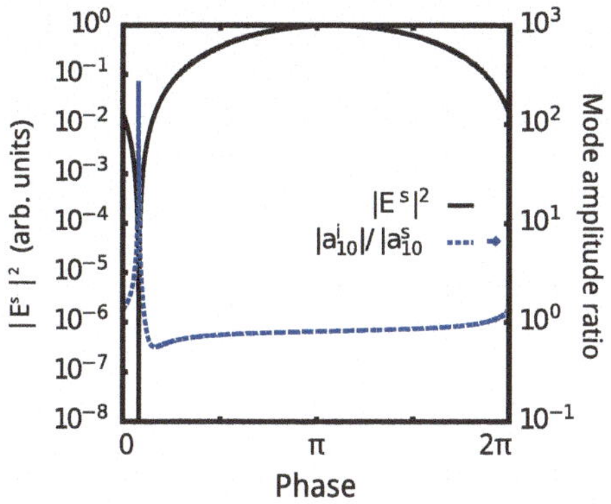

Fig. 1.15 Here and in Fig. 1.16, the pump amplitude, $a^p = 1$. This drives the nonlinearity of a 50 nm radius gold sphere, with dielectric response from the Lorentz-Drude model [71]. The internal and scattered $l = 1, m = 0$ modes are controlled to give perfect absorption of incident light of intensity $|a^c| = 1.47 \times 10^{-12}$ (arbitrary units) and phase of $\pi/2$ compared to the pump beam at $\omega = 563.52$ THz. The scattered intensity, $|E^s|$, is measured in a direction orthogonal to both the pump and control beams. The use of the secondary axis is indicated by arrows in the key

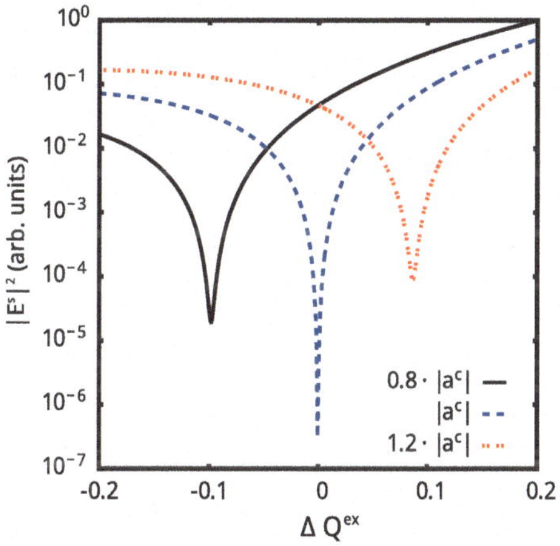

Fig. 1.16 The relative phase and intensity of the pump and control beams are optimised to suppress a s_{10} electric dipole mode at $\Delta\varepsilon^{ex} = 0$, as with Fig. 1.15. The directed scattered intensity is very sensitive to small dielectric changes in the external medium (blue-dashed curve). This enables a sensitive measurement of the surrounding environment. When $|a^c|$ changes by ±20%, since control is no longer optimised for $\Delta\varepsilon^{ex} = 0$, the scattered intensity curve moves (black-solid/red-dotted curves) but sharp features are still observed

amplitude a^c. Amplitudes of the internal and scattered pair, a^i_{lm} and a^s_{lm} are, in this notation,

$$\begin{bmatrix} a^i_{lm} \\ -a^s_{lm} \end{bmatrix} = \begin{bmatrix} i'_{lm} \cdot f^c & i'_{lm} \cdot f^{NL} \\ s'_{lm} \cdot f^c & s'_{lm} \cdot f^{NL} \end{bmatrix} \begin{bmatrix} a^c \\ 1 \end{bmatrix}. \tag{1.29}$$

Figure 1.15 shows control of the relative amplitudes of the i_{10} and s_{10} electric dipole modes. The control is chosen so that amplitudes of these modes can vanish for an appropriate phase ([57] demonstrates absorption control in this way). The phase sensitivity of this control, Fig. 1.15, shows that changes in the position of the particle on the scale $\Delta\lambda/\lambda = \Delta\Phi/2\pi$, where λ is the wavelength of the control beam, can be resolved. Similarly Fig. 1.16 shows how small variations in the surrounding permittivity can be monitored again due to the phase sensitivity.

1.5 Supermodes of Multiple Particles

We now consider a collection of r particles at arbitrary positions. The full Hilbert space of this aggregate is the direct sum of the Hilbert spaces of each particle; we again have two distinct subspaces of solutions of Maxwell's equations, the internal solutions and the external solutions. For each particle we can either use sets of electric and magnetic multipoles, as in Sect. 1.2, or the resulting internal and scattering principal modes of the single particles to expand the solutions of the system. The internal fields of each particle are identically null on the surfaces of the other particles in the system, while scattering fields have singularities only inside the originating particle itself but are non null on all other particles' surfaces. Using the single particle principal modes of each of the particles as a basis, the resulting multi-particle Gram operator is structurally the same as the single particle case, i.e.

$$\begin{bmatrix} 1 & I^\dagger S \\ S^\dagger I & S^\dagger S \end{bmatrix}, \tag{1.30}$$

so we can again define principal modes and resonances. Labelled by particle, the sub-matrices of (1.30) become

$$I^\dagger S = diag\{C_1, \ldots, C_r\} + \begin{bmatrix} 0 & \ldots & I_1^\dagger S_r \\ \vdots & \ddots & \vdots \\ I_r^\dagger S_1 & \ldots & 0 \end{bmatrix}, \tag{1.31}$$

and

$$S^\dagger S = 1 + \begin{bmatrix} \hat{S}_1^\dagger \hat{S}_1 & \ldots & S_1^\dagger S_r \\ \vdots & \ddots & \vdots \\ S_r^\dagger S_1 & \ldots & \hat{S}_r^\dagger \hat{S}_r \end{bmatrix}, \tag{1.32}$$

and analogously for the other blocks. The subscript identifies the particle from which the fields arise, the caret on a matrix indicates integrals are calculated on the surfaces of the system's other particles and C_r is a diagonal block of the particles' principal cosines. In order to implement the principal mode theory, the first step is to transform the matrix in (1.32) into an identity matrix. This can be done using the block structure of the matrix, in analogue to the single particle case of Sect. 1.2. This results in a set of supermodes for the composite system, which are again pairwise coupled at the interface between the external medium and the internal space of *all* of the particles.

Analogous to the problem [72, 73] of treating a sub-region of electronic states of a crystal, one can also define an embedding operator, Σ, for the multiple scattering effects to all orders from the other particles on the ith particle of the system, leading to a matrix problem in the form of (1.9),

$$\begin{bmatrix} 1 & C - \Sigma\hat{S}_i \\ (C - \Sigma\hat{S}_i)^{\dagger} & 1 \end{bmatrix} \begin{bmatrix} \mathbf{a}^i \\ -\mathbf{a}^s \end{bmatrix} = \begin{bmatrix} \Upsilon^{\dagger}\left(f^0 - \Sigma\hat{f}^0\right) \\ \Xi^{\dagger}\left(f^0 - \Sigma\hat{f}^0\right) \end{bmatrix}, \tag{1.33}$$

where again the matrix $C - \Sigma\hat{S}_i$ is diagonal, and the incident field at the particle surface is augmented by the light scattered from the other particles in the system. Similarly, an optical analogue of the Dyson equation for the surface Green function of the composite system can be derived, $G_s = G_s^0 + G_s^0\Sigma' G_s$. Equivalently, using functions to expand the scattered space of the single particle which are themselves exact solutions of Maxwell's equations for an external region which includes all of the other particles in the system, can be applied. Embedding the particle next to a layered material, [74] or a more complex geometry such as a photonic crystal, results in equivalent expressions.

As the distances among the particles increase, the particles become independent from one another. This means that the off diagonal terms in (1.30) and the terms with the hats in (1.31) will vanish, so perturbative schemes for resolvent operators, $(1 + A)^{-1}$, could be used to provide approximate solutions. These are based on the binomial expansion and are convergent if the eigenvalues of A are

$$1 \geq \lambda(A)_1 \geqslant \ldots \lambda(A)_n \geqslant -1.$$

Metallic dimers are particularly interesting from the point of view of sensing applications, as a strong enhancement of the electromagnetic field can be achieved in the gap region between particles. However, a metallic dimer will always be only one of the elements to be considered in a sensing configuration, the other being the molecules to be detected, a point often overlooked. As a matter of fact, any type of sensing in which one uses metallic nanoparticles will detect a signal that results from the optical interaction, induced by the incident field, between the nanoparticle(s) and the molecule, assuming that these are not adsorbed by the particles. For instance, the basic model of an aperture-less SNOM used to study the near field of complex molecules of biological interest near metallic nanoparticles, consists of a dimer of metallic spheres with a nearby dielectric particle. The detected SNOM signal is due to the collective response of all the particles interacting through light. In this section we are interested in studying theoretically how the linear response of the complex system formed by the molecules and the metallic nanoparticles emerges from the optical interaction

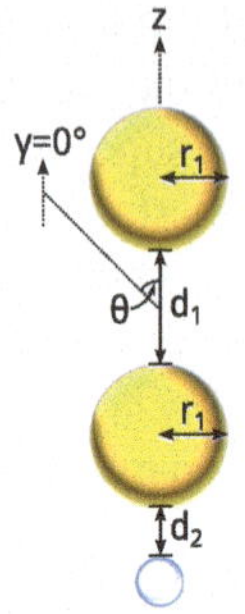

Fig. 1.17 Geometry of the three nanoparticle system, for two gold spheres of radius r_1 and separation d_1 with a smaller dielectric particle d_2 away from the second sphere. The incident orientation and polarisation directions (θ and γ) of the exciting light is also shown

of its constituent parts and depends on their geometrical and material properties. We analyse the simplest configuration relevant for experiments, which consists of a homogeneous medium containing two gold nanospheres and a smaller dielectric sphere representing a simplified small globular protein. The metallic spheres have a radius of 50 nm; the variations of the fields over spheres of this size, also known as retardation effects, are too large for the quasi-static approximation, but for this case could be taken into account exactly using the electromagnetic modes for the spheres, the Mie modes. For a given incident field and configuration of the spheres, internal and scattered fields can then be found using the Generalised Mie Theory [75], with each configuration of the spheres and each incident field requiring a new calculation.

We study instead how the collective response and resonances of this system of particles depends on the frequency of the incident field and the relative positions of the particles by forming, out of the modes of each particle, internal and scattered modes of the whole system that are coupled pairwise over all the surfaces of the particles using the multi-particle Gram matrix defined in (1.30).

The geometry of the system is shown in Fig. 1.17. The extinction and scattering cross sections of the metallic dimer as a function of the frequency for various values of the gap and for both local and non-local permittivity are shown in Fig. 1.18 for light with an incident electric field parallel to the dimer axis (the perpendicular case closely resembles the isolated sphere, since modes associated with the inter-particle gap cannot be strongly excited in this case). With the incident electric field parallel to the axis, extinction and scattering cross sections show a collective dipolar resonance at long wavelengths that moves towards longer wavelengths as the gap increases until it disappears, similar to that seen for the supershapes of Sect. 1.3.2; while the gap has no significant effect for incident electric field perpendicular to the dimer axis. A comparison of the scattering and extinction cross-sections for local and non-local responses is shown in Fig. 1.18b. As already noted by other authors [76, 77], the non-local permittivity affects the behaviour of the dimer most strongly for (very) small gaps when the incident electric field is parallel to the axis. We then add a small (10 nm radius) dielectric sphere of similar size and dielectric properties ($\varepsilon_r = 3.0$) to a globular protein [78], close to the outside of the dimer at a distance of $d_2 \sim$ 1–4 nm. The isolated dielectric particle does not exhibit any resonances in the range of 400–800 nm, and its addition does not visibly change the optical cross sections

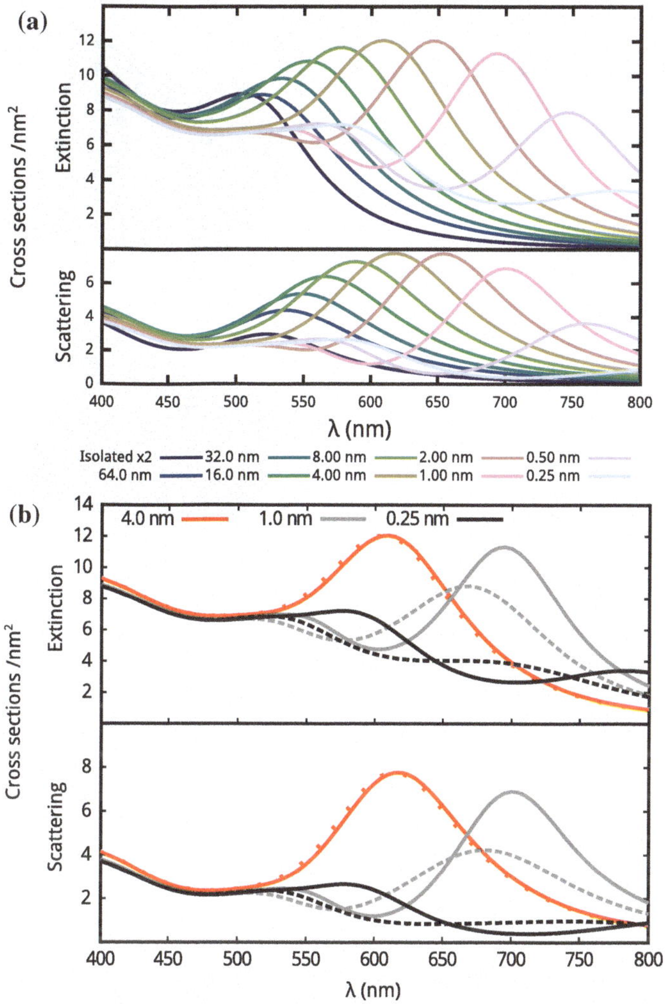

Fig. 1.18 50 nm radius gold sphere dimer in vacuum **a** Optical Spectra of metallic particles for a range of d_1 gaps, and the isolated sphere. **b** Effect of non-locality (dotted lines) on resonant features in the spectra compared to a local dielectric model (solid lines) for a range of inter-particle gaps, d_1

of the resulting three particle system from those of Fig. 1.18b for either local or non-local response models of the gold particles.

Figure 1.19a shows the mode landscape for the principal supermodes of the dimer with incident field consisting of a linearly polarised plane wave with electric field parallel the axis of symmetry and incident at 90° to the axis. From these two figures one can see how the presence of the small dielectric particle modifies the structure of the supermodes, even if the extinction cross section of the dielectric particle on its own is much smaller than the extinction cross section of the metallic dimer: the

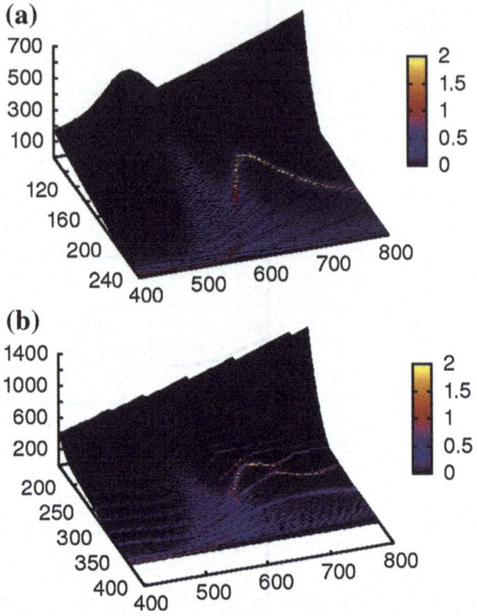

Fig. 1.19 **a** Principle modes of the metallic dimer. **b** Anti-crossing induced by the small neighbouring dielectric particle of Fig. 1.17

dominant supermode of the metallic dimers is replaced by a combination of three supermodes for the three-particle system, two of these modes showing an avoided crossing.

1.6 Conclusion

In this chapter we have reviewed the theory of principal modes that can be derived for any smooth particle starting from a set of distributed electric and magnetic multipoles. We have shown how this theory can be used to identify resonances in nanoparticles and to explain the metamorphosis of SNOM images when the frequency of the incident field is changed. The theory can also be extended to collective principal modes, or supermodes, of several nanoparticles interacting through light. This has shown that supermodes that efficiently transport energy in the far field can emerge from combination of principal modes of different particles that do not transport much energy into the far field individually. This is relevant for all types of microscopy that require the interaction of the object investigated with nearby metallic particles of tips, such as, for instance, SNOM and aperture less SNOM. We have also used principal modes to find how to control the internal and scattered fields on a particle, including those arising from inelastic multiphoton transition when the energy transferred from the pump frequency to the other frequencies by the non linear process is negligible. We are currently extending the numerical codes which implement this theory to deal with particles with lower or no symmetry.

1.7 Appendix

In this appendix we compare the validity of the numerical solutions to the scattering problem calculated via three different methods: the discrete source method using QR decomposition to solve (1.4) in a least squares sense (hereafter referred to as QR); construction of the principal modes by

1. Using QR decompositions to find orthonormal bases corresponding to (1.5), (1.6),
2. Forming the scalar product between the unitary matrices, U^i and U^s,
3. Finding the paired sets of principal modes via a singular value decomposition corresponding to (1.7),
4. Analytically solving for the $\mathbf{a}^i$ and $\mathbf{a}^s$ coefficients,

(this approach being referred to as the QR+SVD method); and finally, purely using singular value decompositions to find the orthonormal bases and then the principal modes (referred to as SVD+SVD). In all cases, standard linear algebra routines were used to perform the decompositions and matrix products [79].

To provide a fair comparison between the algorithms we limit the rank of the output spaces for each method, via regularisation, to be the same for all methods and then study the effect of incrementing this limit. Simulations were run for two distinct particle types, a nanodisc of radius 400 nm and depth 35 nm and a nanorod of length 400 nm and diameter 35 nm, both with rounded edges. Other than their geometries, the two particles differ in the type of sources used to represent the fields. For the rod, multipole sources are distributed along the symmetry axis in the real space, whereas for the disc the sources are located in the complex space—effectively making these

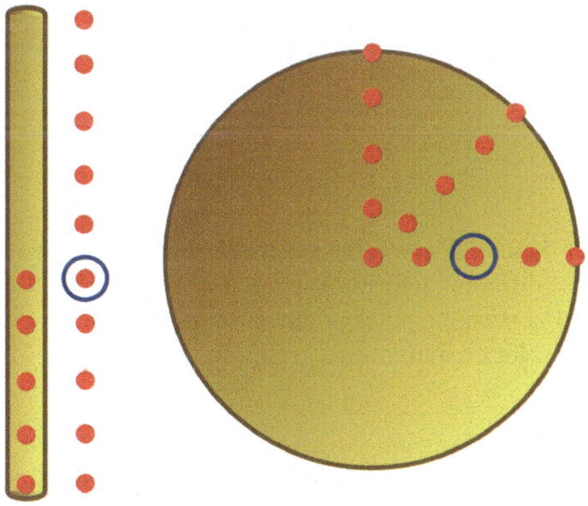

Fig. 1.20 Sampling points of near field excitation for a rounded gold nanodisc and nanorod. The red points indicate the approximate location of the near field source as it was scanned above the gold nanoparticles at a height of 50nm, for a rod with dimensions (l = 400 nm, d = 35 nm) and a disc (d = 800 nm, z = 35 nm). The blue circles indicate the location of the near field source used for the DSCS calculations. There are 15 sampling points for each particle (the centre of the disc was sampled with 3 different polarisations.) All of the following simulations were performed using these particles

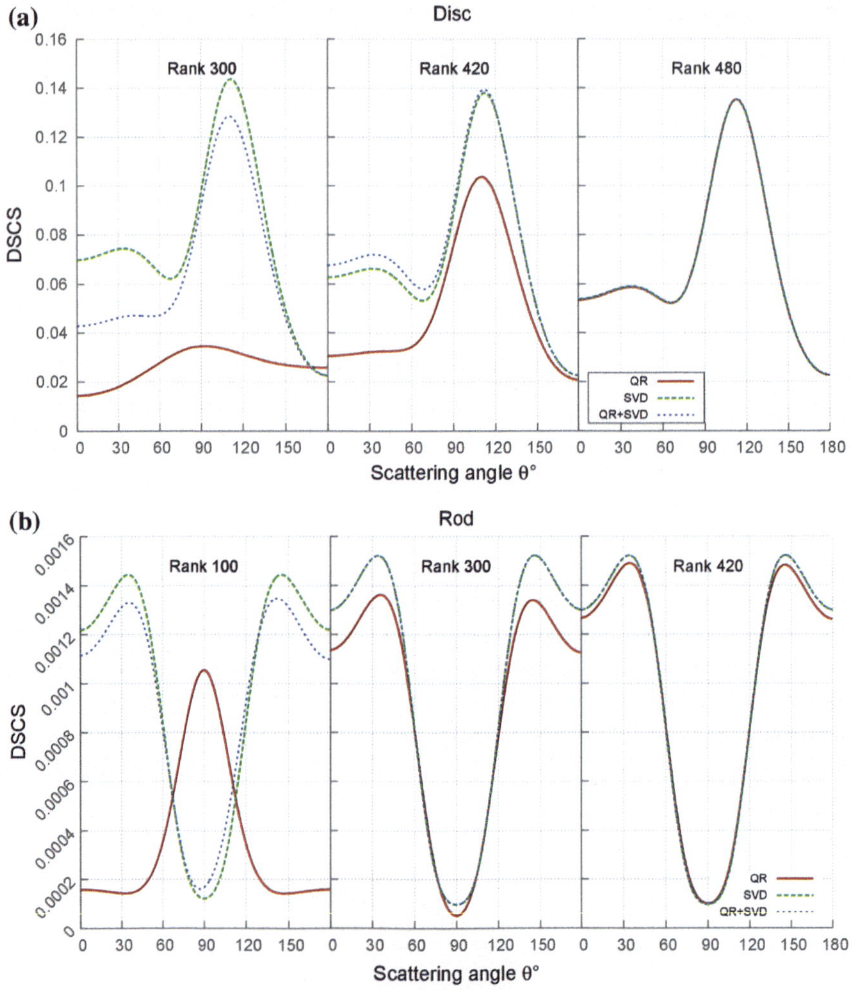

Fig. 1.21 Convergence of the DSCS along the generatrix line with increasing rank. The DSCS, in arbitrary units, for the three different algorithms plotted against the far field angle θ, varied incrementally between 0 and π between the poles of the particle's symmetry axis showing convergence with increasing rank of the solution matrices for a (**a**) disc and (**b**) rod

ring sources distributed concentrically along the particle radius. The particles were illuminated by a near field source of wavelength 720 nm comprised of a combination of electric and magnetic point dipoles located 50 nm above the particle surface. The approximate locations of the near field source, which is moved to obtain average values for some tests by using different locations and polarisations of the source, are highlighted in Fig. 1.20.

Firstly, we compare the convergence of the solutions by plotting the DSCS, i.e. the angular variation of the electric field intensity in the far field, for each of the

Table 1.1 Average total computational time for a full solution of the scattering problem at unrestricted rank for a single wavelength near-field excitation of the disc shaped particle using an AMD Opteron 6344 processor with a 2.6 GHz clock speed. Timings are averaged over 5 runs. For the QR+SVD and purely SVD methods we highlight the time taken for the initial calculation and also the subsequent calculations for the same particle where the principal modes are read back in from disc storage

Algorithm	Time (s)
QR	102
SVD+SVD	260
QR+SVD	194
Using stored modes	36

three methods by increasing the rank from an effective minimum. These results were obtained by calculating the light scattered by the excited particles into the far field along the generatrix line, $\phi = 0$, and sampling θ at equal intervals between the poles of the symmetry axis for the two particles at 0 and π, as shown in Fig. 1.21. We observe that for minimal rank there is an obvious advantage to the principal mode methods, which while not fully converged show the main features of the spectrum at the correct angles. The QR solution however, for both the rod and disc particles, fails to even approximately produce these features of the solution when the rank is minimal. As the rank is increased both Principal Mode methods converge more rapidly than the pure QR solution, which requires the maximum rank considered to show full convergence for the disc, and only an approximate convergence for the rod. Note that with these particular source configurations, the upper bound on the rank obtainable for SVD+SVD and QR+SVD when no limit is imposed is almost half that observed for the QR algorithm.

We have observed that for low rank solutions there is a clear advantage to using a method which splits the space into two subspaces, not only for the extra information about the system which this gives, but also for the accuracy of the calculations. There is also another advantage to using the principal mode methods, due to the sequential way in which the surface fields are calculated using SVD+SVD they can be stored to be used again for a different excitation of the same particle. While, for the initial calculation QR proves to be slightly quicker, as shown in Table 1.1, for multiple calculations the SVD+SVD and QR+SVD methods need only calculate the principal mode fields once and the subsequent calculations are then significantly faster by factors of $\sim$5 for QR+SVD and $\sim$7 for SVD+SVD.

References

1. D. Graham, R. Goodacre, Chem. Soc. Rev. **37**, 883 (2008)
2. T. Aoki, B. Dayan, E. Wilcut, W.P. Bowen, A.S. Parkins, T.J. Kippenberg, K.J. Vahala, H.J. Kimble, Nature **443**, 671 (2006)

3. B. Lukyanchuk, N. Zheludev, S. Maier, N. Halas, P. Nordlander, H. Giessen, C. tow Chong, Nat. Mater. **9**, 707 (2010)
4. N. Liu, L. Langguth, T. Weiss, J. Kstel, M. Fleischhauer, T. Pfau, H. Giessen, Nat. Mater. **8**, 758 (2009)
5. J. Schuller, E. Barnard, W. Cai, Y.C. Jun, J. White, M. Brongersma, Nat. Mater. **9**, 193 (2010)
6. Q. Zhao, J. Zhou, F. Zhang, D. Lippens, Mater. Today **12**(12), 60 (2009)
7. J. Pendry, D. Schuring, D. Smith, Science **312**, 1780 (2006)
8. C. Jordan, Buletin de la Société Mathématique de France **3**, 103 (1875)
9. A. Knyazev, A. Jujusnashvili, M. Argentati, J. Funct. Anal. **259**, 1323 (2010)
10. D. McArthur, B. Hourahine, F. Papoff, Opt. Express **25**(4), 4162 (2017). https://doi.org/10.1364/OE.25.004162
11. A. Doicu, T. Wriedt, Y. Eremin, *Light Scattering by Systems of Particles* (Springer, Berlin, 2006)
12. K. Holms, B. Hourahine, F. Papoff, J. Opt. A: Pure Appl. Opt. **11**, 054009 (2009)
13. A. Aydin, A. Hizal, J. Math. Anal. Appl. **117**, 428 (1986)
14. V.S. Vladimirov, *Equations of Mathematical Physics* (MIR, Moscow, 1984)
15. D.N.G. Roy, L.S. Couchman, *Inverse Problems and Inverse Scattering of Plane Waves* (Academic Press, New York, 2001). (chap. 6)
16. A. Doicu, T. Wriedt, J. Opt. Soc. Am. A **16**(10), 2539 (1999)
17. A. Doicu, T. Wriedt, Opt. Commun. **139**, 85 (1997)
18. T. Rother, M. Kahnert, A. Doicu, J. Wauer, Prog. Electromagn. Res. **38**, 47 (2002)
19. S. Silver, Microwave antenna theory and design. Technical Report 12, M.I.T. Radiation Laboratory Series (1949)
20. C. Müller, *Grundprobleme Der Mathematischen Theorie Elektromagnetischer Schwingungen* (Springer, Berlin, 1957)
21. A. Knyazev, M. Argentati, SIAM J. Sci. Comput. **23**(6), 2008 (2002)
22. D. McArthur, B. Hourahine, F. Papoff, Phys. Sci. Int. J. **4**(4), 564 (2014)
23. E. Hannan, J. Aust. Math. Soc. **2**, 229 (1961/1962)
24. J. Jackson, *Classical Electrodynamics* (Wiley, New York, 1999)
25. B.F. Farrell, P.J. Ioannou, J. Atmos. Sci. **53**, 2025 (1996)
26. B. Hourahine, K. Holms, F. Papoff, J. Phys.: Conf. Ser. **367**, 012010 (2012)
27. G. New, J. Mod. Opt. **42**, 799 (1995)
28. W.J. Firth, A. Yao, Phys. Rev. Lett. **95**, 073903 (2005)
29. F. Papoff, G. D'Alessandro, G.L. Oppo, Phys. Rev. Lett. **100**, 123905 (2008)
30. B. Hourahine, F. Papoff, Opt. Express **21**, 20322 (2013)
31. M.I. Tribelsky, B.S. Lukyanchuk, Phys. Rev. Lett. **97**, 263902 (2006)
32. J.A. Stratton, L.J. Chu, Phys. Rev. **56**, 99 (1939)
33. C.F. Bohren, D.R. Huffman, *Absorption and Scattering of Light by Small Particles* (Wiley, New York, 1998)
34. F. Papoff, B. Hourahine, Opt. Express **19**, 21432 (2011)
35. G. Roll, G. Schweiger, J. Opt. Soc. Am. A **17**, 1301 (2000)
36. G. Mie, Leipzig Ann. Phys. **330**, 377 (1908)
37. Y. Han, Z. Wu, Appl. Opt. **40**, 2501 (2001)
38. M.I. Mishchenko, J.H. Hovernier, L.D. Travis (eds.), *Light Scattering by Nonspherical Particles: Theory, Measurements and Applications* (Academic Press, San Diego, 2000)
39. P.G. Etchegoin, E.C. Le Ru, M. Meyer, J. Chem. Phys. **125**, 164705 (2006)
40. P.G. Etchegoin, E.C. Le Ru, M. Meyer, J. Chem. Phys. **127**, 189901 (2007)
41. L.D. Landau, J. Phys. USSR **10**, 503 (1946). https://doi.org/10.1016/B978-0-08-010586-4.50067-5
42. H. Okamoto, K. Imura, Prog. Surf. Sci. **84**, 199 (2009)
43. J. Gielis, Am. J. Bot. **90**(3), 333 (2003)
44. B. Hourahine, F. Papoff, Meas. Sci. Technol. **23**, 084002 (2012)
45. K. Imura, K. Ueno, H. Misawa, H. Okamoto, D. McArthur, B. Hourahine, F. Papoff, Opt. Express **22**, 12189 (2014)

46. M. Abb, P. Albella, J. Aizpurua, O. Muskens, Nano Lett. **11**, 2457 (2011)
47. A. Kubo, K. Onda, H. Petek, Z. Sun, Y. Jung, H. Kim, Nano Lett. **5**, 1123 (2005)
48. M. Durach, A. Rusina, M. Stockman, Nano Lett. **7**, 3145 (2007)
49. M. Aeschlimann, M. D. Bayer, T. Brixner, F. Garcia de Abajo, W. Pfeiffer, M. Rohmer, C. Spindler, F. Steeb, Nature **446**, 301 (2007)
50. H. Noh, Y. Chomg, A. Stone, H. Cao, Phys. Rev. Lett. **108**, 186805 (2012)
51. R. Pierrat, C. Vandenbem, M. Fink, R. Carminati, Phys. Rev. A **87**, 041801 (2013)
52. J. Jeffers, J. Mod. Opt. **47**, 1819 (2000)
53. M. Shraddha, J. Heitz, T. Roger, N. Westerberg, D. Faccio, Opt. Lett. **39**, 5345 (2014)
54. J. Zhang, K. MacDonald, N. Zheludev, Light: Sci. Appl. **1**, e18 (2012)
55. S.G. Rodrigo, H. Harutyunyan, L. Novotny, Phys. Rev. Lett. **110**, 177405 (2013)
56. S. Roke, M. Bonn, A. Petukhov, Phys. Rev. B **70**, 115106 (2004)
57. F. Papoff, D. McArthur, B. Hourahine, Sci. Rep. **5** (2015). https://doi.org/10.1038/srep12040
58. F.X. Wang, F. Rodríguez, W. Albers, R. Ahorinta, J.E. Sipe, M. Kauranen, Phys. Rev. B **80**, 233402 (2009)
59. J. Butet, G. Bachelier, I. Russier-Antoine, C. Jonin, E. Benichou, P.F. Brevet, Phys. Rev. Lett. **105**, 077401 (2010)
60. A. Melnyk, M.J. Harrison, Phys. Rev. B **2**(4), 835 (1970)
61. R. Ruppin, Phys. Rev. B **11**, 2871 (1975)
62. C. David, F.G. de Abajo, J. Phys. Chem. **115**, 19470 (2011)
63. C. Cirací, R. Hill, J. Mock, Y. Urzhumov, A. Fernández-Domínguez, S. Maier, J. Pendry, A. Chilkoti, D.R. Smith, Science **337**, 1072 (2012)
64. A. Moreau, C. Cirací, R. Smith, Phys. Rev. B **87**, 045401 (2013)
65. T. Christensen, W. Yan, S. Raza, A.P. Jauho, N. Mortensen, M. Wubs, ACS Nano **2**, 1745 (2014)
66. A. Bostrom, G. Kristensson, S. Strom, in *Field Representations and Introduction to Scattering*, ed. by V. Varadan, A. Lakhtakia, V. Varadan (Elsevir Science, Amsterdam, 1991)
67. C. Cirací, E. Poutrina, M. Scalora, D.R. Smith, Phys. Rev. B **86**, 115451 (2012)
68. G. Bachelier, J. Butet, I. Russier-Antoine, C. Jonin, E. Beninchou, P. Brevet, Phys. Rev. B **82**, 235403 (2010)
69. T. Heinz, in *Nonlinear Surface Electromagnetic Phenomena*, ed. by H. Ponath, G. Stegeman (Elsevier, Amsterdam, 1991)
70. J. Dadap, J. Shan, T. Heinz, J. Opt. Soc. Am. B **21**, 1328 (2004)
71. A.D. Rakić, A.B. Djurišić, J.M. Elazar, M.L. Majewski, Appl. Opt. **37**(22), 5271 (1998). https://doi.org/10.1364/AO.37.005271
72. J.E. Inglesfield, J. Phys. C **14**, 3795 (1981)
73. G.A. Baraff, M.M. Schlüter, J. Phys. C **19**, 4383 (1986)
74. J.E. Inglesfield, J. Phys. A **31**, 8495 (1998)
75. G. Gouesbet, G. Grehan, J. Opt. A **1**, 706 (1999)
76. F.J.G. de Abajo, J. Phys. Chem. C **112**(46), 17983 (2008). https://doi.org/10.1021/jp807345h
77. P. Ginzburg, A.V. Zayats, ACS Nano **7**(5), 4334 (2013). https://doi.org/10.1021/nn400842m
78. J.W. Pitera, M. Falta, W.F. van Gunsteren, Biophys. J. **80**, 2546 (2001)
79. E. Anderson, Z. Bai, C. Bischof, S. Blackford, J. Demmel, J. Dongarra, J. Du Croz, A. Greenbaum, S. Hammarling, A. McKenney, D. Sorensen, *LAPACK Users' Guide*, 3rd edn. (Society for Industrial and Applied Mathematics, Philadelphia, PA, 1999)

Chapter 2
The Invariant Imbedding T Matrix Approach

Adrian Doicu and Thomas Wriedt

Abstract The theoretical foundation of the invariant imbedding **T**-matrix method is revised. We present a consistent analysis of the method, show the connection with the superposition **T**-matrix method, and derive new recurrence relations for **T**-matrix calculation. The first recurrence is a numerical method for integrating the Riccati equations by using the Pade approximation to the matrix exponential, while the second one relies on an integral-matrix approach.

2.1 Introduction

At the present time, the invariant imbedding **T**-matrix method seems to be the most efficient tool for analyzing the electromagnetic scattering by large and highly aspherical particles. The method initially proposed by Johnson [4], is based on an electromagnetic volume integral equation in spherical coordinates and iteratively computes the **T** matrix along the radial coordinate. To initialize the iterative procedure, the separation of variables method [1] or the null-field method [2] are used. In the first case, the initial **T** matrix corresponds to a sphere enclosed in the particle, while in the second case, the initial **T** matrix corresponds to a partial volume of the particle generated by the intersection between a sphere and the particle. Because the size parameter of the partial volume of the particle is larger than the size parameter of the sphere enclosed in the particle, the second combined approach is more efficient than the first one. As the volume integral equation is a Fredholm integral equation of the second kind, the method does not suffer from ill-posedness, and its performances are really remarkable. Excellent numerical results have been obtained

A. Doicu
Remote Sensing Technology Institute German Aerospace Centre (DLR), 82234
Oberpfaffenhofen, Germany
e-mail: adrian.doicu@dlr.de

T. Wriedt (✉)
Leibniz-Institut für Werkstofforientierte Technologien—IWT, Bremen, Germany
e-mail: thw@iwt.uni-bremen.de

T. Wriedt and Y. Eremin (eds.), *The Generalized Multipole Technique for Light Scattering*, Springer Series on Atomic, Optical, and Plasma Physics 99,
https://doi.org/10.1007/978-3-319-74890-0_2

for large spheroids and cylinders (size parameters up to 300), cylinders with large aspect ratios (20:1), inhomogeneous particles, and two-particle systems [1]. The invariant imbedding **T**-matrix method is not very efficient because a fine discretization along the radial coordinate has to be considered. To increase the numerical efficiency of the method and to reduce the computational burden, (I) the symmetry relations of the **T** matrix, such as, mirror and point-group symmetries have been exploited, and (II) parallelizations of the code based on the Message Passing Interface (MPI) technology and OPENMP have been used [1, 2].

This chapter which is merely of theoretical nature revised the fundamentals of the invariant imbedding **T**-matrix method. Our goal is to present a consistent analysis of the method, and to derive new recurrence relations for **T**-matrix calculation.

2.2 Mathematical Foundations

Consider a particle which is entirely contained within a sphere of radius R and interior D. Assume that for $\mathbf{r} \in D$, the particle is completely described through the relative refractive index $m_{\mathrm{r}}(\mathbf{r})$, while for $\mathbf{r} \notin D$, we have $m_{\mathrm{r}}(\mathbf{r}) = 1$. The invariant imbedding **T** matrix approach involves the following steps:

1. Derive an ordinary Fredholm integral equation for the radial amplitude vector by making use on a volume integral equation for the electric field and a spherical wave expansion of the free-space dyadic Green's function.
2. Derive a two-terms recurrence relation for the **T** matrix by discretizing the Fredholm integral equation with respect to the radial coordinate and by applying the invariant imbedding procedure to the discretized equation.
3. Derive a matrix Riccati equation for the **T** matrix by passing to the limit $\Delta R \to 0$ in the **T**-matrix recurrence relation, where ΔR is the radial grid spacing.

In the following we will derive the matrix Riccati equation for the **T** matrix by applying the invariant imbedding procedure on the continuous form of the Fredholm integral equation rather than on its discrete form. We will then derive two new recurrence relations for the **T** matrix, by using a numerical scheme for integrating the matrix Riccati equation and an integral-matrix approach.

2.2.1 The Volume Integral Equation in Spherical Coordinates

For the assumed geometry, the total field electric field **E** solves the volume integral equation

$$\mathbf{E}(\mathbf{r}) = \mathbf{E}_0(\mathbf{r}) + \int_D \chi(\mathbf{r}')\overline{\mathbf{G}}(\mathbf{r},\mathbf{r}') \cdot \mathbf{E}(\mathbf{r}')\,\mathrm{d}V', \tag{2.1}$$

where $\mathbf{E}_0$ is the incident field, $\overline{\mathbf{G}}(\mathbf{r}, \mathbf{r}')$ is the free-space dyadic Green's function and $\chi(\mathbf{r}) = k_\mathrm{s}^2(m_\mathrm{r}^2(\mathbf{r}) - 1)$. The dyadic $\overline{\mathbf{G}}$ can be expressed as [3]

$$\overline{\mathbf{G}}(\mathbf{r}, \mathbf{r}') = \overline{\mathbf{G}}_0(\mathbf{r}, \mathbf{r}') - \frac{1}{k_\mathrm{s}^2}\delta(\mathbf{r} - \mathbf{r}')\mathbf{e}_r \otimes \mathbf{e}_r, \tag{2.2}$$

where

$$\overline{\mathbf{G}}_0(\mathbf{r}, \mathbf{r}') = \frac{\mathrm{j}k_\mathrm{s}}{\pi} \sum_\alpha \begin{cases} \mathbf{M}_\alpha^3(k_\mathrm{s}\mathbf{r}) \otimes \mathbf{M}_{\overline{\alpha}}^1(k_\mathrm{s}\mathbf{r}') \\ +\mathbf{N}_\alpha^3(k_\mathrm{s}\mathbf{r}) \otimes \mathbf{N}_{\overline{\alpha}}^1(k_\mathrm{s}\mathbf{r}'), \quad r > r' \\ \\ \mathbf{M}_\alpha^1(k_\mathrm{s}\mathbf{r}) \otimes \mathbf{M}_{\overline{\alpha}}^3(k_\mathrm{s}\mathbf{r}') \\ +\mathbf{N}_\alpha^1(k_\mathrm{s}\mathbf{r}) \otimes \mathbf{N}_{\overline{\alpha}}^3(k_\mathrm{s}\mathbf{r}'), \quad r < r' \end{cases}. \tag{2.3}$$

Here, α is a multiindex incorporating the azimuthal-mode index m and the expansion-order index n, i.e., $\alpha = (m, n)$, while $\overline{\alpha} = (-m, n)$. In the following it is convenient to represent the vector $\mathbf{X} = X_r\mathbf{e}_r + X_\theta\mathbf{e}_\theta + X_\varphi\mathbf{e}_\varphi$ by the column vector $\mathscr{X} = [X_r, X_\theta, X_\varphi]^T$. In this context, the column vector representation of the electric field expansion $\mathbf{E}(\mathbf{r}) = \sum_\alpha a_\alpha\mathbf{M}_\alpha^{1,3}(k\mathbf{r}) + b_\alpha\mathbf{N}_\alpha^{1,3}(k\mathbf{r})$ is $\mathscr{E}(\mathbf{r}) = \sum_\alpha \mathbf{Y}_\alpha(\boldsymbol{\Omega})\mathbf{X}_\alpha^{1,3}(r)\mathbf{c}_\alpha$, where $\boldsymbol{\Omega} = (\theta, \varphi)$, $\mathbf{Y}_\alpha$ is a 3×3 matrix depending on the angular functions, $\mathbf{X}_\alpha^{1,3}$ are 3×2 matrices depending on the spherical Bessel and Hankel functions, and $\mathbf{c}_\alpha = [a_\alpha, b_\alpha]^T$ is a 2×1 matrix depending on the expansion coefficients of the electric field. For regular and radiating spherical vector wave functions, we will write $\mathbf{X}_\alpha^1(r) = \mathbf{J}_\alpha(r)$ and $\mathbf{X}_\alpha^3(r) = \mathbf{H}_\alpha(r)$, respectively, where the expressions of $\mathbf{J}_\alpha$, $\mathbf{H}_\alpha$ and $\mathbf{Y}_\alpha$ can be found in [4]. Inserting (2.2) into (2.1) yields the matrix form representation of the volume integral equation

$$\begin{aligned} \overline{\mathscr{E}}(\mathbf{r}) &= \mathscr{E}_0(\mathbf{r}) \\ &+ \sum_\alpha \mathbf{Y}_\alpha(\boldsymbol{\Omega}) \int_V \chi(\mathbf{r}')\mathbf{g}_\alpha(r, r')\mathbf{Y}_{\overline{\alpha}}^T(\boldsymbol{\Omega}')\mathbf{Z}(\mathbf{r}')\overline{\mathscr{E}}(\mathbf{r}')\,\mathrm{d}V', \end{aligned} \tag{2.4}$$

where $\overline{\mathscr{E}}$ is defined by $\mathscr{E}(\mathbf{r}) = \mathbf{Z}(\mathbf{r})\overline{\mathscr{E}}(\mathbf{r})$ with $\mathbf{Z}(\mathbf{r}) = \mathrm{diag}[1/m_\mathrm{r}^2(\mathbf{r}), 1, 1]$, $\mathscr{E}_0$ and $\mathscr{E}$ are the column vector representation of $\mathbf{E}_0$ and $\mathbf{E}$, respectively, and $\mathbf{g}_\alpha$ is the 3×3 radial Green's function matrix given by [4]

$$\mathbf{g}_\alpha(r, r') = \frac{\mathrm{j}k_\mathrm{s}}{\pi} \begin{cases} \mathbf{H}_\alpha(r)\mathbf{J}_{\overline{\alpha}}^T(r'), \quad r > r' \\ \\ \mathbf{J}_\alpha(r)\mathbf{H}_{\overline{\alpha}}^T(r'), \quad r < r' \end{cases}. \tag{2.5}$$

The column vector representation of the scattered field

$$\mathbf{E}_\mathrm{s}(\mathbf{r}) = \int_D \chi(\mathbf{r}')\overline{\mathbf{G}}_0(\mathbf{r}, \mathbf{r}') \cdot \mathbf{E}(\mathbf{r}')\,\mathrm{d}V' \tag{2.6}$$

reads as

$$\mathscr{E}_s(\mathbf{r}) = \sum_{\alpha} \mathbf{Y}_{\alpha}(\boldsymbol{\Omega})\mathbf{H}_{\alpha}(r)\mathbf{c}_{s\alpha}, \tag{2.7}$$

where, for $\mathscr{E}_0(\mathbf{r}) = \sum_{\alpha} \mathbf{Y}_{\alpha}(\boldsymbol{\Omega})\mathbf{J}_{\alpha}(r)\mathbf{c}_{0\alpha}$, the expansion coefficients of the scattered field

$$\mathbf{c}_{s\alpha} = \frac{jk_s}{\pi} \int_D \chi(\mathbf{r})\mathbf{J}_{\bar{\alpha}}^T(r)\mathbf{Y}_{\bar{\alpha}}^T(\boldsymbol{\Omega})\mathbf{Z}(\mathbf{r})\overline{\mathscr{E}}(\mathbf{r})\,\mathrm{d}V \tag{2.8}$$

are related to the expansion coefficients of the incident field by the **T**-matrix equation

$$\mathbf{c}_{s\alpha} = \sum_{\beta} \mathbf{T}_{\alpha\beta}\mathbf{c}_{0\beta}. \tag{2.9}$$

2.2.2 *An Ordinary Integral Equation*

The volume integral equation (2.4) can be transformed into an ordinary integral equation for the radial amplitude vector

$$\mathbf{F}_{\alpha}(r) = r^2 \int_{\Omega} \chi(\mathbf{r})\mathbf{Y}_{\bar{\alpha}}^T(\boldsymbol{\Omega})\mathbf{Z}(\mathbf{r})\overline{\mathscr{E}}(\mathbf{r})\,\mathrm{d}\Omega, \tag{2.10}$$

which is the analog of the scalar amplitude density function in the quantum mechanical scattering theory [4]. Inserting (2.4) into (2.10) gives the desired equation

$$\begin{aligned} \mathbf{F}_{\alpha}(r) &= \sum_{\beta} \mathbf{U}_{\alpha\beta}(r)\mathbf{J}_{\beta}(r)\mathbf{c}_{0\beta} \\ &+ \int_0^R \sum_{\beta} \mathbf{U}_{\alpha\beta}(r)\mathbf{g}_{\beta}(r, r')\mathbf{F}_{\beta}(r')\,\mathrm{d}r', \end{aligned} \tag{2.11}$$

where the 3×3 matrix $\mathbf{U}_{\alpha\beta}$ is defined by

$$\mathbf{U}_{\alpha\beta}(r) = r^2 \int_{\Omega} \chi(\mathbf{r})\mathbf{Y}_{\bar{\alpha}}^T(\boldsymbol{\Omega})\mathbf{Z}(\mathbf{r})\mathbf{Y}_{\beta}(\boldsymbol{\Omega})\,\mathrm{d}\Omega. \tag{2.12}$$

The scattered field coefficients (2.8) can be also expressed in terms of $\mathbf{F}_{\alpha}$, and the result is

$$\mathbf{c}_{s\alpha} = \frac{jk_s}{\pi} \int_0^R \mathbf{J}_{\bar{\alpha}}^T(r)\mathbf{F}_{\alpha}(r)\,\mathrm{d}r. \tag{2.13}$$

Now, let the 3×2 matrix $\mathbf{F}_{\alpha\beta}$ be the solution of the integral equation

$$\mathbf{F}_{\alpha\beta}(r) = \mathbf{U}_{\alpha\beta}(r)\mathbf{J}_{\beta}(r) + \int_0^R \sum_{\gamma} \mathbf{U}_{\alpha\gamma}(r)\mathbf{g}_{\gamma}(r, r')\mathbf{F}_{\gamma\beta}(r')\,\mathrm{d}r'. \tag{2.14}$$

Obviously, $\mathbf{F}_{\alpha}$ and $\mathbf{F}_{\alpha\beta}$ solve a Fredholm integral equation with the same kernel but with different forcing (matrix) functions. The forcing function of (2.14) is $\mathbf{A}_{\alpha\beta} = \mathbf{U}_{\alpha\beta}\mathbf{J}_{\beta}$, while the forcing function of (2.11) is $\mathbf{A}_{\alpha} = \sum_{\beta} \mathbf{A}_{\alpha\beta}\mathbf{c}_{0\beta}$. Consequently, we have $\mathbf{F}_{\alpha}(r) = \sum_{\beta} \mathbf{F}_{\alpha\beta}(r)\mathbf{c}_{0\beta}$, and from (2.13) we obtain

$$\mathbf{c}_{\mathrm{s}\alpha} = \frac{\mathrm{j}k_{\mathrm{s}}}{\pi} \sum_{\beta} \left[\int_0^R \mathbf{J}_{\bar{\alpha}}^T(r)\mathbf{F}_{\alpha\beta}(r)\,\mathrm{d}r \right] \mathbf{c}_{0\beta}. \tag{2.15}$$

Finally, in view of (2.9) and (2.15), we infer that

$$\mathbf{T}_{\alpha\beta} = \frac{\mathrm{j}k_{\mathrm{s}}}{\pi} \int_0^R \mathbf{J}_{\bar{\alpha}}^T(r)\mathbf{F}_{\alpha\beta}(r)\,\mathrm{d}r. \tag{2.16}$$

Thus, the computation of the 2×2 block matrix elements of the transition matrix requires first, the solution of the Fredholm integral equation (2.14) for $\mathbf{F}_{\alpha\beta}$, and second, the integration of $\mathbf{F}_{\alpha\beta}$ by using (2.16).

Defining the global matrices (or supermatrices) $\mathbf{U} = [\mathbf{U}_{\alpha\beta}]$, $\mathbf{F} = [\mathbf{F}_{\alpha\beta}]$, $\mathbf{J} = [\mathbf{J}_{\alpha}\delta_{\alpha\beta}]$, $\mathbf{H} = [\mathbf{H}_{\alpha}\delta_{\alpha\beta}]$, $\mathbf{g} = [\mathbf{g}_{\alpha}\delta_{\alpha\beta}]$, and $\mathbf{T} = [\mathbf{T}_{\alpha\beta}]$, we express (2.14), (2.16) and (2.5) as

$$\mathbf{F}(r, R) = \mathbf{U}(r)\mathbf{J}(r) + \int_0^R \mathbf{U}(r)\mathbf{g}(r, r')\mathbf{F}(r')\,\mathrm{d}r', \tag{2.17}$$

$$\mathbf{T}(R) = \frac{\mathrm{j}k_{\mathrm{s}}}{\pi} \int_0^R \mathbf{J}^T(r)\mathbf{F}(r, R)\,\mathrm{d}r, \tag{2.18}$$

and

$$\mathbf{g}(r, r') = \frac{\mathrm{j}k_{\mathrm{s}}}{\pi} \begin{cases} \mathbf{H}(r)\mathbf{J}^T(r'), & r > r' \\ \mathbf{J}(r)\mathbf{H}^T(r'), & r < r' \end{cases}, \tag{2.19}$$

respectively, where the transposed matrices $\mathbf{J}^T$ and $\mathbf{H}^T$ should be understood as $\mathbf{J}^T = [\mathbf{J}_{\bar{\alpha}}^T\delta_{\alpha\beta}]$ and $\mathbf{H}^T = [\mathbf{H}_{\bar{\alpha}}^T\delta_{\alpha\beta}]$, respectively.

2.2.3 *The Matrix Riccati Equation*

In (2.17) and (2.18) we indicated the dependency of $\mathbf{F}$ and $\mathbf{T}$ on the length of the integration interval R. The reason is that the Fredholm integral equation (2.17) will be solved by using the invariant imbedding procedure, in which the interval length is parametrized.

Taking the derivative of (2.17) with respect to R, and using $\mathbf{g}(r, R) = (\mathrm{j}k_s/\pi)\,\mathbf{J}(r)\mathbf{H}^T(R)$ gives

$$\frac{\partial \mathbf{F}}{\partial R}(r, R) = \frac{\mathrm{j}k_s}{\pi}\mathbf{U}(r)\mathbf{J}(r)\mathbf{H}^T(R)\mathbf{F}(R, R) + \int_0^R \mathbf{U}(r)\mathbf{g}(r, r')\frac{\partial \mathbf{F}}{\partial R}(r', R)\,\mathrm{d}r'. \tag{2.20}$$

Because the Fredholm integral equations (2.17) and (2.20) have the same kernel but different forcing functions, we deduce that

$$\frac{\partial \mathbf{F}}{\partial R}(r, R) = \frac{\mathrm{j}k_s}{\pi}\mathbf{F}(r, R)\mathbf{H}^T(R)\mathbf{F}(R, R). \tag{2.21}$$

Taking the derivative of (2.18) with respect to R, and using (2.21) yields

$$\frac{\mathrm{d}\mathbf{T}}{\mathrm{d}R}(R) = \frac{\mathrm{j}k_s}{\pi}\mathbf{J}^T(R)\mathbf{F}(R, R) + \frac{\mathrm{j}k_s}{\pi}\mathbf{T}(R)\mathbf{H}^T(R)\mathbf{F}(R, R). \tag{2.22}$$

To find a representation for $\mathbf{F}(R, R)$ we set $r = R$ in (2.17) and use $\mathbf{g}(R, r') = (\mathrm{j}k_s/\pi)\mathbf{H}(R)\mathbf{J}^T(r')$; the result is

$$\mathbf{F}(R, R) = \mathbf{U}(R)\mathbf{J}(R) + \mathbf{U}(R)\mathbf{H}(R)\mathbf{T}(R). \tag{2.23}$$

Combining (2.22) and (2.23) we are led to the following matrix Riccati equation for the $\mathbf{T}$ matrix

$$\frac{\mathrm{d}\mathbf{T}}{\mathrm{d}R}(R) = \mathbf{Q}_{11}(R) + \mathbf{Q}_{12}(R)\mathbf{T}(R) + \mathbf{T}(R)\mathbf{Q}_{21}(R) + \mathbf{T}(R)\mathbf{Q}_{21}(R)\mathbf{T}(R), \tag{2.24}$$

where

$$\begin{aligned}
\mathbf{Q}_{11}(R) &= \frac{\mathrm{j}k_s}{\pi}\mathbf{J}^T(R)\mathbf{U}(R)\mathbf{J}(R),\\
\mathbf{Q}_{12}(R) &= \frac{\mathrm{j}k_s}{\pi}\mathbf{J}^T(R)\mathbf{U}(R)\mathbf{H}(R),\\
\mathbf{Q}_{21}(R) &= \frac{\mathrm{j}k_s}{\pi}\mathbf{H}^T(R)\mathbf{U}(R)\mathbf{J}(R),\\
\mathbf{Q}_{22}(R) &= \frac{\mathrm{j}k_s}{\pi}\mathbf{H}^T(R)\mathbf{U}(R)\mathbf{H}(R).
\end{aligned} \tag{2.25}$$

Substantial attention has been paid in the literature to the numerical integration of Riccati equations. Most of the numerical methods are based on the transformation of the matrix quadratic equation into a system of linear first-order differential equations by using Bernoulli substitution [5–7]. A stable numerical algorithm for this linear differential system is the modified Davison-Maki method of Kenney and Leipnik [6] also known as the Möbius scheme [8]. The algorithm is based on the following two-term recurrence relation with the radial step $\triangle R$:

$$\begin{aligned}\mathbf{T}(R) &= [(\mathbf{I} + \triangle R\mathbf{Q}_{12})\mathbf{T}(R-\triangle R) + \triangle R\mathbf{Q}_{11}] \\ &\quad \times [(\mathbf{I} - \triangle R\mathbf{Q}_{21}) - \triangle R\mathbf{Q}_{22}\mathbf{T}(R-\triangle R)]^{-1}, \end{aligned} \tag{2.26}$$

where the matrices $\mathbf{Q}_{11}$, $\mathbf{Q}_{12}$, $\mathbf{Q}_{21}$ and $\mathbf{Q}_{22}$ are evaluated at $R - \triangle R$.

2.2.4 A Recurrence Relation for the T matrix

A two-term recurrence relation for the $\mathbf{T}$ matrix, which can be regarded as an efficient algorithm for generating numerical solutions to the matrix Riccati equation (2.24), has been proposed by Johnson [4]. In this section we establish this recurrence relation by employing slightly different arguments as in [4]. We choose $\triangle R$ sufficiently small and approximate the integral of a function f over the interval $[R - \triangle R, R]$ by the right-endpoint quadrature formula $\int_{R-\triangle R}^{R} f(r)\,\mathrm{d}r \approx \triangle R f(R)$. For $\mathbf{F}(R, R)$, we have

$$\begin{aligned}\mathbf{F}(R,R) &= \mathbf{U}(R)\mathbf{J}(R) \\ &\quad + \frac{\mathrm{j}k_{\mathrm{s}}}{\pi}\mathbf{U}(R)\mathbf{H}(R)\int_0^{R-\triangle R}\mathbf{J}^T(r)\mathbf{F}(r,R)\,\mathrm{d}r \\ &\quad + \triangle R\mathbf{U}(R)\mathbf{g}(R,R)\mathbf{F}(R,R), \end{aligned} \tag{2.27}$$

and further

$$\mathbf{F}(R,R) = \frac{1}{\triangle R}\mathbf{Q}(R)\left[\mathbf{J}(R) + \mathbf{H}(R)\mathbf{q}(R)\right], \tag{2.28}$$

where the matrices $\mathbf{Q}$ and $\mathbf{q}$ are given by

$$\mathbf{Q}(R) = \triangle R\left[\mathbf{I} - \triangle R\mathbf{U}(R)\mathbf{g}(R,R)\right]^{-1}\mathbf{U}(R) \tag{2.29}$$

and

$$\mathbf{q}(R) = \frac{\mathrm{j}k_{\mathrm{s}}}{\pi}\int_0^{R-\triangle R}\mathbf{J}^T(r)\mathbf{F}(r,R)\,\mathrm{d}r, \tag{2.30}$$

respectively. For $\mathbf{T}(R)$, we proceed similarly and obtain

$$\mathbf{T}(R) = \overline{\mathbf{Q}}_{11}(R) + \left[\mathbf{I} + \overline{\mathbf{Q}}_{12}(R)\right]\mathbf{q}(R), \tag{2.31}$$

where here and in the following, the matrices $\overline{\mathbf{Q}}_{11}$, $\overline{\mathbf{Q}}_{12}$, $\overline{\mathbf{Q}}_{21}$ and $\overline{\mathbf{Q}}_{22}$ are given by (2.25) but with $\mathbf{Q}$ in place of $\mathbf{U}$.

Let us define the matrix $\mathbf{p}$ by the relation

$$\mathbf{p}(R) = \frac{\mathrm{j}k_{\mathrm{s}}}{\pi} \Delta R \mathbf{H}^T(R)\mathbf{F}(R, R), \tag{2.32}$$

which in view of (2.28), can be expressed as

$$\mathbf{p}(R) = \overline{\mathbf{Q}}_{21}(R) + \overline{\mathbf{Q}}_{22}(R)\mathbf{q}(R). \tag{2.33}$$

As $\mathbf{F}(r, R)$ and $\mathbf{F}(r, R - \Delta R)$ solve the same Fredholm integral equation but with different forcing functions, i.e.,

$$\mathbf{F}(r, R) = \mathbf{U}(r)\mathbf{J}(r)\left[\mathbf{I} + \frac{\mathrm{j}k_{\mathrm{s}}}{\pi}\Delta R \mathbf{H}^T(R)\mathbf{F}(R, R)\right] + \int_0^{R-\Delta R} \mathbf{U}(r)\mathbf{g}(r, r')\mathbf{F}(r', R)\,\mathrm{d}r' \tag{2.34}$$

and

$$\mathbf{F}(r, R - \Delta R) = \mathbf{U}(r)\mathbf{J}(r) + \int_0^{R-\Delta R} \mathbf{U}(r)\mathbf{g}(r, r')\mathbf{F}(r' - \Delta R)\,\mathrm{d}r' \tag{2.35}$$

respectively, we infer that

$$\mathbf{F}(r, R) = \mathbf{F}(r, R - \Delta R)\left[\mathbf{I} + \mathbf{p}(R)\right]. \tag{2.36}$$

Multiplying the equation for $\mathbf{T}(R - \Delta R)$,

$$\mathbf{T}(R - \Delta R) = \frac{\mathrm{j}k_{\mathrm{s}}}{\pi} \int_0^{R-\Delta R} \mathbf{J}^T(r)\mathbf{F}(r, R - \Delta R)\,\mathrm{d}r \tag{2.37}$$

from the right by $\mathbf{I} + \mathbf{p}(R)$, and using (2.30) and (2.36) yield

$$\mathbf{T}(R - \Delta R)\left[\mathbf{I} + \mathbf{p}(R)\right] = \mathbf{q}(R). \tag{2.38}$$

Solving (2.33) and (2.38) with respect to $\mathbf{q}$ gives

$$\mathbf{q}(R) = \left[\mathbf{I} - \mathbf{T}(R - \Delta R)\overline{\mathbf{Q}}_{22}(R)\right]^{-1} \times \mathbf{T}(R - \Delta R)\left[\mathbf{I} + \overline{\mathbf{Q}}_{21}(R)\right], \tag{2.39}$$

whence from (2.31), the desired recurrence relation

$$\mathbf{T}(R) = \overline{\mathbf{Q}}_{11} + \left(\mathbf{I} + \overline{\mathbf{Q}}_{12}\right)\left[\mathbf{I} - \mathbf{T}(R - \triangle R)\overline{\mathbf{Q}}_{22}\right]^{-1} \times \mathbf{T}(R - \triangle R)\left(\mathbf{I} + \overline{\mathbf{Q}}_{21}\right) \tag{2.40}$$

readily follows. It should be pointed out that in contrast to (2.26), the matrices $\overline{\mathbf{Q}}_{11}$, $\overline{\mathbf{Q}}_{12}$, $\overline{\mathbf{Q}}_{21}$ and $\overline{\mathbf{Q}}_{22}$ are evaluated at R.

2.2.5 An Integral-Matrix Approach

In this section we describe a combined integral and matrix approach to analyze the scattering by two concentric inhomogeneous spheres. This method, which is similar to the superposition **T**-matrix method, will enable us to derive a recurrence relation for the **T** matrix, without invoking the invariant imbedding procedure. Let us consider two concentric spheres of radii R_1 and $R_2 > R_1$, enclosing an inhomogeneous region. The interior of the sphere of radius R_1 is denoted by D_1, while the domain corresponding to the spherical shell is denoted by D_2. For $D = D_1 \cup D_2$, the matrix form representation of the volume integral equation read as

$$\begin{aligned}\overline{\mathscr{E}}(\mathbf{r}) &= \mathscr{E}_0(\mathbf{r}) \\ &+ \sum_\alpha \mathbf{Y}_\alpha(\boldsymbol{\Omega}) \int_{D_1} \chi(\mathbf{r}')\mathbf{g}_\alpha(r, r')\mathbf{Y}_{\bar{\alpha}}^T(\boldsymbol{\Omega}')\mathbf{Z}(\mathbf{r}')\overline{\mathscr{E}}(\mathbf{r}')\,\mathrm{d}V' \\ &+ \sum_\alpha \mathbf{Y}_\alpha(\boldsymbol{\Omega}) \int_{D_2} \chi(\mathbf{r}')\mathbf{g}_\alpha(r, r')\mathbf{Y}_{\bar{\alpha}}^T(\boldsymbol{\Omega}')\mathbf{Z}(\mathbf{r}')\overline{\mathscr{E}}(\mathbf{r}')\,\mathrm{d}V'.\end{aligned} \tag{2.41}$$

The field scattered by the inhomogeneous spherical shell in D_1 is given by ($\mathbf{r} \in D_1$)

$$\begin{aligned}\mathscr{E}_{\mathrm{s}2}(\mathbf{r}) &= \sum_\alpha \mathbf{Y}_\alpha(\boldsymbol{\Omega}) \int_{D_2} \chi(\mathbf{r}')\mathbf{g}_\alpha(r, r')\mathbf{Y}_{\bar{\alpha}}^T(\boldsymbol{\Omega}')\mathbf{Z}(\mathbf{r}')\overline{\mathscr{E}}(\mathbf{r}')\,\mathrm{d}V' \\ &= \sum_\alpha \mathbf{Y}_\alpha(\boldsymbol{\Omega})\mathbf{J}_\alpha(r)\mathbf{c}_{02\alpha},\end{aligned} \tag{2.42}$$

where

$$\mathbf{c}_{02\alpha} = \frac{\mathrm{j}k_{\mathrm{s}}}{\pi} \int_{R_1}^{R_2} \mathbf{H}_\alpha^T(r)\mathbf{F}_\alpha(r)\,\mathrm{d}r, \tag{2.43}$$

while, the field scattered by the inhomogeneous sphere of radius R_1 in D_2 is given by ($\mathbf{r} \in D_2$)

$$\begin{aligned}\mathscr{E}_{\mathrm{s1}}(\mathbf{r}) &= \sum_{\alpha} \mathbf{Y}_{\alpha}(\boldsymbol{\Omega}) \int_{D_1} \chi(\mathbf{r}')\mathbf{g}_{\alpha}(r,r')\mathbf{Y}^T_{\bar{\alpha}}(\boldsymbol{\Omega}')\mathbf{Z}(\mathbf{r}')\overline{\mathscr{E}}(\mathbf{r}')\,\mathrm{d}V' \\ &= \sum_{\alpha} \mathbf{Y}_{\alpha}(\boldsymbol{\Omega})\mathbf{H}_{\alpha}(r)\mathbf{c}_{\mathrm{s1}\alpha},\end{aligned} \tag{2.44}$$

where

$$\mathbf{c}_{\mathrm{s1}\alpha} = \frac{\mathrm{j}k_{\mathrm{s}}}{\pi} \int_0^{R_1} \mathbf{J}^T_{\alpha}(r)\mathbf{F}_{\alpha}(r)\,\mathrm{d}r. \tag{2.45}$$

By virtue of (2.44), the integral equation (2.41) inside the spherical shell can be written as ($\mathbf{r} \in D_2$)

$$\begin{aligned}\overline{\mathscr{E}}(\mathbf{r}) &= \sum_{\alpha} \mathbf{Y}_{\alpha}(\boldsymbol{\Omega}) \left[\mathbf{J}_{\alpha}(r)\mathbf{c}_{0\alpha} + \mathbf{H}_{\alpha}(r)\mathbf{c}_{\mathrm{s1}\alpha}\right] \\ &+ \sum_{\alpha} \mathbf{Y}_{\alpha}(\boldsymbol{\Omega}) \int_{D_2} \chi(\mathbf{r}')\mathbf{g}_{\alpha}(r,r')\mathbf{Y}^T_{\bar{\alpha}}(\boldsymbol{\Omega}')\mathbf{Z}(\mathbf{r}')\overline{\mathscr{E}}(\mathbf{r}')\,\mathrm{d}V'.\end{aligned} \tag{2.46}$$

It is easy to see that the scattering problem requires the solution of the following integral and matrix equations:

$$\begin{aligned}\mathbf{F}_{\alpha}(r) &= \sum_{\beta} \mathbf{U}_{\alpha\beta}(r) \left[\mathbf{J}_{\beta}(r)\mathbf{c}_{0\beta} + \mathbf{H}_{\beta}(r)\mathbf{c}_{\mathrm{s1}\beta}\right] \\ &+ \int_{R_1}^{R_2} \sum_{\beta} \mathbf{U}_{\alpha\beta}(r)\mathbf{g}_{\beta}(r,r')\mathbf{F}_{\beta}(r')\,\mathrm{d}r', \\ \mathbf{c}_{\mathrm{s1}\alpha} &= \sum_{\beta} \mathbf{T}^1_{\alpha\beta}\left(\mathbf{c}_{0\beta} + \mathbf{c}_{02\beta}\right), \\ \mathbf{c}_{02\alpha} &= \frac{\mathrm{j}k_{\mathrm{s}}}{\pi} \int_{R_1}^{R_2} \mathbf{H}^T_{\alpha}(r)\mathbf{F}_{\alpha}(r)\,\mathrm{d}r,\end{aligned} \tag{2.47}$$

where $\mathbf{T}^1_{\alpha\beta}$ is the transition matrix of the inhomogeneous sphere of radius R_1. The field scattered by the inhomogeneous sphere of radius R_2 is given by

$$\begin{aligned}\mathscr{E}_{\mathrm{s}}(\mathbf{r}) &= \sum_{\alpha} \mathbf{Y}_{\alpha}(\boldsymbol{\Omega})\mathbf{H}_{\alpha}(r)\mathbf{c}_{\mathrm{s1}\alpha} \\ &+ \frac{\mathrm{j}k_{\mathrm{s}}}{\pi} \sum_{\alpha} \mathbf{Y}_{\alpha}(\boldsymbol{\Omega})\mathbf{H}_{\alpha}(r) \int_{R_1}^{R_2} \mathbf{J}^T_{\alpha}(r')\mathbf{F}_{\alpha}(r')\,\mathrm{d}r' \\ &= \sum_{\alpha} \mathbf{Y}_{\alpha}(\boldsymbol{\Omega})\mathbf{H}_{\alpha}(r)\mathbf{c}_{\mathrm{s}\alpha},\end{aligned} \tag{2.48}$$

with

$$\mathbf{c}_{\mathrm{s}\alpha} = \mathbf{c}_{\mathrm{s1}\alpha} + \frac{\mathrm{j}k_{\mathrm{s}}}{\pi} \int_{R_1}^{R_2} \mathbf{J}^T_{\alpha}(r)\mathbf{F}_{\alpha}(r)\,\mathrm{d}r. \tag{2.49}$$

The integral equation in (2.47) describes the scattering by the inhomogeneous spherical shell, while the **T**-matrix equation describes the scattering by the inhomogeneous sphere of radius R_1. The third equation in (2.47) gives the expansion coefficients of the field scattered by the inhomogeneous spherical shell and exciting the inhomogeneous sphere of radius R_1. Finally, the scattered field (2.48) is a superposition of the fields that are scattered from the individual particles (the inhomogeneous sphere of radius R_1 and the spherical shell).

In terms of global matrices, (2.47) and (2.49) read as

$$\begin{aligned}\mathbf{F}(r) &= \mathbf{U}(r)\mathbf{J}(r)\mathbf{c}_0 + \mathbf{U}(r)\mathbf{H}(r)\mathbf{c}_{s1} \\ &\quad + \int_{R_1}^{R_2} \mathbf{U}(r)\mathbf{g}(r, r')\mathbf{F}(r')\,\mathrm{d}r', \\ \mathbf{c}_{s1} &= \mathbf{T}_1\left(\mathbf{c}_0 + \mathbf{c}_{02}\right), \\ \mathbf{c}_{02} &= \frac{\mathrm{j}k_s}{\pi}\int_{R_1}^{R_2} \mathbf{H}^T(r)\mathbf{F}(r)\,\mathrm{d}r,\end{aligned} \tag{2.50}$$

and

$$\mathbf{c}_s = \mathbf{c}_{s1} + \frac{\mathrm{j}k_s}{\pi}\int_{R_1}^{R_2} \mathbf{J}^T(r)\mathbf{F}(r)\,\mathrm{d}r \tag{2.51}$$

respectively, while the **T**-matrix equation takes the form $\mathbf{c}_s = \mathbf{T}\mathbf{c}_0$.

In (2.50) we choose $R_1 = R - \Delta R$ and $R_2 = R$, set $r = R$ and $\mathbf{T}_1 = \mathbf{T}(R - \Delta R)$, and apply the right-endpoint quadrature formula to compute the integrals $\int_{R-\Delta R}^{R} \cdots \mathrm{d}r$. Solving the resulting matrix equations yields the following representation for the **T** matrix of the inhomogeneous sphere of radius R:

$$\begin{aligned}\mathbf{T}(R) &= \mathbf{T}(R - \Delta R) \\ &\quad + \frac{\mathrm{j}k_s}{\pi}\Delta R\left[\mathbf{T}(R-\Delta R)\mathbf{H}^T + \mathbf{J}^T\right] \\ &\quad \times \left[\mathbf{I} - \Delta R\mathbf{U}\mathbf{g} - \frac{\mathrm{j}k_s}{\pi}\Delta R\mathbf{U}\mathbf{H}\mathbf{T}(R-\Delta R)\mathbf{H}^T\right]^{-1} \\ &\quad \times \mathbf{U}\left[\mathbf{J} + \mathbf{H}\mathbf{T}(R - \Delta R)\right].\end{aligned} \tag{2.52}$$

In terms of the matrix **Q** defined in (2.29), the recurrence relation (2.52) becomes

$$\begin{aligned}\mathbf{T}(R) &= \mathbf{T}(R - \Delta R) + \frac{\mathrm{j}k_s}{\pi}\left[\mathbf{T}(R-\Delta R)\mathbf{H}^T + \mathbf{J}^T\right] \\ &\quad \times \left[\mathbf{I} - \frac{\mathrm{j}k_s}{\pi}\mathbf{Q}\mathbf{H}\mathbf{T}(R-\Delta R)\mathbf{H}^T\right]^{-1} \\ &\quad \times \mathbf{Q}\left[\mathbf{J} + \mathbf{H}\mathbf{T}(R - \Delta R)\right].\end{aligned} \tag{2.53}$$

In the following we will show that the recurrences (2.40) and (2.53) are identical, which in turn implies, that the recurrences (2.40) and (2.52) are identical. Setting $\mathbf{T} = \mathbf{T}(R)$, $\mathbf{T}_1 = \mathbf{T}(R - \triangle R)$, and

$$\left(\mathbf{I} - \mathbf{T}_1 \overline{\mathbf{Q}}_{22}\right)^{-1} = \mathbf{I} + \mathbf{A}, \tag{2.54}$$

where

$$\mathbf{A} = \mathbf{T}_1 \overline{\mathbf{Q}}_{22} \left(\mathbf{I} - \mathbf{T}_1 \overline{\mathbf{Q}}_{22}\right)^{-1}, \tag{2.55}$$

and performing the matrix multiplications in (2.40) we obtain

$$\begin{aligned} \mathbf{T} = \mathbf{T}_1 + \overline{\mathbf{Q}}_{11} + \mathbf{T}_1 \overline{\mathbf{Q}}_{21} + \overline{\mathbf{Q}}_{12} \mathbf{T}_1 + \overline{\mathbf{Q}}_{12} \mathbf{T}_1 \overline{\mathbf{Q}}_{21} \\ + \mathbf{A} \mathbf{T}_1 + \mathbf{A} \mathbf{T}_1 \overline{\mathbf{Q}}_{21} + \overline{\mathbf{Q}}_{12} \mathbf{A} \mathbf{T}_1 + \overline{\mathbf{Q}}_{12} \mathbf{A} \mathbf{T}_1 \overline{\mathbf{Q}}_{21}. \end{aligned} \tag{2.56}$$

Similarly, setting $\alpha = jk_s/\pi$, and

$$\left(\mathbf{I} - \alpha \mathbf{Q} \mathbf{H} \mathbf{T}_1 \mathbf{H}^T\right)^{-1} = \mathbf{I} + \mathbf{B}, \tag{2.57}$$

where

$$\mathbf{B} = \alpha \mathbf{Q} \mathbf{H} \mathbf{T}_1 \mathbf{H}^T \left(\mathbf{I} - \alpha \mathbf{Q} \mathbf{H} \mathbf{T}_1 \mathbf{H}^T\right)^{-1}, \tag{2.58}$$

we express the recurrence relation (2.53) as

$$\begin{aligned} \mathbf{T} = \mathbf{T}_1 + \overline{\mathbf{Q}}_{11} + \mathbf{T}_1 \overline{\mathbf{Q}}_{21} + \overline{\mathbf{Q}}_{12} \mathbf{T}_1 + \mathbf{T}_1 \overline{\mathbf{Q}}_{22} \mathbf{T}_1 \\ + \alpha \mathbf{T}_1 \mathbf{H}^T \mathbf{B} \mathbf{Q} \mathbf{J} + \alpha \mathbf{T}_1 \mathbf{H}^T \mathbf{B} \mathbf{Q} \mathbf{H} \mathbf{T}_1 \\ + \alpha \mathbf{J}^T \mathbf{B} \mathbf{Q} \mathbf{J} + \alpha \mathbf{J}^T \mathbf{B} \mathbf{Q} \mathbf{H} \mathbf{T}_1. \end{aligned} \tag{2.59}$$

Using the matrix identity

$$\mathbf{X} \left(\mathbf{I} - \mathbf{Y} \mathbf{X}\right)^{-1} = \left(\mathbf{I} - \mathbf{X} \mathbf{Y}\right)^{-1} \mathbf{X}, \tag{2.60}$$

where $\mathbf{X}$ and $\mathbf{Y}$ are rectangular matrices, but $\mathbf{XY}$ and $\mathbf{YX}$ are square matrices, and taking into account the definition of the matrices $\overline{\mathbf{Q}}_{11}$, $\overline{\mathbf{Q}}_{12}$, $\overline{\mathbf{Q}}_{21}$ and $\overline{\mathbf{Q}}_{22}$, we find the following identities:

$$\begin{aligned} \alpha \mathbf{T}_1 \mathbf{H}^T \mathbf{B} \mathbf{Q} \mathbf{J} &= \mathbf{A} \mathbf{T}_1 \overline{\mathbf{Q}}_{21}, \\ \alpha \mathbf{T}_1 \mathbf{H}^T \mathbf{B} \mathbf{Q} \mathbf{H} \mathbf{T}_1 &= \mathbf{A} \mathbf{T}_1 - \mathbf{T}_1 \overline{\mathbf{Q}}_{22} \mathbf{T}_1, \\ \alpha \mathbf{J}^T \mathbf{B} \mathbf{Q} \mathbf{J} &= \overline{\mathbf{Q}}_{12} \mathbf{T}_1 \overline{\mathbf{Q}}_{21} + \overline{\mathbf{Q}}_{12} \mathbf{A} \mathbf{T}_1 \overline{\mathbf{Q}}_{21}, \\ \alpha \mathbf{J}^T \mathbf{B} \mathbf{Q} \mathbf{H} \mathbf{T}_1 &= \overline{\mathbf{Q}}_{12} \mathbf{A} \mathbf{T}_1. \end{aligned} \tag{2.61}$$

Substituting (2.61) into (2.59) yields (2.56), and the proof is finished.

From a practical point of view, the main results of our analysis are the recurrence relations (2.26), (2.40) and (2.52) for computing the **T** matrix. The recurrences (2.26) and (2.40) have been established by using an integral approach in conjunction with the invariant imbedding procedure, while the recurrence (2.52) has been derived by using an integral-matrix approach. The recurrence (2.40) requires two matrix inversions because the computation of the matrix **Q** demands an additional inversion step, while the recurrences (2.26) and (2.52) requires only one matrix inversion. Thus, (2.52) is more efficient than (2.40) in terms of the computational speed.

2.3 Conclusions

In this chapter, we revised the theoretical foundation of the invariant imbedding **T**-matrix method and established the connection with the superposition **T**-matrix method. Moreover, we derived two new recurrence relations for the **T** matrix: recurrence (2.26) which is essentially the modified Davison-Maki method with Pade approximation to the matrix exponential for solving Riccati differential equations, and recurrence (2.52) which has been obtained by using an integral-matrix approach. Although recurrence (2.52) will not dramatically increase the computation speed (as compared to Johnson's recurrence (2.40)), it appears to be of beneficial use for particles with very large size parameters. As a matter of fact, any numerical method for solving Riccati differential equations can be used to design new recurrence relations. Here we think about methods which transform Riccati differential equation into two coupled nonlinear equations (Chandrasekhar system), methods based on the superposition property of the Riccati solutions, and matrix versions of the ordinary differential equations methods.

References

1. L. Bi, P. Yang, G. Kattawar, M. Mishchenko, J. Quant. Spectrosc. Radiat. Trans. **116**, 169 (2013)
2. L. Bi, P. Yang, G. Kattawar, M. Mishchenko, J. Quant. Spectrosc. Radiat. Trans. **123**, 17 (2013)
3. L. Tsang, J. Kong, R. Shin, *Theory of Microwave Remote Sensing* (Wiley, New York, 1985)
4. B. Johnson, Appl. Opt. **27**, 4861 (1988)
5. E. Davison, M. Maki, IEEE Trans. Autom. Control **18**, 71 (1973)
6. C. Kenney, R. Leipnik, IEEE Trans. Autom. Control **30**, 962 (1985)
7. R. Leipnik, Bull. Aust. Math. Soc. **26**, 355 (1986)
8. J. Schiff, S. Shnider, SIAM J. Numer. Anal. **36**, 1392 (1999)

Chapter 3
Methods for Electromagnetic Scattering by Large Axisymmetric Particles with Extreme Geometries

Adrian Doicu, Yuri Eremin, Dmitry S. Efremenko and Thomas Trautmann

Abstract Several methods for electromagnetic scattering by large axisymmetric particles with extreme geometries are analyzed. These include the discrete sources method and the null-field method with distributed and multiple spherical vector wave functions, as well as, a single spherical coordinate-based null-field method equipped with an analytical approach for computing the elements of the transition matrix. The numerical performances of the methods with distributed and multiple spherical vector wave functions are illustrated through simulations for spheroids and cylinders.

3.1 Introduction

Accurate computation of electromagnetic scattering by large axisymmetric particles with extreme geometries is needed in atmospheric radiative transfer and remote sensing to analyze the parameters of radiation scattered by aerosols, clouds, and precipitations. In the last years, the discrete sources method and the null-field method have become efficient and powerful tools for rigorously computing the electromagnetic scattering by such kind of scatterers.

1. In the discrete sources method, the approximate solution to the scattering problem is written as a superposition of fields of elementary (discrete) sources. The discrete sources, consisting of localized, distributed and multiple spherical vector wave functions, magnetic and electric dipoles, and systems of vector Mie potentials, are placed on a certain support, while the unknown amplitudes of the discrete sources are determined from the boundary conditions [1–4].

A. Doicu (✉) · D. S. Efremenko · T. Trautmann
Remote Sensing Technology Institute, German Aerospace Centre (DLR), 82234 Oberpfaffenhofen, Germany
e-mail: adrian.doicu@dlr.de

Y. Eremin
Lomonosov Moscow State University, Lenin's Hills Moscow, Russia
e-mail: eremin@cs.msu.su

T. Wriedt and Y. Eremin (eds.), *The Generalized Multipole Technique for Light Scattering*, Springer Series on Atomic, Optical, and Plasma Physics 99,
https://doi.org/10.1007/978-3-319-74890-0_3

2. In the null-field method (originally proposed by Watermann [5, 6] and later developed by Barber and Hill [7], and Mishchenko et al. [8]), the transition matrix relating the expansion coefficients of the scattered and incident field is computed. A number of modifications to the conventional null-field method have been proposed, especially to improve the numerical stability in computations for large particles with extreme geometries. These techniques include methods (I) dealing with the numerical stability of the inversion process [6, 9–12], (II) based on accurate computations of the **T**-matrix elements [13–20], (III) relying on formal modifications of the single spherical coordinate-based null-field method [21–24], and (IV) using discrete sources [1, 25]. The fundamentals of the null-field method with discrete sources have been presented in [1], while convergent results for prolate axisymmetric particles with a size parameter of about 100, and oblate axisymmetric particles with a size parameter of about 30 have been reported in [25].

In the first step of our analysis we focus on the discrete sources method and the null-field method with distributed and multiple spherical vector wave functions, and discuss the numerical and theoretical improvements, which are required to handle large and highly aspherical axisymmetric particles. In a second step we present an analytical method for computing the **T**-matrix elements, which may presumably improve the numerical stability of the null-field method with localized spherical vector wave functions.

3.2 Discrete Sources

We consider the scattering by a homogeneous and isotropic particle embedded in a homogeneous, isotropic and nonabsorbing medium. Let D_{i} be the domain occupied by the particle, D_{s} the exterior of D_{i}, S the boundary of D_{i} (the particle surface), and $\mathbf{n}$ the unit normal vector to S directed into D_{s}. The wave number in the free space is k_0, while the wave number in domain D_{t}, t = s, i, is $k_{\mathrm{t}} = k_0\sqrt{\varepsilon_{\mathrm{t}}\mu_{\mathrm{t}}}$. Here, ε_{t} and μ_{t} are the relative permittivity and permeability of the domain D_{t}, respectively. The particle is illuminated by an incident field $\mathbf{E}_0$ and $\mathbf{H}_0$, representing an entire solution to the Maxwell equations. The transmission boundary-value problem to be solved consists in the computation of the vector fields $\mathbf{E}_{\mathrm{s}}$, $\mathbf{H}_{\mathrm{s}}$ and $\mathbf{E}_{\mathrm{i}}$, $\mathbf{H}_{\mathrm{i}}$ satisfying the Maxwell equations

$$\nabla \times \mathbf{E}_{\mathrm{t}} = \mathrm{j}k_0\mu_{\mathrm{t}}\mathbf{H}_{\mathrm{t}}, \quad \nabla \times \mathbf{H}_{\mathrm{t}} = -\mathrm{j}k_0\varepsilon_{\mathrm{t}}\mathbf{E}_{\mathrm{t}}, \tag{3.1}$$

in D_{t}, t = s, i, the two transmission conditions

$$\begin{aligned} \mathbf{n} \times \mathbf{E}_{\mathrm{i}} - \mathbf{n} \times \mathbf{E}_{\mathrm{s}} &= \mathbf{n} \times \mathbf{E}_0, \\ \mathbf{n} \times \mathbf{H}_{\mathrm{i}} - \mathbf{n} \times \mathbf{H}_{\mathrm{s}} &= \mathbf{n} \times \mathbf{H}_0, \end{aligned} \tag{3.2}$$

on S, and the Silver-Miller radiation condition for the scattered field

$$\frac{\mathbf{r}}{r} \times \sqrt{\mu_{\mathrm{s}}}\mathbf{H}_{\mathrm{s}} + \sqrt{\varepsilon_{\mathrm{s}}}\mathbf{E}_{\mathrm{s}} = o\left(\frac{1}{r}\right), \text{ as } r \to \infty, \tag{3.3}$$

uniformly for all directions $\mathbf{r}/r$.

As discrete sources, denoted by$\boldsymbol{\Phi}_{\alpha}^{q}(k\mathbf{r})$ and $\boldsymbol{\Psi}_{\alpha}^{q}(k\mathbf{r})$, with $q = 1, 3$, we consider different types of spherical vector wave functions (SVWF). In particular, $\boldsymbol{\Phi}_{\alpha}^{q}$ and $\boldsymbol{\Psi}_{\alpha}^{q}$, having the properties:

1. $\nabla \times \boldsymbol{\Phi}_{\alpha}^{q} = k\boldsymbol{\Psi}_{\alpha}^{q}$ and $\nabla \times \boldsymbol{\Psi}_{\alpha}^{q} = k\boldsymbol{\Phi}_{\alpha}^{q}$,
2. $\boldsymbol{\Phi}_{\alpha}^{1}$ and $\boldsymbol{\Psi}_{\alpha}^{1}$ are finite at the origin,
3. $\boldsymbol{\Phi}_{\alpha}^{3}$ and $\boldsymbol{\Psi}_{\alpha}^{3}$ satisfy the radiation condition,

stand for

1. the localized SVWF (localized multipoles) $\boldsymbol{\Phi}_{\alpha}^{q}(k\mathbf{r}) = \mathbf{M}_{mn}^{q}(k\mathbf{r})$ and $\boldsymbol{\Psi}_{\alpha}^{q}(k\mathbf{r}) = \mathbf{N}_{mn}^{q}(k\mathbf{r})$, where $\alpha = (m, n)$ for $m \in \mathbb{Z}$ and $n \geq \max(1, |m|)$,
2. the distributed SVWF (lowest-order multipoles) $\boldsymbol{\Phi}_{\alpha}^{q}(k\mathbf{r}) = \mathbf{M}_{m,|m|+l}^{q}(k(\mathbf{r} - z_n\mathbf{e}_{\mathrm{z}}))$ and $\boldsymbol{\Psi}_{\alpha}^{q}(k\mathbf{r}) = \mathbf{N}_{m,|m|+l}^{q}(k(\mathbf{r} - z_n\mathbf{e}_{\mathrm{z}}))$, where $\{z_n \mid n \geq 1\}$ is a dense set of points situated on the z-axis, $\mathbf{e}_{\mathrm{z}}$ is the unit vector in the direction of the z-axis, $l = 1$ if $m = 0$ and $l = 0$ if $m \neq 0$, and $\alpha = (m, n)$ for $m \in \mathbb{Z}$ and $n \geq 1$,
3. the multiple SVWF (multiple multipoles) $\boldsymbol{\Phi}_{\alpha}^{q}(k\mathbf{r}) = \mathbf{M}_{mn}^{q}(k(\mathbf{r} - z_p\mathbf{e}_{\mathrm{z}}))$ and $\boldsymbol{\Psi}_{\alpha}^{q}(k\mathbf{r}) = \mathbf{N}_{mn}^{q}(k(\mathbf{r} - z_p\mathbf{e}_{\mathrm{z}}))$, where $\{z_p \mid p = 1, 2, \ldots, N_{\mathrm{p}}\}$ is a finite set of points (poles) situated on the z-axis, N_{p} is the number of poles, and $\alpha = (m, n)$ for $m \in \mathbb{Z}$ and $n \geq \max(1, |m|)$.

The explicit expressions of the spherical vector wave functions with an orgin at $\widehat{z}$ along the z-axis read as [25]

$$\begin{aligned}\mathbf{M}_{mn}^{1,3}(k(\mathbf{r} - \widehat{z}\mathbf{e}_{\mathrm{z}})) = c_n z_n^{1,3}(kR)\big[\mathrm{j}m\pi_n^{|m|}(\widehat{\theta})(\sin(\theta - \widehat{\theta})\mathbf{e}_r \\ + \cos(\theta - \widehat{\theta})\mathbf{e}_\theta) - \tau_n^{|m|}(\widehat{\theta})\mathbf{e}_\varphi\big]\,\mathrm{e}^{\mathrm{j}m\varphi}\end{aligned} \tag{3.4}$$

and

$$\begin{aligned}\mathbf{N}_{mn}^{1,3}(k(\mathbf{r} - \widehat{z}\mathbf{e}_{\mathrm{z}})) = c_n \Bigg\{ n(n+1)\frac{z_n^{1,3}(kR)}{kR} P_n^{|m|}(\cos\widehat{\theta}) \\ \times (\cos(\theta - \widehat{\theta})\mathbf{e}_r - \sin(\theta - \widehat{\theta})\mathbf{e}_\theta) + \frac{(kR z_n^{1,3}(kR))'}{kR} \\ \times \big[\tau_n^{|m|}(\widehat{\theta})(\sin(\theta - \widehat{\theta})\mathbf{e}_r + \cos(\theta - \widehat{\theta})\mathbf{e}_\theta) \\ + \mathrm{j}m\pi_n^{|m|}(\widehat{\theta})\mathbf{e}_\varphi\big]\Bigg\}\,\mathrm{e}^{\mathrm{j}m\varphi},\end{aligned} \tag{3.5}$$

where $c_n = 1/\sqrt{2n(n+1)}$, z_n^1 and z_n^3 are the spherical Bessel functions j_n and the spherical Hankel functions of the first kind h_n, respectively, $P_n^{|m|}(\cos\theta)$ the normalized associated Legendre functions, $\tau_n^{|m|}(\theta) = \mathrm{d}P_n^{|m|}(\cos\theta)/\mathrm{d}\theta$, $\pi_n^{|m|}(\theta) = P_n^{|m|}(\cos\theta)/\sin\theta$, $(\mathbf{e}_r, \mathbf{e}_\theta, \mathbf{e}_\varphi)$ the unit vectors in spherical coordinates, (r, θ, φ) and

(ρ, φ, z) the spherical and the cylindrical coordinates of the field point $\mathbf{r}$, respectively, and

$$R^2 = \rho^2 + (z - \widehat{z})^2, \quad \sin\widehat{\theta} = \frac{\rho}{R}, \quad \cos\widehat{\theta} = \frac{z - \widehat{z}}{R}.$$

The localized SVWF correspond to $\widehat{z} = 0$ in (3.4) and (3.5), in which case, $R = r$ and $\widehat{\theta} = \theta$.

3.2.1 *Discrete Sources Method for the Transmission Boundary-Value Problem*

The discrete sources method uses the completeness of the systems of tangential vector functions

$$\left\{ \begin{pmatrix} \mathbf{n} \times \boldsymbol{\Phi}_{\alpha}^{q_{\mathrm{t}}}(k_{\mathrm{t}}\mathbf{r}) \\ -\mathrm{j}\sqrt{\frac{\varepsilon_{\mathrm{t}}}{\mu_{\mathrm{t}}}}\mathbf{n} \times \boldsymbol{\Psi}_{\alpha}^{q_{\mathrm{t}}}(k_{\mathrm{t}}\mathbf{r}) \end{pmatrix}, \begin{pmatrix} \mathbf{n} \times \boldsymbol{\Psi}_{\alpha}^{q_{\mathrm{t}}}(k_{\mathrm{t}}\mathbf{r}) \\ -\mathrm{j}\sqrt{\frac{\varepsilon_{\mathrm{t}}}{\mu_{\mathrm{t}}}}\mathbf{n} \times \boldsymbol{\Phi}_{\alpha}^{q_{\mathrm{t}}}(k_{\mathrm{t}}\mathbf{r}) \end{pmatrix} \right\}$$

on the surface S, i.e., for any tangential field $(\mathbf{e}_0, \mathbf{h}_0)$ and any $\delta > 0$, there exists $N_0 = N_0(\delta)$, such that for all $N > N_0$,

$$\left\| \mathbf{e}_0 + \mathbf{e}_{\mathrm{s}}^N - \mathbf{e}_{\mathrm{i}}^N \right\|_{2S} + \left\| \mathbf{h}_0 + \mathbf{h}_{\mathrm{s}}^N - \mathbf{h}_{\mathrm{i}}^N \right\|_{2S} \le \delta, \tag{3.6}$$

where $\mathbf{e}_0 = \mathbf{n} \times \mathbf{E}_0$, $\mathbf{h}_0 = \mathbf{n} \times \mathbf{H}_0$, and

$$\begin{pmatrix} \mathbf{e}_{\mathrm{t}}^N(\mathbf{r}) \\ \mathbf{h}_{\mathrm{t}}^N(\mathbf{r}) \end{pmatrix} = \sum_{\alpha=1}^{N} a_{\mathrm{t}\alpha} \begin{pmatrix} \mathbf{n} \times \boldsymbol{\Phi}_{\alpha}^{q_{\mathrm{t}}}(k_{\mathrm{t}}\mathbf{r}) \\ -\mathrm{j}\sqrt{\frac{\varepsilon_{\mathrm{t}}}{\mu_{\mathrm{t}}}}\mathbf{n} \times \boldsymbol{\Psi}_{\alpha}^{q_{\mathrm{t}}}(k_{\mathrm{t}}\mathbf{r}) \end{pmatrix} + b_{\mathrm{t}\alpha} \begin{pmatrix} \mathbf{n} \times \boldsymbol{\Psi}_{\alpha}^{q_{\mathrm{t}}}(k_{\mathrm{t}}\mathbf{r}) \\ -\mathrm{j}\sqrt{\frac{\varepsilon_{\mathrm{t}}}{\mu_{\mathrm{t}}}}\mathbf{n} \times \boldsymbol{\Phi}_{\alpha}^{q_{\mathrm{t}}}(k_{\mathrm{t}}\mathbf{r}) \end{pmatrix}, \tag{3.7}$$

with $\mathbf{e}_{\mathrm{t}} = \mathbf{n} \times \mathbf{E}_{\mathrm{t}}$, $\mathbf{h}_{\mathrm{t}} = \mathbf{n} \times \mathbf{H}_{\mathrm{t}}$, $\mathrm{t} = \mathrm{s}, \mathrm{i}$, $q_{\mathrm{s}} = 3$ and $q_{\mathrm{i}} = 1$. In view of (3.6), the amplitudes of the discrete sources, encapsulated in the $4N$-dimensional vector $\mathbf{x} = [a_{\mathrm{i}\alpha}, b_{\mathrm{i}\alpha}, a_{\mathrm{s}\alpha}, b_{\mathrm{s}\alpha}]^T$, can be computed by solving the minimization problem

$$\mathbf{x} = \arg\min \left(\left\| \mathbf{e}_0 + \mathbf{e}_{\mathrm{s}}^N - \mathbf{e}_{\mathrm{i}}^N \right\|_{2S}^2 + \left\| \mathbf{h}_0 + \mathbf{h}_{\mathrm{s}}^N - \mathbf{h}_{\mathrm{i}}^N \right\|_{2S}^2 \right). \tag{3.8}$$

However, this procedure leads to a normal system of equations which is fundamentally unstable for amplitudes determination. To deal with this problem, the point matching method is used instead. Essentially, for a set of P matching points on the particle surface $\{\mathbf{r}_p\}_{p=1}^{P}$, with $P > N$, we compute the least squares solution $\mathbf{x} = \mathrm{argmin}_{\mathbf{y}} \|\mathbf{A}\mathbf{y} - \mathbf{b}\|^2$, where $\mathbf{A}$ is a $4P \times 4N$ overdetermined matrix, and $\mathbf{b}$ is a $4P$-dimensional vector specified by the values of the incident field at the matching points.

3.2.2 Null-Field Method with Discrete Sources for the Transmission Boundary-Value Problem

In the null-field method with discrete sources, the internal surface fields $\mathbf{e}_{\mathrm{i}}$ and $\mathbf{h}_{\mathrm{i}}$ are approximated by (3.7), while the expansion coefficients $a_{\mathrm{i}\alpha}$ and $b_{\mathrm{i}\alpha}$ are computed from the null-field equations

$$\frac{\mathrm{j}k_{\mathrm{s}}^2}{\pi}\int_S \left[(\mathbf{e}_{\mathrm{i}}^N(\mathbf{r}) - \mathbf{e}_0(\mathbf{r}))\cdot\begin{pmatrix}\boldsymbol{\Psi}_{\bar{\alpha}}^3(k_{\mathrm{s}}\mathbf{r})\\ \boldsymbol{\Phi}_{\bar{\alpha}}^3(k_{\mathrm{s}}\mathbf{r})\end{pmatrix}\right.$$
$$\left. +\mathrm{j}\sqrt{\frac{\mu_{\mathrm{s}}}{\varepsilon_{\mathrm{s}}}}(\mathbf{h}_{\mathrm{i}}^N(\mathbf{r}) - \mathbf{h}_0(\mathbf{r}))\cdot\begin{pmatrix}\boldsymbol{\Phi}_{\bar{\alpha}}^3(k_{\mathrm{s}}\mathbf{r})\\ \boldsymbol{\Psi}_{\bar{\alpha}}^3(k_{\mathrm{s}}\mathbf{r})\end{pmatrix}\right]\,\mathrm{d}S(\mathbf{r}) = 0, \tag{3.9}$$

for $\bar{\alpha} = (-m, n)$ and $\bar{\alpha} = 1, 2, \ldots, N$. Making use on the vector spherical wave expansion of the incident field,

$$\mathbf{E}_0(\mathbf{r}) = \sum_{\alpha=1}^{N} a_{0\alpha}\mathbf{M}_{\alpha}^1(k_{\mathrm{s}}\mathbf{r}) + b_{0\alpha}\mathbf{N}_{\alpha}^1(k_{\mathrm{s}}\mathbf{r}), \tag{3.10}$$

yields

$$\mathbf{Q}^{31}(k_{\mathrm{s}}, k_{\mathrm{i}})\begin{bmatrix}a_{\mathrm{i}\beta}\\ b_{\mathrm{i}\beta}\end{bmatrix} = -\mathbf{Q}_0^{31}(k_{\mathrm{s}}, k_{\mathrm{s}})\begin{bmatrix}a_{0\beta}\\ b_{0\beta}\end{bmatrix}, \tag{3.11}$$

where the entries of the matrix $\mathbf{Q}^{31}(k_{\mathrm{s}}, k_{\mathrm{i}})$,

$$\mathbf{Q}^{31}(k_{\mathrm{s}}, k_{\mathrm{i}}) = \begin{bmatrix}(Q^{31})_{\alpha\beta}^{11} & (Q^{31})_{\alpha\beta}^{12}\\ (Q^{31})_{\alpha\beta}^{21} & (Q^{31})_{\alpha\beta}^{22}\end{bmatrix} \tag{3.12}$$

are given by

$$\begin{aligned}(Q^{31})_{\alpha\beta}^{11} &= \frac{\mathrm{j}k_{\mathrm{s}}^2}{\pi}\int_S \left[(\mathbf{n}\times\boldsymbol{\Phi}_{\beta}^1(k_{\mathrm{i}}\mathbf{r}))\cdot\boldsymbol{\Psi}_{\bar{\alpha}}^3(k_{\mathrm{s}}\mathbf{r})\right.\\ &\quad\left.+\mathrm{j}m_{\mathrm{r}}(\mathbf{n}\times\boldsymbol{\Psi}_{\beta}^1(k_{\mathrm{i}}\mathbf{r}))\cdot\boldsymbol{\Phi}_{\bar{\alpha}}^3(k_{\mathrm{s}}\mathbf{r})\right]\,\mathrm{d}S(\mathbf{r}),\\ (Q^{31})_{\alpha\beta}^{12} &= \frac{\mathrm{j}k_{\mathrm{s}}^2}{\pi}\int_S \left[(\mathbf{n}\times\boldsymbol{\Psi}_{\beta}^1(k_{\mathrm{i}}\mathbf{r}))\cdot\boldsymbol{\Psi}_{\bar{\alpha}}^3(k_{\mathrm{s}}\mathbf{r})\right.\\ &\quad\left.+\mathrm{j}m_{\mathrm{r}}(\mathbf{n}\times\boldsymbol{\Phi}_{\beta}^1(k_{\mathrm{i}}\mathbf{r}))\cdot\boldsymbol{\Phi}_{\bar{\alpha}}^3(k_{\mathrm{s}}\mathbf{r})\right]\,\mathrm{d}S(\mathbf{r}),\end{aligned} \tag{3.13}$$

$$(Q^{31})^{21}_{\alpha\beta} = \frac{\mathrm{j}k_{\mathrm{s}}^2}{\pi}\int_S \left[(\mathbf{n}\times\boldsymbol{\Phi}^1_{\beta}(k_{\mathrm{i}}\mathbf{r}))\cdot\boldsymbol{\Phi}^3_{\bar{\alpha}}(k_{\mathrm{s}}\mathbf{r})\right.$$
$$\left.+\mathrm{j}m_{\mathrm{r}}(\mathbf{n}\times\boldsymbol{\Psi}^1_{\beta}(k_{\mathrm{i}}\mathbf{r}))\cdot\boldsymbol{\Psi}^3_{\bar{\alpha}}(k_{\mathrm{s}}\mathbf{r})\right]\,\mathrm{d}S(\mathbf{r}),$$
$$(Q^{31})^{22}_{\alpha\beta} = \frac{\mathrm{j}k_{\mathrm{s}}^2}{\pi}\int_S \left[(\mathbf{n}\times\boldsymbol{\Psi}^1_{\beta}(k_{\mathrm{i}}\mathbf{r}))\cdot\boldsymbol{\Phi}^3_{\bar{\alpha}}(k_{\mathrm{s}}\mathbf{r})\right.$$
$$\left.+\mathrm{j}m_{\mathrm{r}}(\mathbf{n}\times\boldsymbol{\Phi}^1_{\beta}(k_{\mathrm{i}}\mathbf{r}))\cdot\boldsymbol{\Psi}^3_{\bar{\alpha}}(k_{\mathrm{s}}\mathbf{r})\right]\,\mathrm{d}S(\mathbf{r}), \tag{3.14}$$

with $m_{\mathrm{r}} = \sqrt{\varepsilon_{\mathrm{i}}/\varepsilon_{\mathrm{s}}}$ being the relative refractive index of the particle. The matrix $\mathbf{Q}_0^{31}$ has the same structure as the matrix $\mathbf{Q}^{31}$, but it contains as columns the vectors $\mathbf{M}^1_{\beta}(k_{\mathrm{s}}\mathbf{r})$ and $\mathbf{N}^1_{\beta}(k_{\mathrm{s}}\mathbf{r})$ in place of the vectors $\boldsymbol{\Phi}^1_{\beta}(k_{\mathrm{i}}\mathbf{r})$ and $\boldsymbol{\Psi}^1_{\beta}(k_{\mathrm{i}}\mathbf{r})$, respectively. The expansion coefficients of the scattered field

$$\mathbf{E}_{\mathrm{s}}(\mathbf{r}) = \sum_{\alpha=1}^{N} a_{\mathrm{s}\alpha}\mathbf{M}^3_{\alpha}(k_{\mathrm{s}}\mathbf{r}) + b_{\mathrm{s}\alpha}\mathbf{N}^3_{\alpha}(k_{\mathrm{s}}\mathbf{r}), \tag{3.15}$$

are computed from Huygens principle in conjunction with the surface fields approximation (3.7). The result is

$$\begin{bmatrix} a_{\mathrm{s}\alpha} \\ b_{\mathrm{s}\alpha} \end{bmatrix} = \mathbf{Q}^{11}(k_{\mathrm{s}}, k_{\mathrm{i}}) \begin{bmatrix} a_{\mathrm{i}\beta} \\ b_{\mathrm{i}\beta} \end{bmatrix}, \tag{3.16}$$

where the matrix $\mathbf{Q}^{11}$ has the same structure as the matrix $\mathbf{Q}^{31}$, but it contains as rows the vectors $\mathbf{M}^1_{\bar{\alpha}}(k_{\mathrm{s}}\mathbf{r})$ and $\mathbf{N}^1_{\bar{\alpha}}(k_{\mathrm{s}}\mathbf{r})$ in place of the vectors $\boldsymbol{\Phi}^3_{\bar{\alpha}}(k_{\mathrm{s}}\mathbf{r})$ and $\boldsymbol{\Psi}^3_{\bar{\alpha}}(k_{\mathrm{s}}\mathbf{r})$, respectively. Combining (3.11) and (3.16) we find that the transition matrix $\mathbf{T}$, relating the scattered field coefficients to the incident field coefficients, is given by

$$\mathbf{T} = -\mathbf{Q}^{11}(k_{\mathrm{s}}, k_{\mathrm{i}})(\mathbf{Q}^{31}(k_{\mathrm{s}}, k_{\mathrm{i}}))^{-1}\mathbf{Q}_0^{31}(k_{\mathrm{s}}, k_{\mathrm{s}}). \tag{3.17}$$

For localized SVWF, $\mathbf{Q}_0^{31}$ is the identity matrix, and (3.17) gives the standard form representation of the transition matrix.

3.2.3 Algorithm Details

For axisymmetric particles, the scattering problem is solved independently for each azimuthal modes m. The system of SVWF $\{\boldsymbol{\Phi}^q_{mn}(k\mathbf{r}), \boldsymbol{\Psi}^q_{mn}(k\mathbf{r})\}$ is truncated at an appropriate expansion order N_{rank} with the following meaning:

1. for the localized SVWF

$$\{\mathbf{M}^q_{mn}(k\mathbf{r}), \mathbf{N}^q_{mn}(k\mathbf{r}) \mid n = \max(1, |m|), \ldots, N_{\mathrm{rank}}\},$$

N_{rank} is the order of the localized pole,

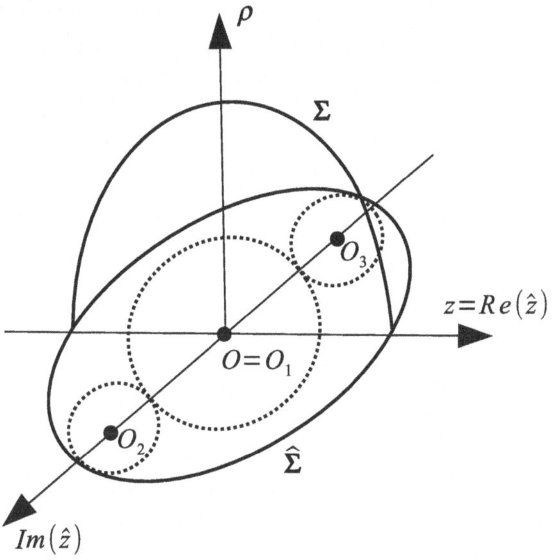

Fig. 3.1 The complex plane $(\mathrm{Re}\widehat{z}, \mathrm{Im}\widehat{z})$, the generatrix Σ of an oblate spheroid, and the image $\widehat{\Sigma}$ of Σ in the complex plane. The dotted circles represent the areas of influence of the poles O_1, O_2, and O_3 situated on the imaginary axis

2. for the distributed SVWF

$$\{\mathbf{M}^q_{m,|m|+l}(k(\mathbf{r} - z_n\mathbf{e}_z)), \mathbf{N}^q_{m,|m|+l}(k(\mathbf{r} - z_n\mathbf{e}_z)) \\ \mid n = 1, \ldots, N_{\mathrm{rank}}\},$$

   N_{rank} is number of lowest-order multipoles (discrete sources),
3. for the multiple SVWF

$$\{\mathbf{M}^q_{mn}(k(\mathbf{r} - z_p\mathbf{e}_z)), \mathbf{N}^q_{mn}(k(\mathbf{r} - z_p\mathbf{e}_z)) \mid p = 1, \ldots, N_{\mathrm{p}}, \\ n = \max(1, |m|), \ldots, N_{\mathrm{rank}p}\},$$

   $N_{\mathrm{rank}} = \sum_p N_{\mathrm{rank}p}$, where $N_{\mathrm{rank}p}$ is the order of the pole p.

The poles are distributed in the complex plane $(\mathrm{Re}\widehat{z}, \mathrm{Im}\widehat{z})$, which is the dual of the azimuthal plane $\varphi = \mathrm{const}$, i.e., (ρ, z) with $\rho \geq 0$ and $z \in \mathbb{R}$ (Fig. 3.1). In the discrete sources method, only lowest-order multipoles are used. For oblate particles, the poles for internal field representation are distributed on the imaginary axis in the interior and exterior of $\widehat{\Sigma}$ (the image of the generatrix Σ in the complex plane), while the poles for scattered field representation are distributed on both the real and imaginary axis in the interior of $\widehat{\Sigma}$. In the null-field method, all types of spherical vector wave functions are considered, and the same poles are used for representing the radiating and the regular system of vector functions in (3.14). For oblate particles, the pole with the largest order is placed at the origin and is called the dominant pole, while the rest of the poles are distributed on the imaginary axis in the interior of $\widehat{\Sigma}$. In both methods, a uniform distribution of the poles along the real axis is used for prolate particles.

In the discrete sources method, the least squares problem $\mathbf{x} = \text{argmin}_{\mathbf{y}} \|\mathbf{Ay} - \mathbf{b}\|^2$ is solved by means of Tikhonov regularization using (I) the QR factorization, (II) the identity matrix as regularization matrix, and (III) an a priori regularization parameter choice method based on the size parameter and particle eccentricity. In the null-field method, the matrix inversion is performed by the Gauss elimination method with back-substitution [10], or alternatively, by the block matrix inversion method [11]. To increase the accuracy and efficiency of the Gauss elimination method, the routine has been modified to work in multiple-precision arithmetic with the Multiprecision System (MPFUN90) package [26], and parallelized with OpenMP API.

3.2.4 Convergence Analysis

In the null-field method we choose the incident direction along the axis of symmetry of the particle and perform a convergence test over the expansion order $N_{\mathtt{rank}}$. Essentially, we solve the scattering problem for a reference and a lower-order system of SVWF, and check the convergence of the differential scattering cross-sections at a number of scattering angles [7]. The lower-order system of SVWF is chosen as follows: (I) for localized SVWF, the lower-order system is the reference system in which the order of the localized pole is reduced from $N_{\mathtt{rank}}$ to $N_{\mathtt{rank}} - 1$, (II) for distributed SVWF, the lower-order system is the reference system in which the pole placed at the origin is omitted, and finally, (III) for multiple SVWF, the lower-order system is the reference system in which the order of the dominant pole is reduced from $N_{\mathtt{rank1}}$ to $N_{\mathtt{rank1}} - 1$. In the discrete sources method we estimate the residual field at the particle surface for a given configuration of poles. This test which does not require the solution of two scattering problems can be regarded an "internal convergence criterion".

In our analysis we consider spheroids and cylinders, and in order to reduce the numerical instability, we perform the computations using extended- instead of double-precision floating-point variables. For spheroids, we denote by a and b the polar radius and the equatorial radius, respectively, while for cylinders a and b stand for the half-length and the cylinder radius, respectively.

In the first test case we consider a prolate spheroid and a prolate cylinder with a size parameter of $k_{\mathrm{s}}a = 80$ and an aspect ratio of $a/b = 8$. The normalized differential scattering cross-sections are illustrated in Fig. 3.2. In the null-field method with multiple SVWF, the parameters of calculation are $N_{\mathrm{p}} = 31$, $N_{\mathtt{rank}} = 101$, $N_{\mathtt{rank1}} = 11$ and $N_{\mathtt{rank}p} = 3$ for $p \neq 1$ in the case of the spheroid, and $N_{\mathrm{p}} = 31$, $N_{\mathtt{rank}} = 131$, $N_{\mathtt{rank1}} = 11$ and $N_{\mathtt{rank}p} = 4$ for $p \neq 1$ in the case of the cylinder, while in the null-field method with distributed SVWF, the parameters of calculation are $N_{\mathtt{rank}} = 101$ for the spheroid, and $N_{\mathtt{rank}} = 131$ for the cylinder. As it can be seen from Fig. 3.2, the agreement between the discrete sources method and the null-field method with discrete sources is excellent.

In a second test case we consider an oblate spheroid and an oblate cylinder with a size parameter of $k_{\mathrm{s}}b = 50$ and an aspect ratio of $a/b = 1/5$. Because the

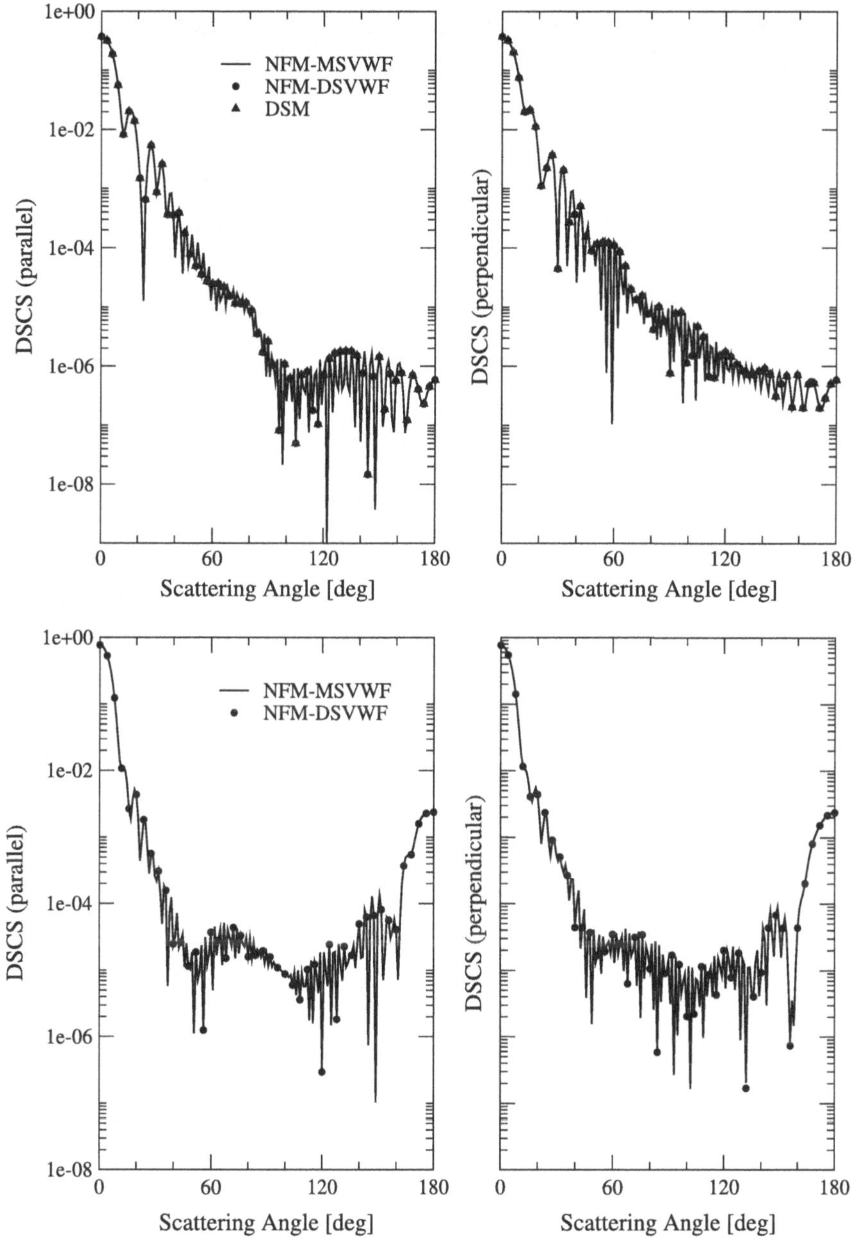

Fig. 3.2 Normalized differential scattering cross-sections computed with the discrete sources method (DSM), and the null-field method with multiple (MSVWF), and distributed (DSVWF) spherical vector wave functions. The results correspond to a prolate spheroid (top panels) and a prolate cylinder (bottom panels) with $m_{\rm r} = 1.5 + 0.02{\rm j}$, $k_{\rm s}a = 80$ and $k_{\rm s}b = 10$

extended-precision versions of null-field method with localized and distributed SVWF fail to converge for $k_s b > 30$, the results reported here correspond to the discrete sources method, the null-field method with multiple SVWF, and a multiple-precision version of the null-field method with localized SVWF. In the latter case, the calculation of the **Q**-matrix elements and the inversion are performed with multiple-precision arithmetic. The normalized differential scattering cross-sections are illustrated in Fig. 3.3. In the null-field method with multiple SVWF, the parameters of calculation are $N_{\mathrm{p}} = 3$, $N_{\mathrm{rank}} = 96$, $N_{\mathrm{rank1}} = 60$, $N_{\mathrm{rank2,3}} = 18$, $k_s \mathrm{Im} z_1 = 0.0$, and $k_s \mathrm{Im} z_{2,3} = \pm 42$ for the spheroid, and $N_{\mathrm{p}} = 3$, $N_{\mathrm{rank}} = 130$, $N_{\mathrm{rank1}} = 90$, $N_{\mathrm{rank2,3}} = 20$, $k_s \mathrm{Im} z_1 = 0.0$, and $k_s \mathrm{Im} z_{2,3} = \pm 45$ for the cylinder. From Fig. 3.3 we may conclude that the agreement between the different methods is quite good. Coming to the computational efficiency, we mention that for the oblate cylinder, (I) the computational time of the discrete sources method is 7:51 (min:s), (II) the computational times of the null-field method with multiple SVWF are 3:42 (min:s) for the Gauss elimination routine working in extended precision, 3:16 (min:s) for the block matrix inversion routine working in extended precision, and 5:24 (min:s) for the Gauss elimination routine working in multiple precision, and finally, (III) the computational time of the multiple-precision version of the null-field method with localized SVWF is 58:33 (min:sec). Thus, the use of the multiple-precision Gauss elimination routine increases the computational time by a factor of 2, while the multiple-precision version of the null-field method with localized SVWF is extremely inefficient.

In the final test case we consider an oblate spheroid with a size parameter of $k_s b = 80$ and an aspect ratio of $a/b = 1/8$, as well as an oblate cylinder with a size parameter of $k_s b = 70$ and an aspect ratio of $a/b = 1/7$. The results in Fig. 3.4 correspond to the null-field method with multiple SVWF working with a multiple-precision version of the Gauss elimination routine. It should be pointed out that for this test case only this method converges. The parameters of calculation are $N_{\mathrm{p}} = 7$, $N_{\mathrm{rank}} = 264$, $N_{\mathrm{rank1}} = 112$, $N_{\mathrm{rank2,3}} = 40$, $N_{\mathrm{rank4,5}} = 20$, $N_{\mathrm{rank6,7}} = 16$, $k_s \mathrm{Im} z_1 = 0.0$, $k_s \mathrm{Im} z_{2,3} = \pm 55$, $k_s \mathrm{Im} z_{4,5} = \pm 65$, and $k_s \mathrm{Im} z_{6,7} = \pm 75$ for the spheroid, and $N_{\mathrm{p}} = 7$, $N_{\mathrm{rank}} = 268$, $N_{\mathrm{rank1}} = 108$, $N_{\mathrm{rank2,3}} = 40$, $N_{\mathrm{rank4,5}} = 20$, $N_{\mathrm{rank6,7}} = 20$, $k_s \mathrm{Im} z_1 = 0.0$, $k_s \mathrm{Im} z_{2,3} = \pm 45$, $k_s \mathrm{Im} z_{4,5} = \pm 55$, and $k_s \mathrm{Im} z_{6,7} = \pm 65$ for the cylinder. The use of the multiple-precision version of the Gauss elimination routine leads to a relatively high computational times: 86 min for the spheroid and 93 min for the cylinder.

The conclusion of our numerical analysis is that the null-field method with multiple SVWF is superior to the null-field method with distributed SVWF considered in [25]. Another method for improving the numerical stability of the conventional null-field method relies on an analytic computation of the **Q**-matrix elements. This method, which is rather technical, is described in the next section.

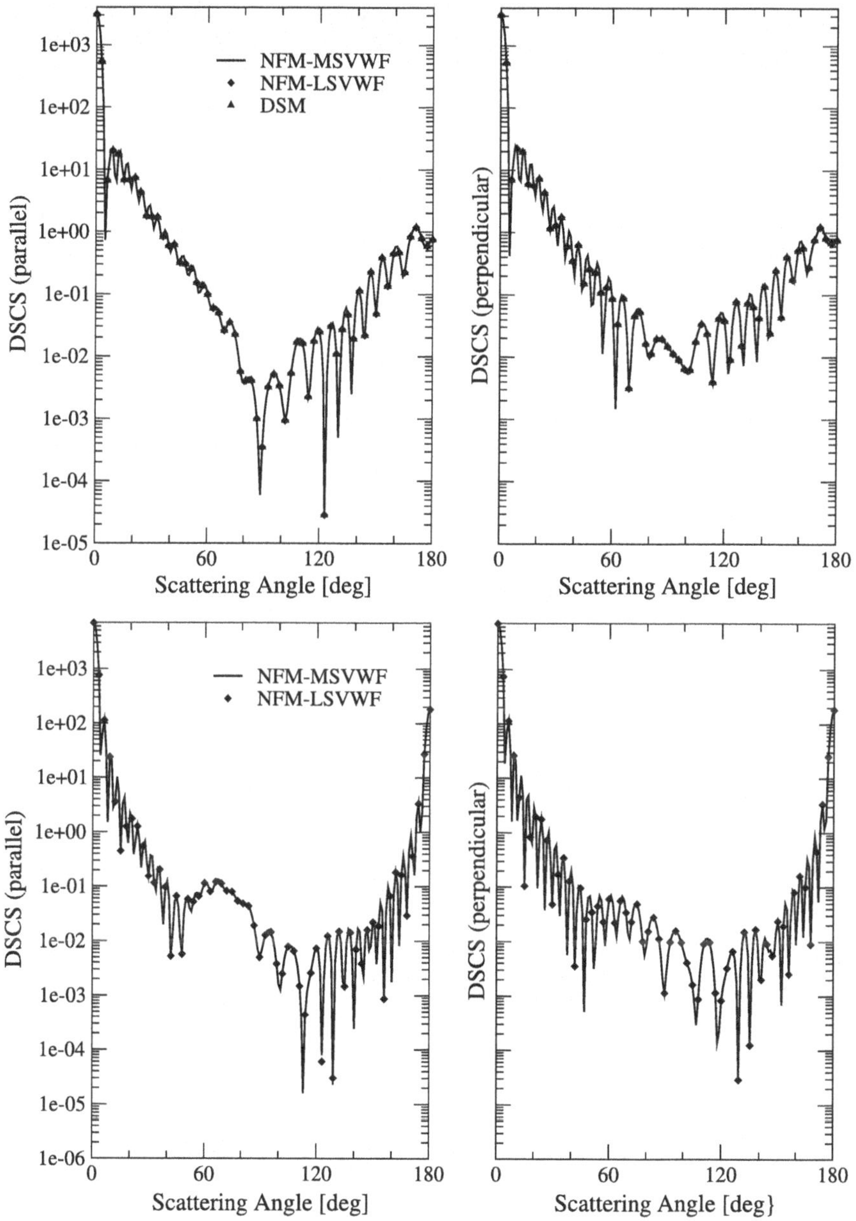

Fig. 3.3 Normalized differential scattering cross-sections computed with the discrete sources method (DSM), and the null-field method with multiple (MSVWF) and localized (LSVWF) spherical vector wave functions. The results correspond to an oblate spheroid (top panels) and an oblate cylinder (bottom panels) with $m_{\mathrm{r}} = 1.5 + 0.02\mathrm{j}$, $k_{\mathrm{s}}a = 10$ and $k_{\mathrm{s}}b = 50$

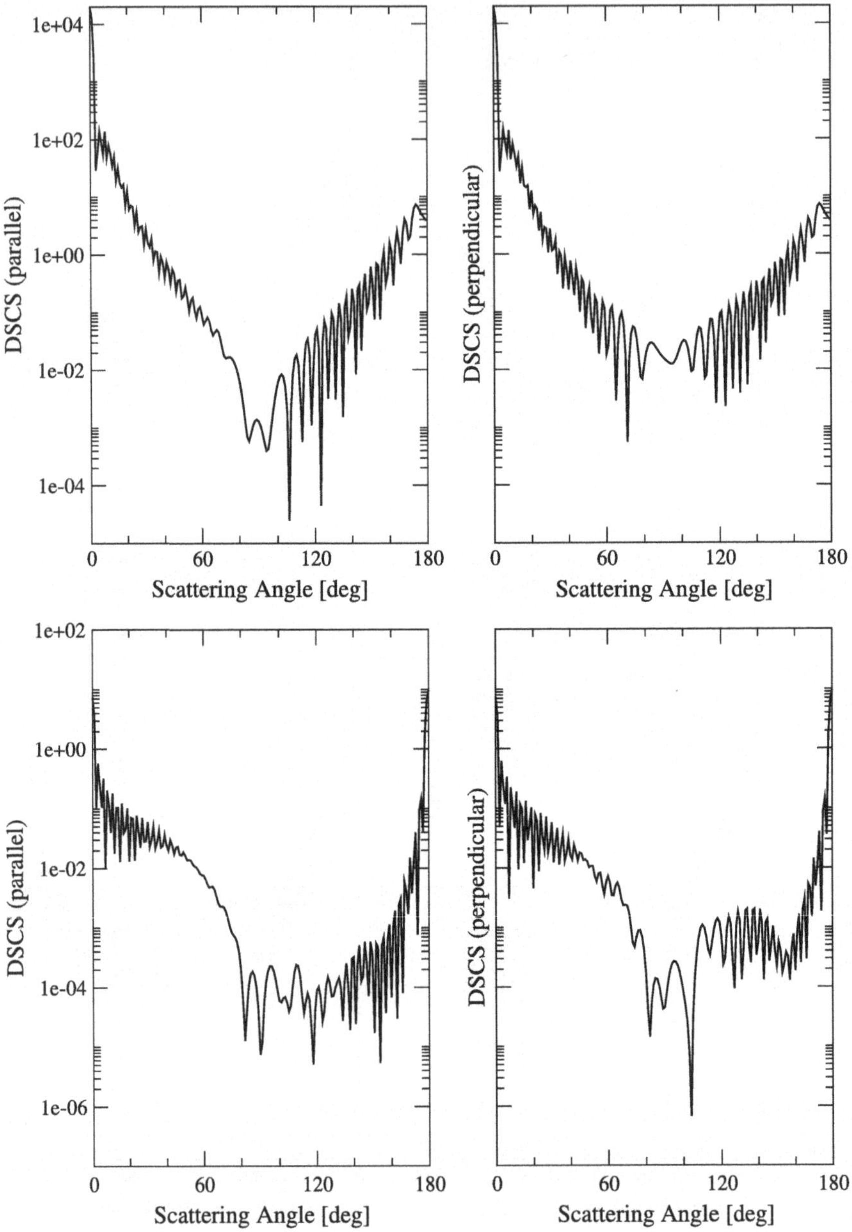

Fig. 3.4 Normalized differential scattering cross-sections computed with the null-field method with multiple spherical vector wave functions. The results correspond to an oblate spheroid with $m_{\mathrm{r}} = 1.5 + 0.02\mathrm{j}$, $k_{\mathrm{s}}a = 10$ and $k_{\mathrm{s}}b = 80$ (top panels) and an oblate cylinder with $m_{\mathrm{r}} = 1.311$, $k_{\mathrm{s}}a = 10$ and $k_{\mathrm{s}}b = 70$ (bottom panels)

3.3 An Analytical Method for Computing the Q-Matrix Elements

In this section we sketch the main ideas of an analytical method for computing the **Q**-matrix elements in the framework of the null-field method with localized SVWF.

For axisymmetric particles, Somerville et al. [27] have shown that the **Q**-matrix elements can be expressed in terms of six types of integrals. The integrands depend on the regular Riccati-Bessel functions $\psi_n = \psi_n(m_{\mathrm{r}}x) = m_{\mathrm{r}}xj_n(m_{\mathrm{r}}x)$ and their derivatives ψ_n', the Riccati-Hankel functions of the first kind $\xi_n = \xi_n(x) = xh_n(x)$ and their derivatives ξ_n', the unnormalized associated Legendre functions $P_n^{|m|} = P_n^{|m|}(\cos\theta)$, satisfying

$$\int_0^\pi P_n^{|m|}(\cos\theta)P_{n'}^{|m|}(\cos\theta)\sin\theta\,\mathrm{d}\theta = \frac{2}{2n+1}\frac{(n+|m|)!}{(n-|m|)!}\delta_{nn'}, \tag{3.18}$$

and the angular functions $\tau_n^{|m|} = \tau_n^{|m|}(\theta)$. The argument of the Riccati functions is $x = x(\theta) = k_{\mathrm{s}}r(\theta)$, where $r(\theta)$ describes the generatrix in polar coordinates. Assuming that x can be written as $x = x(z(\theta))$, where $z = \cos\theta$, using the decomposition $\xi_n = \psi_n + \mathrm{j}\chi_n$, where $\chi_n(x) = xy_n(x)$ are the irregular Riccati-Bessel functions, and the representations

$$\sin\theta\,\tau_n^{|m|} = \frac{n(n-|m|+1)}{2n+1}P_{n+1}^{|m|} - \frac{(n+1)(n+|m|)}{2n+1}P_{n-1}^{|m|}, \tag{3.19}$$

and

$$X_n' = \frac{n+1}{2n+1}X_{n-1} - \frac{n}{2n+1}X_{n+1}, \tag{3.20}$$

where X_n stands for ψ_n and χ_n, it can be shown that the computation of the six integrals simplifies to the computation of the generic integrals

$$A_{lnkn'k'} = \int_{-1}^{1} \chi_{n'}\psi_{k'}P_n^{|m|}P_k^{|m|}x_z^l\,\mathrm{d}z, \tag{3.21}$$

$$B_{lnkn'k'} = \int_{-1}^{1} \psi_{n'}\psi_{k'}P_n^{|m|}P_k^{|m|}x_z^l\,\mathrm{d}z, \tag{3.22}$$

where $x_z = \mathrm{d}x/\mathrm{d}z$. The indices n' and k' are defined by $n' = n + \overline{n} \geq 0$ and $k' = k + \overline{k} \geq 0$, respectively, and we have (1) $l = 0, 1$, (2) $\overline{n}, \overline{k} \in \{-2, -1, 0, 1, 2\}$, $\overline{n} + \overline{k}$ is odd, and (3) $n, k = \max(1, |m|), \ldots, N_{\mathtt{rank}} + 2$.

A dramatic loss of precision may occur during the computation of the integral $A_{lnkn'k'}$ by a numerical scheme [13]. For large size parameters and large values of $n - k$, the integrand oscillates around zero and its magnitude varies significantly across the range of integration. Consequently, severe cancellations can occur during the summation step of a Gauss–Legendre quadrature method. To overcome this problem

1. we may perform the computation with increased precision, or we may use the algorithm for summation of floating point numbers as in K-fold working precision developed by Ogita et al. [28],
2. we may remove analytically the terms causing the source of cancellation [14], or finally,
3. we may compute analytically the integral $A_{lnkn'k'}$.

In the following we will be concerned with an analytical computation of the integral $A_{lnkn'k'}$ by using the Gaunt formula for the integral of triple products of associated Legendre functions as suggested in [13]. Other techniques for computing the **Q**-matrix elements can be found in [16–19].

Assuming the representation $x = X\overline{x}$, where X is the size parameter, and making use on the power series representations of the Riccati–Bessel functions

$$\psi_n(x) = \sum_{s=0}^{\infty} \frac{1}{s!} \left(-\frac{1}{2}\right)^s \alpha_{sn} x^{2s+n+1} \tag{3.23}$$

and

$$\chi_n(x) = -\sum_{s=0}^{\infty} \frac{1}{s!} \left(-\frac{1}{2}\right)^s \beta_{sn} x^{2s-n}, \tag{3.24}$$

where

$$\alpha_{sn} = \frac{1}{(2n+2s+1)!!} \tag{3.25}$$

and

$$\beta_{sn} = \begin{cases} (-1)^s\,(2n-2s-1)!! & \text{for } s \le n \\ (-1)^n/(2s-2n-1)!! & \text{for } s > n \end{cases}, \tag{3.26}$$

we obtain

$$\begin{aligned} A_{lnkn'k'} &= -m_{\mathrm{r}}^{k'+1} \sum_{q=0}^{\infty} \gamma_{qn'k'}(m_{\mathrm{r}}) \frac{1}{q!} \left(-\frac{1}{2}\right)^q \\ &\times X^{2q+k'-n'+l+1} \int_{-1}^{1} P_n^{|m|} P_k^{|m|} \overline{x}^{2q+k'-n'+1} \overline{x}_z^l \, \mathrm{d}z, \end{aligned} \tag{3.27}$$

with

$$\gamma_{qnk}(m_{\mathrm{r}}) = \sum_{s=0}^{q} \frac{q!}{s!(q-s)!} \alpha_{q-s,k} \beta_{sn} m_{\mathrm{r}}^{2(q-s)}. \tag{3.28}$$

An efficient method for computing the coefficients γ_{qnk} without loss of precision has been described in [14]. The method uses a recursion scheme and avoids severe cancellations which may occur when the relative refractive index m_{r} is close to 1. Setting $s = 2q + k' - n' + 1$ in (3.27) yields

$$A_{lnk'k'} = -m_{\mathrm{r}}^{k'+1} \sum_{s=k'-n'+1;2}^{\infty} \gamma_{qn'k'}(m_{\mathrm{r}}) \frac{1}{q!} \left(-\frac{1}{2}\right)^q X^{s+l} I_{lsnk}, \tag{3.29}$$

with $q = q(s)$ and

$$I_{lsnk} = \int_{-1}^{1} P_n^{|m|} P_k^{|m|} \overline{x}^s \, \overline{x}_z^l \, \mathrm{d}z. \tag{3.30}$$

In (3.29), the notation $\sum_{s=s_0;2}^{s_1}$ means that the index s increases from s_0 to s_1 in steps of 2. To compute $A_{lnk'k'}$ we split the infinite series into a finite series involving negative powers of $\overline{x}$, and an infinite series involving positive powers of $\overline{x}$, i.e.,

$$\sum_{s=k'-n'+1;2}^{\infty} = \sum_{s=k'-n'+1;2}^{s_0} + \sum_{s=s_0+2}^{\infty}, \tag{3.31}$$

where $s_0 = -1$ if $k' - n'$ is even, and $s_0 = -2$ if $k' - n'$ is odd.

The integrals I_{lsnk} depend only on the particle shape, and therefore, they need to be computed only once for a class of particles of similar shapes, but having different sizes and refractive indices. The connection to the shape matrix approach, which is based on power series representations of the spherical Bessel and Neumann functions, and the multiplication theorem

$$\begin{aligned} j_n(X\overline{x}) &= X^n \sum_{k=0}^{\infty} \frac{(-1)^k (X^2-1)^k}{k!} \left(\frac{\overline{x}}{2}\right)^k j_{n+k}(\overline{x}), \\ y_n(X\overline{x}) &= \frac{1}{X^{n+1}} \sum_{k=0}^{\infty} \frac{(X^2-1)^k}{k!} \left(\frac{\overline{x}}{2}\right)^k y_{n-k}(\overline{x}), \end{aligned} \tag{3.32}$$

is apparent. Although, from a theoretical point of view, the design of the shape matrix approach by using the multiplication theorem is elegant and ingenious, the final computational relations for the **Q**-matrix elements are identical.

The key point in computing I_{lsnk} is the integral of triple products of associated Legendre functions

$$\int_{-1}^{1} P_n^m P_k^m P_p \, \mathrm{d}z = (-1)^m \frac{2}{2p+1} \overline{a}(m, n, k, p), \tag{3.33}$$

where the coefficients $\overline{a}(m, n, k, p)$ are proportional to the Gaunt coefficients

$$\begin{aligned} &a(m,n,m',k,p) \\ &= \frac{2p+1}{2}\frac{(p-m-m')!}{(p+m+m')!}\int_{-1}^{1} P_n^m P_k^{m'} P_p^{m+m'}\,\mathrm{d}z \end{aligned} \tag{3.34}$$

in the particular case $m' = -m$ and $m \geq 0$, i.e.,

$$\overline{a}(m,n,k,p) = \frac{(k+m)!}{(k-m)!}a(m,n,-m,k,p). \tag{3.35}$$

The coefficients $\overline{a}(m,n,k,p)$ are nonzero for $p = |n-k|, |n-k|+2, \ldots, n+k$, and can be computed by using the downward recurrence relation

$$\varsigma_{p+1}\overline{a}(\cdot,p) - \left(4m^2 + \varsigma_{p+2} + \varsigma_{p+3}\right)\overline{a}(\cdot,p+2) + \varsigma_{p+4}\overline{a}(\cdot,p+4) = 0, \tag{3.36}$$

with the starting values

$$\begin{aligned} \overline{a}(\cdot,n+k) &= \frac{(2n-1)!!(2k-1)!!}{(2n+2k-1)!!}\frac{(n+k)!}{(n-m)!(k-m)!}, \\ \overline{a}(\cdot,n+k-2) &= \frac{(2n+2k-3)}{(2n-1)(2k-1)(n+k)} \\ &\quad\times\left[nk - m^2(2n+2k-1)\right]\overline{a}(\cdot,n+k), \end{aligned} \tag{3.37}$$

where $\overline{a}(\cdot,p)$ stands for $\overline{a}(m,n,k,p)$ and

$$\varsigma_p = \frac{\left[4p^2 - (n+k+1)^2\right]\left[p^2 - (n-k)^2\right]}{4p^2-1}. \tag{3.38}$$

The three-term recurrence formula (3.36) is due to Bruning and Lo [29], and it provides accurate numerical results for all low- and high-degree coefficients.

Let us assume that $\overline{x}$ can be expressed as a finite Legendre series, which approximates a real shape function $\overline{x}_{\mathrm{r}}$ with a truncation error ε, i.e., $\overline{x}_{\mathrm{r}} \approx \overline{x} = \sum_{p=0}^{N_{01}} a_{01p}P_p$, where

$$a_{01p} = \frac{2p+1}{2}\int_{-1}^{1}\overline{x}_{\mathrm{r}}(z)P_p(z)\,\mathrm{d}z, \tag{3.39}$$

and

$$\varepsilon = \int_{-1}^{1}(\overline{x}(z) - \overline{x}_{\mathrm{r}}(z))^2\mathrm{d}z \Big/ \int_{-1}^{1}\overline{x}_{\mathrm{r}}(z)^2\mathrm{d}z. \tag{3.40}$$

In other words, we replace the real shape $\overline{x}_{\mathrm{r}}$ by a pseudo-shape $\overline{x}$, which we then use to compute the **Q**-matrix elements. The same technique has been employed by Petrov et al. in [19]. In Table 3.1 we list the number of terms N_{01} in the finite Legendre series for the superellipse

Table 3.1 The number of terms N_{01} in the finite Legendre series for a superellipse of eccentricity e and exponent n. The results correspond to the truncation errors $\varepsilon = 10^{-6}$

	e		
n	0.5	0.25	0.125
2	10	20	42
4	14	24	48
6	16	32	64
∞	96	166	306

$$\overline{x}_r(\theta) = (\cos^n\theta + e^n \sin^n\theta)^{-1/n}$$

and the truncation error $\varepsilon = 10^{-6}$. The case $n = 2$ corresponds to an ellipse, the cases $n = 4$ and $n = 6$ correspond to a rectangle with rounded corners, and the case $n \to \infty$ corresponds to a rectangle. From this table we see that N_{01} increases with increasing the eccentricity e and the exponent n, and that the method is especially efficient for particles with smooth surfaces.

The computation of I_{lsnk} depends on the sign of s, and we distinguish two cases:

A. Case $s < 0$. As any continuous function defined in the interval $[-1, 1]$ can be expanded in a Legendre series, we set, for any negative s in the finite range $k' - n' + 1, k' - n' + 3, \ldots, s_0$,

$$\overline{x}^s\,\overline{x}_z^l = \sum_{p=0}^{\infty} a_{lsp} P_p, \tag{3.41}$$

where

$$a_{lsp} = \frac{2p+1}{2}\int_{-1}^{1} \overline{x}^s(z)\overline{x}_z^l(z) P_p(z)\,\mathrm{d}z, \tag{3.42}$$

and $\overline{x}_z = \sum_{p=0}^{N_{01}} a_{01p} P'_p$. We then obtain

$$I_{lsnk} = \sum_{p=|n-k|;2}^{n+k} (-1)^{|m|}\frac{2}{2p+1} a_{lsp}\overline{a}(|m|, n, k, p). \tag{3.43}$$

From (3.43) we see that the vanishing property of Gaunt coefficients implies that only the first $n + k$ terms in expansion (3.41) are required for computing I_{lsnk}.

B. Case $s \geq 0$. Setting $f_{01} = \overline{x}$, and making use on the representation of $\overline{x}$, it is apparent that the product $f_{ls} = \overline{x}^s\,\overline{x}_z^l$ can be expressed as a finite Legendre expansion of the form

$$f_{ls} = \sum_{p=0}^{N_{ls}} a_{lsp} P_p, \tag{3.44}$$

yielding

$$I_{lsnk} = \sum_{p=|n-k|;2}^{\min(N_{ls},n+k)} (-1)^{|m|} \frac{2}{2p+1} a_{lsp} \overline{a}(|m|, n, k, p). \quad (3.45)$$

The process of computing the expansion coefficients of f_{ls} is based on the following results:

1. The derivative of the finite Legendre series

$$x(z) = \sum_{p=0}^{N} a_p P_p(z)$$

is given by $x_z(z) = \sum_{p=0}^{N-1} b_p P_p(z)$, where in view of the recurrence relation $P'_{p+1} - P'_{p-1} = (2p+1)P_p$, the coefficients b_p are computed as $b_{2p-1} = (4p-1)\sum_{k=p}^{K} a_{2k}$, $p = 1, \ldots, K$ for $N = 2K$ and $N = 2K+1$, and as $b_{2p} = (4p+1)\sum_{k=p}^{K-1} a_{2k+1}$, $p = 0, \ldots, K-1$ for $N = 2K$, and $b_{2p} = (4p+1) \sum_{k=p}^{K} a_{2k+1}$, $p = 0, \ldots, K$ for $N = 2K+1$.
2. The product of the finite Legendre series $x = \sum_{n=0}^{N} a_n P_n$ and $y = \sum_{k=0}^{K} b_k P_k$ is given by $xy = \sum_{p=0}^{N+K} c_p P_p$, where in view of (3.33) and the orthogonality relation of the Legendre polynomials, the coefficients c_p are computed as $c_p = \sum_{n=0}^{N} \sum_{k=|p-n|}^{n+p} a_n b_k \overline{a}(0, n, k, p)$.

The computational process is then organized as follows:

1. compute the expansion coefficients of $\overline{x}_z = f_{10} = \mathrm{d} f_{01}/\mathrm{d}z = \sum_{p=0}^{N_{10}} a_{10p} P_p$, with $N_{10} = N_{01} - 1$, by applying the derivative rule to f_{01},
2. compute the expansion coefficients of f_{0s}, $s \geq 2$, by using the product rule and the recurrence $f_{0s} = f_{0,s-1} f_{01}$, and the expansion coefficients of f_{1s}, $s \geq 1$, by using the product rule and the recurrence $f_{1s} = f_{1,s-1} f_{01}$.

Note that in the case $s = 0$ and $l = 0$ (the case $f_{00} = 1$), we have, $N_{00} = 0$ and $a_{000} = 1$. The numbers of terms in the finite series are $N_{0s} = sN_{01}$ and $N_{1s} = (s+1)N_{01} - 1$.

In the proposed method, the set of expansion coefficients defined by (3.39), i.e., $\{a_{01p} | p = 0, \ldots, N_{01}\}$, and the set of expansion coefficients defined by (3.42), i.e.,

$$\{a_{lsp} | l = 0, 1;\ s = -(N_{\mathtt{rank}} + 3), \ldots, -1;\ p = 0, \ldots, 2N_{\mathtt{rank}} + 4\},$$

describe the particle geometry. They can be computed in the preprocessing step with a desired accuracy, or they can be stored in a database and used as input parameters of the algorithm.

In summary, the analytical method for computing the **Q**-matrix elements involves the following steps:

1. Express the **Q**-matrix elements as integrals of products of Riccati-Bessel functions, and use the power series representations of these functions to obtain a Laurent series representation of the integrands.
2. Approximate the real shape function $\overline{x}_{\mathrm{r}}$ by a finite Legendre series representing a pseudo-shape function $\overline{x}$.
3. For negative s, expand the product $\overline{x}^s\,\overline{x}_z^l$ in an infinite series of Legendre functions, while for non-negative s, linearize $\overline{x}^s\,\overline{x}_z^l$ by using the product and derivative rule for finite Legendre series.
4. Integrate the linearized terms by using the Gaunt formula for the integral of triple products of associated Legendre functions.

Other techniques for computing I_{lsnk} are summarized below:

1. Assuming that a Taylor expansion of the product $\overline{x}^s(z)\overline{x}_z^l(z)$ around $z = 0$ is available, we may compute the integrals of the form $\int_{-1}^{1} P_n^m(z)P_k^m(z)z^s\,\mathrm{d}z, s \geq 0$, by applying the recurrence relation

$$zP_n^m(z) = \frac{n-m+1}{2n+1}P_{n+1}^m + \frac{n+m}{2n+1}P_{n-1}^m \tag{3.46}$$

and the orthogonality relation of the associated Legendre functions, or alternatively, by using the representation

$$z^s = \sum_{p=s,s-2,\ldots} \frac{(2p+1)s!}{2^{(s-p)/2}\left((s-p)/2\right)!(s+p+1)!!}P_p(z) \tag{3.47}$$

and the integral of triple products of associated Legendre functions.
2. If $r(\theta)$ has a simple representation in terms of trigonometric functions, we may use the relation

$$P_n^m(\cos\theta) = \sum_{k=0}^{n-m}(-1)^k\frac{n!(n+m)!}{k!(n-k)!(n-m-k)!(m+k)!} \times \left(\cos\frac{\theta}{2}\right)^{2n-2k-m}\left(\sin\frac{\theta}{2}\right)^{2k+m}, \quad m \geq 0, \tag{3.48}$$

to reduce the computation of I_{lsnk} to the computation of trigonometric integrals. It seems that this technique has been used by Petrov et al. in [16, 17].

The same approach can be used to compute the **Q**-matrix elements of a non-axisymmetric particle. In this case, the Gaunt coefficients of any degree and order can be computed by using, for example, the recursion scheme developed by Xu [30].

3.4 Conclusions

In the first part of this chapter we proved through a numerical analysis that the discrete sources method and the null-field method with distributed and multiple SVWF can be applied to electromagnetic scattering by large axisymmetric particles with extreme geometries. By an appropriate distribution of the poles in the complex plane, the condition number of the matrix to be inverted is decreased, and so, the stability of the computational scheme is increased. In particular, the null-field method with multiple SVWF enabled us to compute the scattering characteristics of oblate particles with a size parameter of 80 and an aspect ratio of $1/8$. A drawback of the method is its reduced efficiency; the multiple-precision version of the Gauss elimination routine increases the computational time by a factor of 2 as compared to an extended-precision version. In this regard, we plan to analyze the applicability of an inversion algorithm using floating-point format and a multiplicative correction as described in [31].

In the second part of this chapter we sketched the main ideas of an analytical method for computing the **Q**-matrix elements in the framework of the null-field method with localized SVWF. In summary, the **Q**-matrix elements are expressed as integrals of products of Riccati-Bessel functions, the power series representations of these functions are used to obtain a Laurent series representation of the integrands, and finally, the integrals are computed by using the Gaunt formula for the integral of triple products of associated Legendre functions. Although not yet proved, we expect that this method will avoid the dramatic loss of precision when the integrals in the expressions of the **Q**-matrix elements are computed by a Gauss-Legendre quadrature method.

References

1. A. Doicu, Y. Eremin, T. Wriedt, *Acoustic and Electromagnetic Scattering Analysis Using Discrete Sources* (Academic Press, London, 2000)
2. Y. Eremin, A. Sveshnikov, *The Discrete Sources Method in Electromagnetic Diffraction Problems* (Moscow State University Press, Moscow, 1992)
3. Y. Eremin, Sov. Phys. Dokl. **28**, 451 (1983)
4. C. Hafner, *The Generalized Multipole Technique for Computational Electromagnetics* (Artech House, Boston, 1990)
5. P. Waterman, Proc. IEEE **53**, 805 (1965)
6. P. Waterman, Phys. Rev. D **3**, 825 (1971)
7. P. Barber, S. Hill, *Light Scattering by Particles: Computational Methods* (World Scientific, Singapore, 1990)
8. M. Mishchenko, L. Travis, A. Lacis, *Scattering, Absorption and Emission of Light by Small Particles* (Cambridge University Press, Cambridge, 2002)
9. A. Lakhtakia, V. Varadan, V. Varadan, Appl. Opt. **23**, 3502 (1984)
10. A. Moroz, Appl. Opt. **44**, 3604 (2005)
11. D. Petrov, Y. Shkuratov, G. Videen, Opt. Lett. **32**, 1168 (2007)
12. M. Kahnert, T. Rother, Opt. Express **19**, 11138 (2011)
13. W. Somerville, B. Auguié, E. Le Ru. J. Quant. Spectrosc. Radiat. Trans. **113**, 524 (2012)

14. W. Somerville, B. Auguié, E. Le Ru. J. Quant. Spectrosc. Radiat. Trans. **123**, 153 (2013)
15. D. Petrov, E. Synelnyk, Y. Shkuratov, G. Videen, J. Quant. Spectrosc. Radiat. Trans. **102**, 85 (2006)
16. D. Petrov, G. Videen, Y. Shkuratov, M. Kaydash, J. Quant. Spectrosc. Radiat. Trans. **108**, 81 (2007)
17. D. Petrov, Y. Shkuratov, G. Videen, J. Quant. Spectrosc. Radiat. Trans. **109**, 1474 (2008)
18. D. Petrov, Y. Shkuratov, G. Videen, J. Opt. **12** (2010). http://dx.doi.org/10.1088/2040-8978/12/9/095701
19. D. Petrov, Y. Shkuratov, G. Videen, J. Quant. Spectrosc. Radiat. Trans. **112**, 1636 (2011)
20. M. Mishchenko, L. Travis, Opt. Commun. **109**, 16 (1994)
21. M. Iskander, A. Lakhtakia, C. Durney, IEEE Trans. Antennas Propagat. **31**, 317 (1983)
22. A. Lakhtakia, V. Varadan, V. Varadan, J. Opt. Soc. Am. A **76**, 906 (1984)
23. R. Bates, D. Wall, Phil. Trans. R. Soc. Lond. **287**, 45 (1977)
24. R. Hackman, J. Acoust. Soc. Am. **75**, 35 (1984)
25. A. Doicu, T. Wriedt, Y. Eremin, *Light Scattering by Systems of Particles* (Springer, Berlin, 2006)
26. D. Bailey, ACM Trans. Math. Softw. **21**, 379 (1995)
27. W. Somerville, B. Auguié, E. Le Ru. Opt. Lett. **36**, 3482 (2011)
28. T. Ogita, S. Rump, S. Oishi, SIAM J. Sci. Comput. **26**, 1955 (2005)
29. J. Bruning, Y. Lo, IEEE Trans. Antennas Propagat. **19**, 378 (1971)
30. Y. Xu, J. Comput. Appl. Math. **85**, 53 (1971)
31. S. Rump, Jpn. J. Indust. Appl. Math. **26**, 249 (2009)

Chapter 4
Fictitious Particle Approach for Light Scattering Investigation from the Line Features of a Substrate Based on the Discrete Sources Method

Yuri Eremin and Thomas Wriedt

Abstract Computer simulation of light scattering from features on plane interfaces is of interest in the semiconductor and nanoelectronic industry. Submicrometer defects on silicon substrates are detected and characterized by in-line laser scanning surface inspection systems. A reliable computer model for predicting this light scattering would provide a flexible and efficient tool for efficient surface features detection and discrimination. Based on the Discrete Sources Method (DSM) a new fictitious particle concept has been elaborated and realized. Based on this conception an updated DSM computer model enables to analyze light scattering from line defects of a plane silicon substrate such as a line bump and a pit. Computer simulation results corresponding to the Scattering Cross-Section (SCS) for P/S polarized excitation are presented. It was found that the Total Scattering Cross-Section (TSC) can be change by an order of magnitude depending on the orientation of the linear feature with respect to the plane wave incident direction.

4.1 Introduction

Many practical applications in nanotechnology require considering a particle embedded in an infinite stratified medium. Biosensors, optical antennas, solar cells, UV lithography and silicon wafer purification are examples of the topics of interest [1–6]. In the semiconductor industry, submicrometer particle contaminants and defects on silicon wafers are detected and characterized by optical inspection systems. These systems use a focused laser beam striking a surface and thus creating a field of scattered light which is registrated by a photodetector. The resulting scattered field is sampled and quantified with the use of a system of photodetectors, which

Y. Eremin
Lomonosov Moscow State University, Lenin's Hills Moscow, Russia
e-mail: eremin@cs.msu.su

T. Wriedt (✉)
Leibniz-Institut für Werkstofforientierte Technologien—IWT, Bremen, Germany
e-mail: thw@iwt.uni-bremen.de

T. Wriedt and Y. Eremin (eds.), *The Generalized Multipole Technique for Light Scattering*, Springer Series on Atomic, Optical, and Plasma Physics 99,
https://doi.org/10.1007/978-3-319-74890-0_4

enables to detect the defects and the contaminants. As design sizes of structures on silicon wafers continue to decrease, the importance of proper detection and characterization of such detects grows. The design of more sensitive inspection systems require a fundamental predictive capability for light scattering from nanosized structures [6]. A reliable computer model for predicting this light scattering would provide a flexible and efficient way to understand scattering processes and would increase the resolution ability of the inspection tools. So the importance in a sophisticated computer model which is based on rigorous mathematical statement of the scattering problem and which enables to predict light scattering from nanofeatures of silicon substrate seems to be highly required.

Various technique have been used so far for light scattering simulation. Direct methods solve Maxwell's equations in their differential form. They rely on a volume discretization in a finite computational domain. A first approach consists of directly discretizing Maxwell's equations in time and space using finite differences: this is the core of the so called Finite Difference Time Domain (FDTD) method [7] The FDTD method is one of the most popular methods in nanooptics because of its ability to handle a large variety of the scattering problems. FDTD is a simple technique, which can be effectively implemented on a computer or a *graphics processing unit* (*GPU*). Unfortunately, these models might not be accurate enough in some interesting cases [8]. Besides, the conventional FDTD numerical scheme does not incorporate an infinite interface in its theoretical model and it is required to apply special near to far field transformation to take into account it [9].

The Finite Element Method (FEM) [10] consists of expanding the electromagnetic fields as local functions in elements, which results in higher accuracy in the frequency domain. Their advantage as compared to FDTD is their higher accuracy, which comes at the price of a more challenging implementation and generally higher computational cost. Application of the FEM to surface features of a plane interface can create problems related to a truncation of the simulation domain [11]. Besides, FDTD and FEM have the disadvantage that they require both the local feature and the surrounding space to be discretized, leading to higher volume of the simulation domain.

For our particular case the most suitable and fast numerical approach seems to be a semi-analytical method. Such a method applies the Green's theorem to the system of Maxwell equations [12] to reduce the scattering problem formulated in the whole of 3D space to the impurity domain. Incorporation of the Green Tensor of the half-space enables to account for multiple scattering interactions between impurity and the plane interface [13]. Let us mainly focus on the methods that enable to treat an inhomogeneity located near an infinite plane interface.

There are volume-based methods, similar to the Discrete Dipole Approximation (DDA) [14, 15] and the Volume Integral Equation (VIE) [16, 17], which are suitable for modeling of light scattering by arbitrary impurities. Additionally the surface based methods, such as the T-matrix method [18], the Surface Integral Equation (SIE) [19], the Multiple MultiPole Technique (MMP) [20] and the Discrete Sources Method (DSM) [21] are applicable. While VIE and DDA can handle any kind of inhomogeneity, they are pretty time consuming, especially if it is required to account for strong scattering interaction between cells of mesh attaching the interface and the infinite interface itself. Besides, the volume-based methods might require different discretization for different parts of an embedded particle because of the essential difference in the refractive indexes of the upper and lower half spaces. This results into switching from a propagating plane wave in the upper half space to a damped wave inside the substrate. So surface-based methods seem to be more appropriate for the treatment of a homogeneous particle partially embedded into a plane infinite substrate. A detail description of the current status of the main numerical methods can be found in reviews [22, 23] as well.

Among the other surface-based methods mentioned above, the MMP and the DSM have several advantages. First of all they are semi-analytical meshless methods that do not require any mesh generation or integration procedure over the impurity domain. They enables to obtain the far field directly for multiple external excitations and will account for the interaction between the impurity and the plane interface in an analytical way. Besides, MMP and DSM provide the unique opportunity for a reliable validation of the numerical results obtained because the errors of the solution obtained can be evaluated explicitly by computation of the surface residual over the surface of the impurity [24, 25]. While in MMP models, dipoles and multipoles are placed on an auxiliary surface inside a scatterer [24] DSM can use complex valued discrete sources (DS) for each DS coordinate [25]. Employing complex DS allows to get a compromise between a resultant matrix condition number and the accuracy of the results in accordance with the so called "uncertainty principle" [26]. The uncertainty principle can be formulated in a brief way as "the worse the condition number the better the results".

In this paper we present a new scheme of the DSM [27, 28] that enables investigation of polarized light scattering from a penetrable particle partially embedded into a plane substrate. In contrast to the axial-symmetric case examined in [7], the new DSM scheme allows to consider particles, which together with the substrate, does not form an axial-symmetric structure. The DSM developed is a semi-analytical surface based meshless method, which requires neither mesh generation, nor an integration procedure over the particle surface. In the present realization of the DSM the scattered field everywhere outside a particle is represented as a finite linear combination of electromagnetic fields originated by electric dipoles distributed over an auxiliary surface deposited inside the particle. The field of each dipole is constructed incorporating the Green's Tensor of half-space. Thus, the DSM solution analytically satisfies Maxwell's equations, the radiation or attenuation conditions in the far zone and the transmission conditions established at the infinite plane interface. The unknown amplitudes of the DS are to be determined from the boundary conditions enforced at the particle surface only. For the examination of scattering by nano-dimensional line pit and line bump we employ the "fictitious particle" concept which already has successfully been used for the investigation of axial-symmetrical features on a plane substrate [7]. A fictitious particle means a virtual particle that is partially embedded into a substrate that has a refractive index which coincides with the air in case of a scratch or with the substrate index in case of a bump.

The paper is organized as follows. In the next section we consider the statement of the scattering problem and then describe the new mathematical formalism of the DSM. In the following section a detailed description of the corresponding numerical scheme of the DSM is provided. In the final section of the paper some numerical results are presented and discussed.

4.2 Discrete Sources Method for Non-axial Symmetric Case

Let us consider the case of a penetrable particle partially embedded into an infinite plane substrate. Choose a Cartesian coordinate system with its origin O at the plane interface - Ξ $(z = 0)$ assuming that the Oz axis is perpendicular to Ξ and directed into the upper half-space. Consider the substrate domain being D_1 (half-space $z < 0$), and the rest of R^3 space corresponds to air D_0 $(z > 0)$. We denote the particle's internal domain as D_i and will refer to its surface as ∂D_i. The external excitation $\{E^0,\ H^0\}$ is the electromagnetic field of linearly polarized plane wave propagating at an angle of $\pi - \theta_0$ with respect to the Oz axis and the plane of wave incidence is determined by the azimuth angle φ_0. Then the mathematical statement of the scattering problem can be written in the form including the time harmonic Maxwell equations
$\nabla \times H_\zeta = jk\varepsilon_\zeta E_\zeta; \quad \nabla \times E_\zeta = -jk\mu_\zeta H_\zeta \quad \text{in } D_\zeta, \quad \zeta = 0, 1, i,$
the transmission conditions enforced at the particle surface and plane interface

$$\begin{aligned} n_p \times \left(E_i(p) - E_{0,1}(p)\right) &= 0, \\ n_p \times \left(H_i(p) - H_{0,1}(p)\right) &= 0, \end{aligned} \quad p \in \partial D_i, \quad \begin{aligned} e_z \times (E_0(p) - E_1(p)) &= 0, \\ e_z \times (H_0(p) - H_1(p)) &= 0, \end{aligned} \quad p \in \Xi, \tag{4.1}$$

the Silver–Muller radiation conditions for all radial directions defined by the $e_r vector$

$$\lim_{r\to\infty} r \cdot \left(\sqrt{\varepsilon_0} E_0^s \times e_z - \sqrt{\mu_0} H_0^s\right) = 0, \quad r = |M| \to \infty, \quad z > 0,$$

attenuation conditions inside the substrate

$$\left(\left|E_1^s\right|, \left|H_1^s\right|\right) = o\left(\exp\{-\left|\mathrm{Im}(k_1)\right| r\}\right), \quad z < 0.$$

Let $\{E_\zeta,\ H_\zeta\}$ stand for the total field in the domains D_ζ, $\zeta = 0, 1,$ and $\{E_\zeta^s,\ H_\zeta^s\}$ - for the scattered field, $k = \omega/c, n_p$ is the outward unit normal to the surface ∂D_i, e_z is the unit vector directed along the Oz axis, $k_\zeta = k\sqrt{\varepsilon_\zeta \mu_\zeta}$. $|M|$ is denoted as a distance from the origin O to M point. Let the particle surface ∂D_i belong to the Hölder space $\partial D_i \subset C^{(2,\upsilon)}$, the relative permittivity and permeability are chosen such, that the following conditions are valid: $\mathrm{Im}(\varepsilon_0), \mathrm{Im}(\mu_0) = 0, \quad \mathrm{Im}(\varepsilon_1), \mathrm{Im}(\mu_1) < 0$. Time dependence is expected to be $\exp\{j\omega t\}$. Then, the boundary scattering problem (4.1) has a unique solution.

Following the basic DSM scheme described in [7] we construct the external excitation field $\{E_\zeta^0,\ H_\zeta^0\}$ in each domain D_ζ, $\zeta = 0, 1$ solving the reflecting and transmiting problem for the exciting plane wave $\{E^0,\ H^0\}$ at the plane interface Ξ [29]. The constructed field should satisfy the transmission conditions have been installed at the plane interface Ξ. Let us introduce the following notations

$$
\begin{aligned}
\psi_\zeta^{\pm} &= \exp\left\{-jk_\zeta\left(x\sin\theta_\zeta\cos\varphi_0 + y\sin\theta_\zeta\sin\varphi_0 \pm z\cos\theta_\zeta\right)\right\}\\
e_\zeta^{\pm,P} &= \left(\mp e_x\cos\theta_\zeta\cos\varphi_0 \mp e_y\cos\theta_\zeta\sin\varphi_0 + e_z\sin\theta_\zeta\right)\\
h_\zeta^{\pm,P} &= \left(e_x\sin\varphi_0 - e_y\cos\varphi_0\right) \qquad \zeta = 0, 1. \qquad (4.2)\\
e_\zeta^{\pm,S} &= \left(-e_x\sin\varphi_0 + e_y\cos\varphi_0\right)\\
h_\zeta^{\pm,S} &= \left(\mp e_x\cos\theta_\zeta\cos\varphi_0 \mp e_y\cos\theta_\zeta\sin\varphi_0 + e_z\sin\theta_\zeta\right)
\end{aligned}
$$

where e_x, e_y, e_z are Cartesian basis, θ_1 is the angle at which the transmitted plane wave penetrates into the substrate, according to Snell's law. For P-polarized excitation the incoming, outgoing and transmitting plane waves are

$$E_\zeta^{P(\pm)} = e_\zeta^{\pm,P}\cdot\psi_\zeta^{\pm};\quad H_\zeta^{P(\pm)} = h_\zeta^{\pm,P}\cdot n_\zeta\cdot\psi_\zeta^{\pm}$$

and similar to for S-polarization

$$E_\zeta^{S(\pm)} = e_\zeta^{\pm,S}\cdot\psi_\zeta^{\pm};\quad H_\zeta^{S(\pm)} = h_\zeta^{\pm,S}\cdot n_\zeta\cdot\psi_\zeta^{\pm}.$$

Here $n_\zeta = \sqrt{\varepsilon_\zeta\mu_\zeta}$ is the refractive index in the corresponding half-space $D_{0,1}$. Then the fields of the external excitation can be written as

$$E_0^{0(P,S)} = E_0^{P,S(-)} + R_{P,S}\cdot E_0^{P,S(+)};\quad H_0^{0(P,S)} = H_0^{P,S(-)} + R_{P,S}\cdot H_0^{P,S(+)};\quad z \ge 0$$

$$E_1^{0(P,S)} = T_{P,S}\cdot E_1^{P,S(-)};\quad H_1^{0(P,S)} = T_{P,S}\cdot H_1^{P,S(-)};\quad z < 0 \qquad (4.3)$$

and the corresponding Fresnel coefficients $R_{P,S}$; $T_{P,S}$ [29], are given by

$$R_P = \frac{n_1\cos\theta_0 - n_0\cos\theta_1}{n_1\cos\theta_0 + n_0\cos\theta_1};\quad T_P = \frac{2n_0\cos\theta_0}{n_1\cos\theta_0 + n_0\cos\theta_1}$$

$$R_S = \frac{n_0\cos\theta_0 - n_1\cos\theta_1}{n_0\cos\theta_0 + n_1\cos\theta_1};\quad T_S = \frac{2n_0\cos\theta_0}{n_0\cos\theta_0 + n_1\cos\theta_1}$$

while $E_0^{(-)} = E^0$; $H_0^{(-)} = H^0$ coincide with the original exciting plane wave $\{E^0,\ H^0\}$. Thus, we have obtained the field of external excitation in entire space (4.3) which satisfies the transmission conditions at the plane interface.

Let us now specify the definition of the scattered field. We define in each domainD_ζ, $\zeta = 0, 1$ the scattered field as $E_\zeta^s = E_\zeta - E_\zeta^0$, $H_\zeta^s = H_\zeta - H_\zeta^0$. The scattered field should satisfy the infinity conditions and the transmission conditions enforced at the plane interface Ξ. Thus, the transmission conditions imposed at the particle surface accept the following form

$$n_p \times (E_i - E^s_{0,1}) = n_p \times E^0_{0,1}; \quad n_p \times (H_i - H^s_{0,1}) = n_p \times H^0_{0,1} \tag{4.4}$$

We construct an approximate solution of problem (4.1) based on the DSM concept [30]. The main idea of the DSM is to represent the scattered field as a finite linear combination of elementary DS, which in this particular case are electric dipoles distributed over an auxiliary surface, located inside the particle. To construct the dipoles field satisfying the transmission conditions at the infinite plane interface we incorporate the Green Tensor (GT) of the half-space. Such a representation should satisfy Maxwell equations, the infinity conditions and transmission conditions at interface Ξ. In this case, the solution to the scattering problem (4.1) is reduced to the problem of approximating the fields of external excitation (4.3) on the surface of the embedded particle ∂D_i. Thus, the unknown amplitudes of the dipoles will be determined from the matching transmission conditions imposed on the particle surface ∂D_i (4.4).

To construct the fields of DS, which fit the transmission conditions at the plane Ξ we incorporate the GT. This enables to analytically fulfill the transmission conditions enforced at the plane ($z = 0$). The electric GT can be written in the following form [31]

$$\overset{\leftrightarrow e}{G}(M, M_0) = \begin{bmatrix} G^e_{11} & 0 & 0 \\ 0 & G^e_{11} & 0 \\ \partial G_{31}/\partial x_M & \partial G_{31}/\partial y_M & G^h_{11} \end{bmatrix}$$

The components of the tensor can be represented in the form of Sommerfeld integrals:

$$G^{e,h}_{11}(M, M_0) = \int_0^\infty J_0(\lambda r)\, v^{e,h}_{11}(\lambda, z, z_0)\lambda d\lambda, \; G_{31}(M, M_0) = \int_0^\infty J_0(\lambda r)\, v_{31}(\lambda, z, z_0)\lambda d\lambda. \tag{4.5}$$

Here $R^2_{MM_0} = r^2 + (z - z_0)^2, r^2 = \rho^2 + \rho_0^2 - 2\rho\rho_0\cos(\varphi - \varphi_0)$, $J_0(.)$ is the Bessel function of zero order, and (ρ_0, φ_0, z_0) are the cylindrical coordinates of a DS located at M_0. The corresponding spectral functions $v^{e,h}_{11}$, v_{31} ensure the satisfaction of the transmission conditions at the plane interface $z = 0$. For this case, they are written as

$$v^{e,h}_{11}(\lambda, z, z_0) = \begin{cases} \frac{\exp\{-\eta_0|z-z_0|\}}{\eta_0} + A^{e,h}_{11}(\lambda) \cdot \frac{\exp\{-\eta_0(z+z_0)\}}{\eta_0}, & z_0 > 0,\; z \geq 0, \\ B^{e,h}_{11}(\lambda) \cdot \frac{\exp\{\eta_1 z - \eta_0 z_0\}}{\eta_0}, & z_0 > 0,\; z \leq 0, \\ C^{e,h}_{11}(\lambda) \cdot \frac{\exp\{-\eta_0 z + \eta_1 z_0\}}{\eta_1}, & z_0 < 0,\; z \geq 0, \\ \frac{\exp\{-\eta_1|z-z_0|\}}{\eta_1} + D^{e,h}_{11}(\lambda) \cdot \frac{\exp\{\eta_1(z+z_0)\}}{\eta_1}, & z_0 < 0,\; z \leq 0, \end{cases} \tag{4.6}$$

Let us emphasize that the first term in the first and last lines of (4.6) are associated with singular part of the GT corresponding to DS location inside $D_{0,1}$.

$$v_{31}(\lambda, z, z_0) = \begin{cases} A_{31}(\lambda) \cdot \exp\{-\eta_0(z+z_0)\}, & z_0 > 0,\ z \geq 0, \\ B_{31}(\lambda) \cdot \exp\{\eta_1 z - \eta_0 z_0\}, & z_0 > 0,\ z \leq 0, \\ C_{31}(\lambda) \cdot \exp\{-\eta_0 z + \eta_1 z_0\}, & z_0 < 0,\ z \geq 0, \\ D_{31}(\lambda) \cdot \exp\{\eta_1(z+z_0)\}, & z_0 < 0,\ z \leq 0, \end{cases} \tag{4.7}$$

here $\eta_\zeta^2 = \lambda^2 - k_\zeta^2$. The spectral coefficients A, B, C, and D are determined from the following jump conditions enforced at $z = 0$

$$[v_{11}^e] = \left[\frac{1}{\mu}\frac{\partial v_{11}^e}{\partial z}\right] = 0; \left[\frac{1}{\mu} v_{31}^e\right] = 0; \quad \left[\frac{1}{\varepsilon\mu}\frac{\partial v_{31}^e}{\partial z}\right] = -\left[\frac{1}{\varepsilon\mu}\right] v_{11}^e$$

$$[v_{11}^h] = \left[\frac{1}{\varepsilon}\frac{\partial v_{11}^h}{\partial z}\right] = 0; \left[\frac{1}{\varepsilon} v_{31}^h\right] = 0; \quad \left[\frac{1}{\varepsilon\mu}\frac{\partial v_{31}^h}{\partial z}\right] = -\left[\frac{1}{\varepsilon\mu}\right] v_{11}^h$$

Using these conditions leads to the following expressions for the corresponding coefficients A, B, C, D

$$A_{11}^{e,h}(\lambda) = \frac{\chi_0^{e,h} - \chi_1^{e,h}}{\chi_0^{e,h} + \chi_1^{e,h}}; \; B_{11}^{e,h}(\lambda) = \frac{2\chi_0^{e,h}}{\chi_0^{e,h} + \chi_1^{e,h}};$$

$$C_{11}^{e,h}(\lambda) = \frac{2\chi_1^{e,h}}{\chi_0^{e,h} + \chi_1^{e,h}}; \; D_{11}^{e,h}(\lambda) = \frac{\chi_1^{e,h} - \chi_0^{e,h}}{\chi_0^{e,h} + \chi_1^{e,h}}.;$$

$$A_{31}(\lambda) = \frac{2\delta}{\left(\chi_0^e + \chi_1^e\right)\left(\chi_0^h + \chi_1^h\right)}; \; B_{31}(\lambda) = \frac{\mu_1}{\mu_0} A_{31}(\lambda);$$

$$C_{31}(\lambda) = \frac{\mu_0}{\mu_1} A_{31}(\lambda); \; D_{31}(\lambda) = A_{31}(\lambda).$$

where $\chi_\zeta^e = \frac{\eta_\zeta}{\mu_\zeta}$, $\chi_\zeta^h = \frac{\eta_\zeta}{\varepsilon_\zeta}$, $\delta = \frac{1}{\varepsilon_0\mu_0} - \frac{1}{\varepsilon_1\mu_1}$.

We start to construct the approximate solution for the scattered field in $D_{0,1}$ based on the system of electric dipoles localized on the set of points $\{M_n\}_{n=1}^{\infty}$. Let the points M_n belong to an auxiliary surface $S_0 \in C^{(1,\upsilon)}$, located inside the particle D_i. Assume that in each point M_n there are three orthogonal dipoles $\{e_m\}_{m=1}^{3}$ and they are directed along the basic vectors of the cylindrical coordinate system (ρ, φ, z). Then the vector potential of the DS system can be written as

$$A_n^1 = \left\{G_{11}^e(M, M_n)\cos(\varphi - \varphi_n); \quad -G_{11}^e \sin(\varphi - \varphi_n); \right.$$
$$\left. \frac{\partial G_{31}(M, M_n)}{\partial \rho}\cos(\varphi - \varphi_n) - \frac{1}{\rho}\frac{\partial G_{31}}{\partial \varphi}\sin(\varphi - \varphi_n)\right\}$$

$$A_n^2 = \left\{G_{11}^e(M, M_n)\sin(\varphi - \varphi_n);\ \ G_{11}^e\cos(\varphi - \varphi_n);\ \frac{\partial G_{31}(M, M_n)}{\partial \rho}\sin(\varphi - \varphi_n) + \frac{1}{\rho}\frac{\partial G_{31}}{\partial \varphi}\cos(\varphi - \varphi_n)\right\} \quad (4.8)$$

$$A_n^3 = \{0;\ 0;\ G_{11}^h(M, M_n)\}$$

Here A_n^1 corresponds to dipoles located in M_n with coordinates (ρ_n, φ_n, z_n) directed along the basis vector e_ρ, $A_n^2 - e_\varphi$ and $A_n^3 - e_z$. Thus, for the scattered field in the domains $D_{0,1}$ satisfying the transmission conditions at Ξ the following representation holds

$$E_\zeta^N(M) = \sum_{n=1}^{N_0}\sum_{m=1}^{3} p_{nm}^0\ \nabla \times \nabla \times\ A_n^m(M),$$

$$H_\zeta^N(M) = \frac{j}{k\mu_\zeta}\nabla \times E_\zeta^N(M),\ \ \zeta = 0, 1;\quad M \in D_{0,1} \quad (4.9)$$

We would like to emphasize that the electromagnetic fields both in D_0 and in D_1 are defined by the unified set of amplitudes $\left\{\{p_{nm}^0\}_{n=1}^{N_0}\right\}_{m=1}^{3}$. This is a consequence of incorporating the GT for the scattered field representation. The main difference between the representation (4.9) and the field representation for axial symmetric structures [7] consists in that (4.9) is valid for non-axial symmetric geometries and for any kind of excitation regardless of its polarization.

Let us construct the representation for the field inside the penetrable particle D_i. Choose an auxiliary surface $S_1 \in C^{(1,\upsilon)}$, enclosing the scattering surface ∂D_i and consider the following vector potential system

$$A_{ni}^m(M) = h_0^{(2)}(k_i R_{MM_n^1})e_m;\ \ M_n^1 \in S_1\ . \quad (4.10)$$

Here $h_0^{(2)}$ is spherical Hankel function and e_m - are basic vectors of the cylindrical coordinate system. Then the representation for the total field inside the particle accepts the following form

$$E_i^N(M) = \sum_{n=1}^{N_i}\sum_{m=1}^{3} p_{nm}^i\ \nabla \times \nabla \times A_{ni}^m(M),\ \ H_i^N(M) = \frac{j}{k\mu_i}\nabla \times E_i^N(M),\ \ M \in D_i \quad (4.11)$$

The approximate solutions (4.9)–(4.11) satisfy all conditions of the scattering problem (4.1) except the transmission conditions enforced at the particle surface (4.4). Therefore by matching these conditions one can determine the set of DS amplitudes $\left\{\{p_{nm}^{0,i}\}_{m=1}^{3}\right\}_{n=1}^{N_{0,i}}$.

The completeness of the system of the distributed electric multipoles used in (4.8), (4.10) distributed over the auxiliary surfaces $S_{0,1}$ guarantees the convergence of the approximate solutions (4.9) and (4.11) to the exact ones [30]. Thus, the following result can be formulated:

Theorem *Let the penetrable particle D_i with smooth surface $\partial D_i \subset C^{(2,\upsilon)}$ be embedded into a plane substrate and the DS origins $\{M_n, M_n^1\}_{n=1}^{\infty}$ form dense sets at $S_{0,1}$, then for any external excitation $\{E_{0,1}^0, H_{0,1}^0\} \in L_\tau^2(\partial D_i)$ and any $\delta > 0$ there exist such numbers $N_{0,i}$ and the sets of amplitudes $\left\{\{p_{nm}^{0,i}\}_{m=1}^{3}\right\}_{n=1}^{N_{0,i}}$ that the following result holds*

$$\left\| n_p \times \begin{Bmatrix} E_i^N - E_e^N - E_{0,1}^0 \\ H_i^N - H_e^N - H_{0,1}^0 \end{Bmatrix} \right\|_{L_\tau^2(\partial D_i)} < \delta.$$

The latter result means convergence of the approximate solution to the exact one in any closed subset in D_0, that is

$$\lim_{N_{0,i}\to\infty} \{E_0^N, H_0^N\} = \{E_0^s, H_0^s\}, \text{ in } D_0.$$

4.3 Numerical Scheme of the DSM

In this section the numerical scheme of the DSM will be described in detail. As outlined above the representations (4.9), (4.11) satisfy all conditions of the scattering problem (4.1) except the transmission conditions (4.4) at the particle's surface ∂D_i. These last conditions (4.4) are used to determine the unknown amplitudes of the DS $\left\{\{p_{nm}^{0,i}\}_{m=1}^{3}\right\}_{n=1}^{N_{0,i}}$. The computational algorithm is based on the Generalized Point Matching Technique (GPMT) [32]. Choose a set of matching points (MPs) distributed over the whole particle surface $\{Q_l\}_{l=1}^{L} \in \partial D_i$, then the DS amplitudes can be evaluated as

$$p_{nm}^{0,i} := \arg\min \left\| n_p \times \begin{Bmatrix} E_i^N - E_e^N - E_{0,1}^0 \\ H_i^N - H_e^N - H_{0,1}^0 \end{Bmatrix} \right\|_{L_\tau^2(\partial D_i)} \tag{4.12}$$

Application of GPMT is followed by the pseudo solution of the corresponding over-determined system of linear algebraic equations [33]. Tikhonov's regularization in the l_2 sense is applied [34]. Adjusting the regularization parameter to the computational errors of the Sommerfeld integrals ensures stable and convergent results with increasing number of MPs and DS. As mentioned above, calculation of the surface residual of the boundary conditions (4.4) at intermediate points with respect to the MPs, we obtain a posteriori error estimate for the approximate solution (4.9), (4.11).

The unknown vector of amplitudes $\left\{\{p_{nm}^{0,i}\}_{m=1}^{3}\right\}_{n=1}^{N_{0,i}}$ of dimension $3(N_0 + N_i)$ can be computed as a pseudo solution of the following over-determined system of linear equation

$$\Phi p = f \tag{4.13}$$

Here Φ is a rectangular matrix having a dimension of $4L \times 3(N_0 + N_i)$, $4L \geq 3(N_0 + N_i)$ and the vector f: (4.13) can be represented as the following *4L* vector: $f = \left(E_\zeta^0 \cdot \tau^1,\ E_\zeta^0 \cdot \tau^2,\ H_\zeta^0 \cdot \tau^1,\ H_\zeta^0 \cdot \tau^2\right)^T$, where L is the number of the MPs and $\tau^{1,2}$ are orthogonal vectors tangential to the surface ∂D_i corresponding to each MP $Q_l \in \partial D_i$ and $\left\{E_\zeta^0, H_\zeta^0\right\}$, $\zeta = 0, 1$.

Let MPs and tangential vectors, which are defined by their Cartesian coordinates as $Q_l = (\rho_l \cos\varphi_l,\ \rho_l sin\varphi_l,\ z_l)$; $\tau_l^{1,2} = (\alpha_l^{1,2},\ \beta_l^{1,2},\ \gamma_l^{1,2})$. Then

$$\begin{aligned}
&\psi_\zeta^\pm(P_l) = \exp\left\{-jk_\zeta\left(\rho_l \sin\theta_\zeta \cos(\varphi_l - \varphi_0) \pm z_l \cos\theta_\zeta\right)\right\}\\
&e_\zeta^{\pm,P} \cdot \tau_l^{1,2} = \left(\mp\alpha_l^{1,2} \cos\theta_\zeta \cos\varphi_0 \mp \beta_l^{1,2} \cos\theta_\zeta \sin\varphi_0 + \gamma_l^{1,2} \sin\theta_\zeta\right)\\
&h_\zeta^{\pm,P} \cdot \tau_l^{1,2} = \left(\alpha_l^{1,2} \sin\varphi_0 - \beta_l^{1,2} \cos\varphi_0\right)
\end{aligned}$$

$$\begin{aligned}
&e_\zeta^{\pm,S} \cdot \tau_l^{1,2} = \left(-\alpha_l^{1,2} \sin\varphi_0 + \beta_l^{1,2} \cos\varphi_0\right)\\
&h_\zeta^{\pm,S} \cdot \tau_l^{1,2} = \left(\mp\alpha_l^{1,2} \cos\theta_\zeta \cos\varphi_0 \mp \beta_l^{1,2} \cos\theta_\zeta \sin\varphi_0 + \gamma_l^{1,2} \sin\theta_\zeta\right)
\end{aligned}$$

here $\left\{e_\zeta^{\pm,P,S};\ h_\zeta^{\pm,P,S}\right\}$ - notations introduced in (4.2), Following the notations above the components of vector f relevant to P/S polarization accept the following form

$$E_\zeta^{0(P,S)} \cdot \tau_l^{1,2} = \left[\begin{array}{l} \left(e_0^{-(P,S)} \cdot \tau_l^{1,2}\right) \psi_0^-(Q_l) + R_P \cdot \left(e_0^{+(P,S)} \cdot \tau_l^{1,2}\right) \psi_0^+(Q_l),\quad z \geq 0 \\ \qquad T_P \cdot \left(e_1^{-(P,S)} \cdot \tau_l^{1,2}\right) \psi_1^-(Q_l),\quad z < 0 \end{array}\right.$$

$$H_\zeta^{0(P,S)} \cdot \tau_l^{1,2} = \left[\begin{array}{l} \left(h_0^{-(P,S)} \cdot \tau_l^{1,2}\right) \psi_0^-(Q_l) + R_P \cdot \left(h_0^{+(P,S)} \cdot \tau_l^{1,2}\right) \psi_0^+(Q_l),\quad z \geq 0 \\ \qquad T_P \cdot \left(h_1^{-(P,S)} \cdot \tau_l^{1,2}\right) \psi_1^-(Q_l),\quad z < 0 \end{array}\right. \tag{4.14}$$

The solution of the linear system (4.13) is performed for P and S polarized excitation and all incident planes φ_0 and incident angles θ_0 at once. First we make a QR factorization of the Φ matrix and then calculate pseudo solutions for the whole set of the right hand parts [32].

Three columns of the Φ matrix (4.13) relating to the scattered field representation (4.9) can be written as

$$\vec{\Phi}_\zeta^m = \left\{W_\zeta^m,\ Y_\zeta^m,\ \frac{j}{k\mu_\zeta} V_\zeta^m,\ \frac{j}{k\mu_\zeta} U_\zeta^m\right\}^T, m = 1, 2, 3 \tag{4.15}$$

The rows associate to the following sequence of the field's components $E \cdot \tau^1$, $E \cdot \tau^2$, $H \cdot \tau^1$, $H \cdot \tau^2$. So for the field the matrix elements can be evaluated as $W \Rightarrow \tau^1 \cdot \nabla \times \nabla \times$, $Y \Rightarrow \tau^2 \cdot \nabla \times \nabla \times$, and for **H** field as follows $V \Rightarrow \tau^1 \cdot \nabla \times$, $U \Rightarrow \tau^2 \cdot \nabla \times$.

In each MP there are three electric dipoles oriented in accordance with cylindrical coordinate system. Specific expressions for the matrix elements corresponding to the scattered fields are:

$$\left[W_\zeta^m(Q_l)\right] = \alpha_l^1(w_{\zeta,\rho}^m \cos\varphi_l - w_{\zeta,\varphi}^m \sin\varphi_l) + \beta_l^1(w_{\zeta,\rho}^m \sin\varphi_l + w_{\zeta,\varphi}^m \cos\varphi_l) + \gamma_l^1 w_{\zeta,z}^m;$$

$$\left[Y_\zeta^m(Q_l)\right] = \alpha_l^2(w_{\zeta,\rho}^m \cos\varphi_l - w_{\zeta,\varphi}^m \sin\varphi_l) + \beta_l^2(w_{\zeta,\rho}^m \sin\varphi_l + w_{\zeta,\varphi}^m \cos\varphi_l) + \gamma_l^2 w_{\zeta,z}^m;$$

$$\left[V_\zeta^m(Q_l)\right] = \alpha_l^1(v_{\zeta,\rho}^m \cos\varphi_l - v_{\zeta,\varphi}^m \sin\varphi_l) + \beta_l^1(v_{\zeta,\rho}^m \sin\varphi_l + v_{\zeta,\varphi}^m \cos\varphi_l) + \gamma_l^1 v_{\zeta,z}^m;$$

$$\left[U_\zeta^m(Q_l)\right] = \alpha_l^2(v_{\zeta,\rho}^m \cos\varphi_l - v_{\zeta,\varphi}^m \sin\varphi_l) + \beta_l^2(v_{\zeta,\rho}^m \sin\varphi_l + v_{\zeta,\varphi}^m \cos\varphi_l) + \gamma_l^2 v_{\zeta,z}^m; \quad (4.16)$$

where the following notations have been introduced

$$w_{\zeta,\rho}^1 = k_\zeta^2 \cos(\varphi_l - \varphi_n) \int_0^\infty J_0(\lambda r)\, v_{11}^e \lambda d\lambda - \frac{\cos(\varphi_l - \varphi_n)}{r} \int_0^\infty J_1(\lambda r)\, \lambda \left[v_{11}^e + \frac{\partial v_{31}}{\partial z}\right] \lambda d\lambda -$$
$$- \frac{\rho_n - \rho_l \cos(\varphi_l - \varphi_n)}{r} \frac{\rho_l - \rho_n \cos(\varphi_l - \varphi_n)}{r} \int_0^\infty \left[\frac{2J_1(\lambda r)\lambda}{r} - J_0(\lambda r)\,\lambda^2\right]\left[v_{11}^e + \frac{\partial v_{31}}{\partial z}\right] \lambda d\lambda;$$

$$w_{\zeta,\varphi}^1 = -k_\zeta^2 \sin(\varphi_l - \varphi_n) \int_0^\infty J_0(\lambda r)\, v_{11}^e \lambda d\lambda + \frac{\sin(\varphi_l - \varphi_n)}{r} \int_0^\infty J_1(\lambda r)\, \lambda \left[v_{11}^e + \frac{\partial v_{31}}{\partial z}\right] \lambda d\lambda$$
$$- \frac{\rho_n - \rho_l \cos(\varphi_l - \varphi_n)}{r} \frac{\rho_n \sin(\varphi_l - \varphi_n)}{r} \int_0^\infty \left[\frac{2J_1(\lambda r)\lambda}{r} - J_0(\lambda r)\,\lambda^2\right]\left[v_{11}^e + \frac{\partial v_{31}}{\partial z}\right] \lambda d\lambda;$$

$$w_{\zeta,z}^1 = \frac{\rho_0 - \rho\cos(\varphi_l - \varphi_n)}{r} \int_0^\infty J_1(\lambda r)\,\lambda \left[\frac{\partial v_{11}^e}{\partial z} + v_{31}\lambda^2\right] \lambda d\lambda;$$

$$w_{\zeta,\rho}^2 = k_\zeta^2 \sin(\varphi_l - \varphi_n) \int_0^\infty J_0(\lambda r)\, v_{11}^e \lambda d\lambda - \frac{\sin(\varphi_l - \varphi_n)}{r} \int_0^\infty J_1(\lambda r)\, \lambda \left[v_{11}^e + \frac{\partial v_{31}}{\partial z}\right] \lambda d\lambda$$
$$+ \frac{\rho_l \sin(\varphi_l - \varphi_n)}{r} \frac{\rho_l - \rho_n \cos(\varphi_l - \varphi_n)}{r} \int_0^\infty \left[\frac{2J_1(\lambda r)\lambda}{r} - J_0(\lambda r)\,\lambda^2\right]\left[v_{11}^e + \frac{\partial v_{31}}{\partial z}\right] \lambda d\lambda;$$

$$w_{\zeta,\varphi}^2 = k_\zeta^2 \cos(\varphi_l - \varphi_n) \int_0^\infty J_0(\lambda r)\, v_{11}^e \lambda d\lambda - \frac{\cos(\varphi_l - \varphi_n)}{r} \int_0^\infty J_1(\lambda r)\, \lambda \left[v_{11}^e + \frac{\partial v_{31}}{\partial z}\right] \lambda d\lambda -$$
$$+ \frac{\rho_l \sin(\varphi_l - \varphi_n)}{r} \frac{\rho_n \sin(\varphi_l - \varphi_n)}{r} \int_0^\infty \left[\frac{2J_1(\lambda r)\lambda}{r} - J_0(\lambda r)\,\lambda^2\right]\left[v_{11}^e + \frac{\partial v_{31}}{\partial z}\right] \lambda d\lambda; \quad (4.17)$$

$$w_{\zeta,z}^2 = -\frac{\rho_l \sin(\varphi_l - \varphi_n)}{r} \int_0^\infty J_1(\lambda r)\,\lambda \left[\frac{\partial v_{11}^e}{\partial z} + v_{31}\lambda^2\right] \lambda d\lambda$$

$$w_{\zeta,\rho}^3 = -\frac{\rho_l - \rho_n \cos(\varphi_l - \varphi_n)}{r} \int_0^\infty J_1(\lambda r)\,\lambda \frac{\partial v_{11}^h}{\partial z} \lambda d\lambda$$

$$w^3_{\zeta,\varphi} = -\frac{\rho_l - \rho_n \sin\left(\varphi_l - \varphi_n\right)}{r} \int\limits_0^\infty J_1\left(\lambda r\right) \lambda \frac{\partial v^h_{11}}{\partial z} \lambda d\lambda,$$

$$w^3_{\zeta,z} = \int\limits_0^\infty J_0\left(\lambda r\right) v^h_{11} \lambda^2 \lambda d\lambda$$

and

$$\begin{aligned}
v^1_{\zeta,\rho} &= \sin\left(\varphi_l - \varphi_n\right) \int_0^\infty J_0\left(\lambda r\right) \tfrac{\partial v^e_{11}}{\partial z} \lambda d\lambda + \tfrac{\sin(\varphi_l - \varphi_n)}{r} \int_0^\infty J_1\left(\lambda r\right) \lambda v_{31} \lambda d\lambda - \\
&- \tfrac{\rho_n - \rho_l \sin(\varphi_l - \varphi_n)}{r} \tfrac{\rho_n \sin(\varphi_l - \varphi_n)}{r} \int_0^\infty \left[\tfrac{2J_1(\lambda r)\lambda}{r} - J_0\left(\lambda r\right)\lambda^2\right] v_{31} \lambda d\lambda \\
v^1_{\zeta,\varphi} &= \cos\left(\varphi_l - \varphi_n\right) \int_0^\infty J_0\left(\lambda r\right) \tfrac{\partial v^e_{11}}{\partial z} \lambda d\lambda + \tfrac{\cos(\varphi_l - \varphi_n)}{r} \int_0^\infty J_1\left(\lambda r\right) \lambda v_{31} \lambda d\lambda + \\
&+ \tfrac{\rho_n - \rho_l \cos(\varphi_l - \varphi_n)}{r} \tfrac{\rho_l - \rho_n \cos(\varphi_l - \varphi_n)}{r} \int_0^\infty \left[\tfrac{2J_1(\lambda r)\lambda}{r} - J_0\left(\lambda r\right)\lambda^2\right] v_{31} \lambda d\lambda \\
v^1_{\zeta,z} &= \rho_l \frac{\sin\left(\varphi_l - \varphi_n\right)}{r} \int\limits_0^\infty J_1\left(\lambda r\right) \lambda v^e_{11} \lambda d\lambda \\
v^2_{\zeta,\rho} &= -\cos\left(\varphi_l - \varphi_n\right) \int_0^\infty J_0\left(\lambda r\right) \tfrac{\partial v^e_{11}}{\partial z} \lambda d\lambda - \tfrac{\cos(\varphi_l - \varphi_n)}{r} \int_0^\infty J_1\left(\lambda r\right) \lambda v_{31} \lambda d\lambda + \\
&+ \tfrac{\rho_l \sin(\varphi_l - \varphi_n)}{r} \tfrac{\rho_n \sin(\varphi_l - \varphi_n)}{r} \int_0^\infty \left[\tfrac{2J_1(\lambda r)\lambda}{r} - J_0\left(\lambda r\right)\lambda^2\right] v_{31} \lambda d\lambda \\
v^2_{\zeta,\varphi} &= \sin\left(\varphi_l - \varphi_n\right) \int_0^\infty J_0\left(\lambda r\right) \tfrac{\partial v^e_{11}}{\partial z} \lambda d\lambda + \tfrac{\sin(\varphi_l - \varphi_n)}{r} \int_0^\infty J_1\left(\lambda r\right) \lambda v_{31} \lambda d\lambda - \\
&- \tfrac{\rho_l \sin(\varphi_l - \varphi_n)}{r} \tfrac{\rho_l - \rho_n \cos(\varphi_l - \varphi_n)}{r} \int_0^\infty \left[\tfrac{2J_1(\lambda r)\lambda}{r} - J_0\left(\lambda r\right)\lambda^2\right] v_{31} \lambda d\lambda
\end{aligned} \tag{4.18}$$

$$v^2_{\zeta,z} = \frac{\rho_n - \rho_l \cos\left(\varphi_l - \varphi_n\right)}{r} \int\limits_0^\infty J_1\left(\lambda r\right) \lambda v^e_{11} \lambda d\lambda,$$

$$v^3_{\zeta,\rho} = -\frac{\rho_n \sin\left(\varphi_l - \varphi_n\right)}{r} \int\limits_0^\infty J_1\left(\lambda r\right) \lambda v^h_{11} \lambda d\lambda$$

$$v^3_{\zeta,\varphi} = -\frac{\rho - \rho_0 \cos\left(\varphi_l - \varphi_n\right)}{r} \int\limits_0^\infty J_1\left(\lambda r\right) \lambda v^h_{11} \lambda d\lambda, \; v^3_{\zeta,z} = 0$$

We emphasize that we removed the singular part from the first and last lines of the spectral functions $v^{e,h}_{11}$ representation (4.6). Next we would like to consider the evaluation of the GT components (4.5). There are two possible singularities associated with the first line in (4.6)

$$\frac{\exp\{-\eta_0 |z - z_0|\}}{\eta_0} + A_{11}^{e,h}(\lambda) \cdot \frac{\exp\{-\eta_0 (z + z_0)\}}{\eta_0}, \; z_0 > 0, \; z \geq 0, \; .$$

under $\lambda = k_0$. The first singularity corresponds to the GT of the ambient space and it can be represented in clear analytic form. The other one is associated with the second term, which can be evaluated in the following way:

1. extract the vicinity of the singularity $\lambda = k_0$ and consider the integral $I(\delta) = \int\limits_{k_0-\delta}^{k_0+\delta} \frac{F(\lambda)d\lambda}{\sqrt{\lambda^2 - k_0^2}}$. The latter integral can be estimated as $I(\delta) \approx \sqrt{\frac{2\delta}{k_0}} F(k_0), \;\; \delta \to 0.$
2. for the residual of the Sommerfeld integral in the infinite domain we compute using double-exponential code [35] that enables to evaluate the integral with any desired accuracy degree. When this accuracy has been assigned as ε one can estimate the δ value so that $I(\delta) \leq \varepsilon$. We prefer not use the integral path transition to a complex plane because DS coordinates M_n can be complex valued. DSM can also be applied to analyze light scattering by a particle located at a stratified medium (see for example [36]).

The matrix elements related to the internal field representation (4.11) look similar to (4.16) where

$$\begin{aligned} w_{\zeta,\rho}^1 &= k_\zeta^2 \cos(\varphi_l - \varphi_n) X_\zeta^0 - k_\zeta \frac{\cos(\varphi_l - \varphi_n)}{r} X_\zeta^1 \\ &- \frac{\rho_n - \rho_l \cos(\varphi_l - \varphi_n)}{r} \frac{\rho_l - \rho_n \cos(\varphi_l - \varphi_n)}{r} X_\zeta^2; \end{aligned}$$

$$\begin{aligned} w_{\zeta,\varphi}^1 &= -k_\zeta^2 \sin(\varphi_l - \varphi_n) X_\zeta^0 + k_\zeta \frac{\sin(\varphi_l - \varphi_n)}{r} X_\zeta^1 \\ &- k_\zeta^2 \frac{\rho_n - \rho_l \cos(\varphi_l - \varphi_n)}{r} \frac{\sin(\varphi_l - \varphi_n)}{r} X_\zeta^2; \end{aligned}$$

$$w_{\zeta,z}^1 = -k_\zeta^2 \frac{\rho_n - \rho_l \cos(\varphi_l - \varphi_n)}{r} \frac{z_l - z_n}{r} X_\zeta^2;$$

$$\begin{aligned} w_{\zeta,\rho}^2 &= k_\zeta^2 \sin(\varphi_l - \varphi_n) X_\zeta^0 - k_\zeta \frac{\sin(\varphi_l - \varphi_n)}{r} X_\zeta^1 \\ &+ k_\zeta^2 \frac{\rho_l - \rho_n \cos(\varphi_l - \varphi_n)}{r} \frac{\sin(\varphi_l - \varphi_n)}{r} \rho_l X_\zeta^2; \end{aligned}$$

$$w_{\zeta,\varphi}^2 = k_\zeta^2 \cos(\varphi_l - \varphi_n) X_\zeta^0 - k_\zeta \frac{\cos(\varphi_l - \varphi_n)}{r} X_\zeta^1 + k_\zeta^2 \left(\frac{\sin(\varphi_l - \varphi_n)}{r} \right)^2 \rho_l \rho_n X_\zeta^2; \tag{4.19}$$

$$w^2_{\zeta,z} = k^2_\zeta \frac{\sin(\varphi_l - \varphi_n)}{r} \frac{z_l - z_n}{r} \rho_l X^2_\zeta;$$

$$w^3_{\zeta,\rho} = k^2_\zeta \frac{\rho_l - \rho_n \cos(\varphi_l - \varphi_n)}{r} \frac{z_l - z_n}{r} X^2_\zeta;$$

$$w^3_{\zeta,\varphi} = k^2_\zeta \frac{\sin(\varphi_l - \varphi_n)}{r} \frac{z_l - z_n}{r} \rho_n X^2_\zeta;$$

$$w^3_{\zeta,z} = k^2_\zeta X^0_\zeta - k_\zeta \frac{1}{r} X^1_\zeta + k^2_\zeta \left(\frac{z_l - z_n}{r}\right)^2 X^2_\zeta;$$

and

$$v^1_{\zeta,\rho} = -k_\zeta \frac{z_l - z_n}{r} \sin(\varphi_l - \varphi_n) X^1_\zeta;$$

$$v^1_{\zeta,\varphi} = -k_\zeta \frac{z_l - z_n}{r} \cos(\varphi_l - \varphi_n) X^1_\zeta;$$

$$v^1_{\zeta,z} = k_\zeta \frac{\sin(\varphi_l - \varphi_n)}{r} \rho_l X^1_\zeta;$$

$$v^2_{\zeta,\rho} = k_\zeta \frac{z_l - z_n}{r} \cos(\varphi_l - \varphi_n) X^1_\zeta;$$

$$v^2_{\zeta,\varphi} = -k_\zeta \frac{z_l - z_n}{r} \sin(\varphi_l - \varphi_n) X^1_\zeta; \qquad (4.20)$$

$$v^2_{\zeta,z} = k_\zeta \frac{\rho_n - \rho_l \cos(\varphi_l - \varphi_n)}{r} X^1_\zeta;$$

$$v^3_{\zeta,\rho} = -k_\zeta \frac{\sin(\varphi_l - \varphi_n)}{r} \rho_n X^1_\zeta;$$

$$v^3_{\zeta,\varphi} = k_\zeta \frac{\rho_l - \rho_n \cos(\varphi_l - \varphi_n)}{r} X^1_\zeta;$$

$$v^3_{\zeta,z} = 0.$$

where $\zeta = i, \quad M^1_n = (\rho_n, \varphi_n, z_n) \in S_1$.

The singular part of the matrix elements related to the scattered field representation (4.9) can written in the same manner as (4.19) and (4.20) assuming that $X^m_\zeta = \frac{k_\zeta}{j} h^{(2)}_m(k_\zeta R_{Q_l M_n}), \quad \zeta = 0, 1, \quad M_n \in S_0$ are used.

The main differences between the present numerical scheme and the one used for the axial-symmetric case [7] consist in following:

1. The approximate solution (4.9), (4.11) does not depend on the polarization of the external excitation and can be directly applied for all available polarizations at once.
2. The numerical scheme enables rotation of the incident plane by varying φ_0.
3. The representation for the scattered field (4.9) contains dipoles DS only in contrast to the axial-symmetric scheme which involves multipoles as well. This circumstance makes the computation of Sommerfeld integrals (4.5) much easier, less time consuming and more accurate.

To compute the scattered field intensity, we need the corresponding far field pattern $F(\theta, \varphi)$ [12]. In the case under consideration, the far field pattern is determined only in the upper half space D_0 as

$$E(r)/\left|E^0(z=0)\right| = \frac{\exp\{-jk_0 r\}}{r} F(\theta, \varphi) + O(1/r^2), \quad r = |M| \to \infty, \quad z > 0. \tag{4.21}$$

The far field pattern in the frame of the DSM is obtained using the asymptotic representations of the Sommerfeld integrals as described in [27, 37]. Then, the (θ, φ)-components of the scattering diagram, can be written in the following form

$$F_\theta(\theta_0, \theta; \varphi, \varphi_0) = jk \sum_{n=1}^{N_e} \{p_{n1}^0 f_{\theta 1} + p_{n2}^0 f_{\theta 2} - p_{n3}^0 f_{\theta 3}\}$$

$$F_\varphi(\theta_0, \theta; \varphi, \varphi_0) = jk \sum_{n=1}^{N_e} \{-p_{n1}^0 f_{\varphi 1} + p_{n2}^0 f_{\varphi 2}\} \tag{4.22}$$

where

$$f_{\theta 1} = \left\{\psi^+ \delta_{0\zeta} + [\bar{v}_{11}(\theta) + \bar{v}_{31}(\theta) \sin^2\theta] \cdot \psi^-\right\} \cos\theta \cos(\varphi - \varphi_n)$$

$$f_{\theta 2} = \left\{\psi^+ \delta_{0\zeta} + [\bar{v}_{11}(\theta) + \bar{v}_{31}(\theta) \sin^2\theta] \cdot \psi^-\right\} \cos\theta \sin(\varphi - \varphi_n);$$

$$f_{\theta 3} = \left\{\psi^+ \delta_{0\zeta} + \bar{v}_{33}(\theta) \cdot \psi^-\right\} \sin\theta \tag{4.23}$$

$$f_{\varphi 1} = \left\{\psi^+ \delta_{0\zeta} + \bar{v}_{11}(\theta) \cdot \psi^-\right\} \sin(\varphi - \varphi_n);$$

$$f_{\varphi 2} = \left\{\psi^+ \delta_{0\zeta} + \bar{v}_{11}(\theta) \cdot \psi^-\right\} \cos(\varphi - \varphi_n)$$

Here (ρ_n, φ_n, z_n) are cylindrical coordinates of $M_n \in S_0$, $\delta_{0\zeta}$ - Kronecker delta $\zeta = 0, 1$, $\psi^\pm$ and spectral functions accept the following expressions

$$\psi^+ = \exp\{ik_0(\rho_n \sin\theta \cos(\varphi - \varphi_n) + z_n \cos\theta)\};$$

$$\psi^- = \begin{bmatrix} \exp\{jk_0(\rho_n \sin\theta\cos(\varphi-\varphi_n) - z_n\cos\theta)\}, & z_n \geq 0 \\ \exp\{jk_0(\rho_n \sin\theta\cos(\varphi-\varphi_n) + z_n\kappa(\theta))\}, & z_n < 0 \end{bmatrix}$$

$$\overline{v}_{11}(\theta) = \begin{bmatrix} \frac{\mu_1\cos\theta-\mu_0\kappa(\theta)}{\mu_1\cos\theta+\mu_0\kappa(\theta)}, & z_n \geq 0 \\ \frac{2\mu_0\cos\theta}{\mu_1\cos\theta+\mu_0\kappa(\theta)}, & z_n < 0 \end{bmatrix}, \overline{v}_{33}(\theta) = \begin{bmatrix} \frac{\varepsilon_1\cos\theta-\varepsilon_0\kappa(\theta)}{\varepsilon_1\cos\theta+\varepsilon_0\kappa(\theta)}, & z_n \geq 0 \\ \frac{2\varepsilon_0\cos\theta}{\varepsilon_1\cos\theta+\varepsilon_0\kappa(\theta)}, & z_n < 0 \end{bmatrix}, \tag{4.24}$$

$$\overline{v}_{31}(\theta) = \frac{2(\mu_1\varepsilon_1 - \mu_0\varepsilon_0)}{(\mu_1\cos\theta + \mu_0\kappa(\theta))\cdot(\varepsilon_1\cos\theta+\varepsilon_0\kappa(\theta))}, \quad \kappa(\theta) = \sqrt{\varepsilon_1 - \sin^2\theta}.$$

Thus, the components of far field pattern do not contain any integrals. After determining of the unknown DS amplitudes $\left\{\{p^0_{nm}\}_{n=1}^{N_0}\right\}_{m=1}^{3}$ the scattering characteristics are obtained by computing finite linear combinations of elementary functions (4.23)–(4.24). As a result, we can analyze all scattering characteristics in detail, such as the Differential Scattering Cross-Section (DSC)

$$DSC(\theta_0, \theta; \varphi, \varphi_0) = |F_\theta(\theta_0, \theta; \varphi, \varphi_0)|^2 + \left|F_\varphi(\theta_0, \theta; \varphi, \varphi_0)\right|^2 \tag{4.25}$$

where $F^{P,S}_{\theta,\varphi}(\theta_0, \theta, \varphi)$ are the corresponding components of the scattering diagram (4.22), or the Total Scattering Cross-Section (TSC), i.e., the intensity of the scattered field integrated over a given solid angle Ω (response):

$$\sigma(\theta_0; \varphi_0) = \int_\Omega DSC\ (\theta_0, \theta; \varphi, \varphi_0)\ d\omega, \tag{4.26}$$

Via the definition of far field pattern (4.21), the unit of DSC and TSC is μm^2. In our case we will use $\Omega = \{89° \geq \theta \geq 0°;\ \ 360° \geq \varphi \geq 0°\}$ which belongs to the upper unit hemi-sphere.

Summarising, the computation scheme of the non-axial symmetric version of the DSM can be described as consisting of the following steps:

1. Estimation of the numbers of matching points L distributed over the whole scattering surface ∂D_i accounting for the value of the exciting wavelength λ, corresponding to the surface area of the particle and the particle refractive index;
2. The number of DS needed for the internal N_i and external fields N_0 depends on the refractive indices of the particle and the substrate, their numbers should follow the rule $3(N_0 + N_i) \leq 4L$;
3. Complete matrix Φ: (4.13) computation and extension of the matrix by applying regularization in the $l_2$sense and adding an associating regularization parameter adjusted to the errors of Sommerfeld integrals evaluation;
4. Extended matrix factorization using the QR algorithm;
5. Computing the corresponding right hand parts according to (4.14) and evaluation of the pseudo solution for all the incident planes $0 \div \varphi_0$, incident angles $0 \div \theta_0$ and both polarizations P/S simultaneously after the matrix QR resolution was determined;

6. Determination of the DS amplitudes for the all incident planes, angles and both polarizations.
7. For some particular incidence, checking the relative surface residual at the testing points distributed over the whole particle surface ∂D_i employing transmission conditions (4.4).
8. Computation of the scattering characteristics required: DSC (4.25) and TSC (4.26).

4.4 Simulation Results

The verification of the non-axial symmetric version of the DSM computer model has been performed in [38]. In all our computations the relative surface residual corresponding to (4.4) did not exceed 0.6% in mean square norm with respect to the incident field and the convergence test showed that this is enough to guarantee two valid digits of the computed DSC.

Similar to [38] we consider a P/S-polarized plane wave of wavelength $\lambda = 196$ nm as an external excitation. For this wavelength, the corresponding refractive index of Si substrate was chosen equal to $n_1 = 1.004 - 2.14j$ [39]. For simulation of the line bump and line pit we choose a prolate spheroid as a fictitious particle which has larger axis a = 120 nm parallel to the 0x axis and its aspect ratio is a/b = 4. This fictitious spheroid was embedded into the substrate with a depth of 15 nm. So the length of the feature was 120 nm, width 30 nm and height 15 nm. To simulate scattering different kinds of the line features we vary the corresponding selecting refractive index of the substrate for the bump and air refractive index for the pit.

The behavior of P/S-polarized TSC: (4.26) - $\sigma^{P,S}(\theta_0, \varphi_0)$ versus the incident angle θ_0 in two incident planes $\varphi_0 = 90°; 180°$ are depicted in Fig. 4.1 for the line bump and in Fig. 4.2 for the line pit. At the same figure the TSC corresponding to a

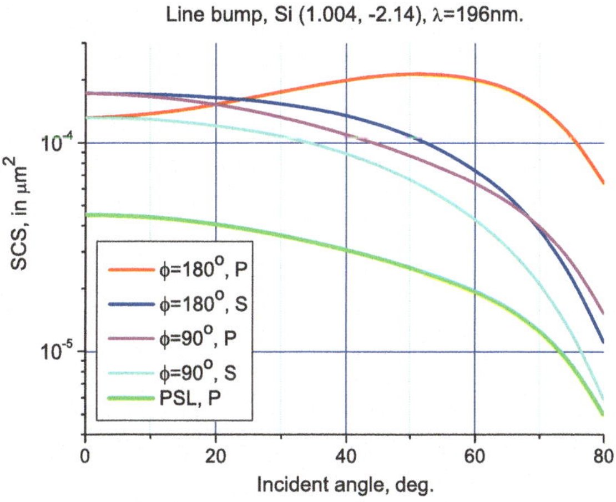

Fig. 4.1 TSC: - $\sigma^{P,S}(\theta_0, \varphi_0)$: (4.26) for the line bump versus incident angle θ_0 in two incident planes $\varphi_0 = 90°; 180°$. Results for PSL sphere D = 35 nm, P-polarized excitation are shown as well

Fig. 4.2 Similar results: $\sigma^{P,S}(\theta_0, \varphi_0)$ versus θ_0 for the line pit

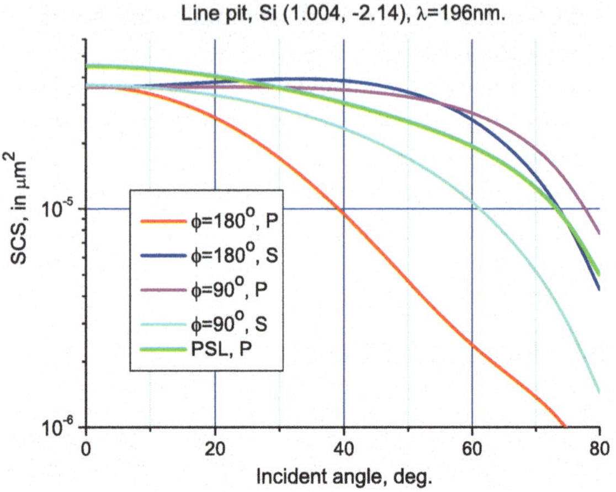

Fig. 4.3 $\sigma^{P,S}(\theta_0, \varphi_0)$ versus incident plane φ_0 for the line bump under oblique incidence $\theta_0 = 50°$. σ^P for PSL D = 35 nm is depicted as a level of defects visibility

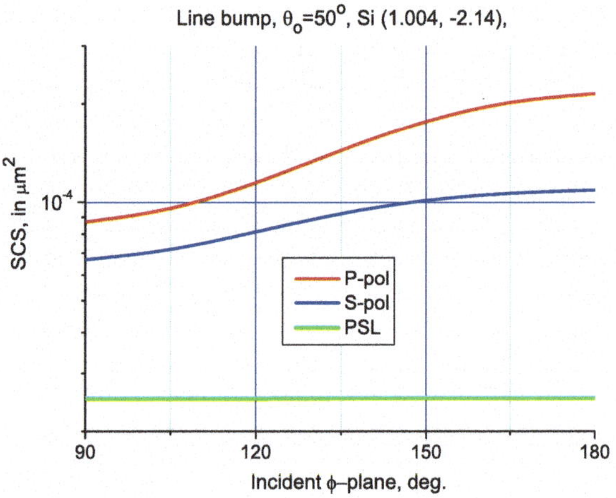

Polystyrene Latex (PSL) sphere ($n_i = 1.62$) of diameter D = 35 nm is shown. This PSL sphere diameter corresponds to the limit of resolution ability of some surface scanners was are used for substrate examination [40]. We remind that the larger axis of the spheroid belongs to the plane $\varphi_0 = 180°$ and that the $\varphi_0 = 90°$ plane is perpendicular to the larger axis. The behavior of the curves in Fig. 4.1 demonstrates that almost all TSCs decrease versus the incident angle except $\sigma^P(\theta_0, 180°)$ for a line bump which increases. Nevertheless the TSCs for the line bump exceed the PSL TSC in the whole range of incident angles. Similar style of the TSCs corresponding to the line pit can be observed in Fig. 4.2. But in contrast to Fig. 4.1 all curves drop down versus the incident angle. Besides some TSCs for the line pit exceed the TSC for PSL only at the incidence behind 30°.

Figure 4.3 demonstrates TSCs for the line bump under the incident angle $\theta_0 = 50°$ versus the incident plane φ_0 variation from 90° to 180°. As it is clear that the integral

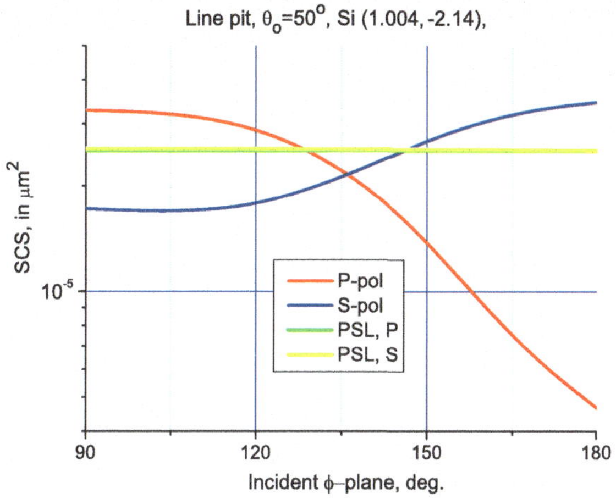

Fig. 4.4 Similar results: $\sigma^{P,S}(\theta_0, \varphi_0)$ versus φ_0 for the line pit

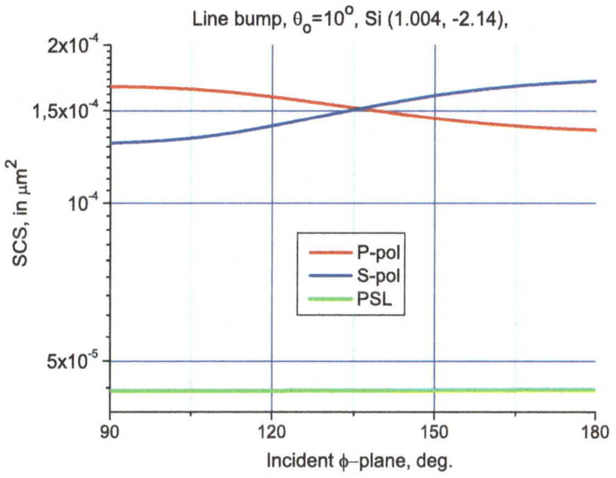

Fig. 4.5 $\sigma^{P,S}(\theta_0, \varphi_0)$ versus incident plane φ_0 for a line bump under small angle of incidence $\theta_0 = 10°$. σ^P for PSL D = 35 nm is depicted as a level of defects visibility

responses both for P and S-polarized light surpass the response for the PSL sphere in the whole range of the incident planes. Similar results are depicted in Fig. 4.4 for the line pit. In contrast to the line bump for line pit P-polarized response decreases versus the incident plane while S-polarized increases. Besides the most parts of the integral responses for P/S polarized excitation corresponding to the incident directions are less then the responses for the PSL particle. The latter means that it is possible to miss a line pit defect during the substrate examination via the optical scanner employing [40].

In Figs. 4.5 and 4.6 one can see the results for $\sigma^{P,S}(\theta_0, \varphi_0)$ versus the incident plane corresponding to an incident angle $\theta_0 = 10°$. If for the line bump the TSCs exceed the integral response for PSL in the whole range of incidences for the line pit the situation it is just the opposite. The PSL response is much higher than the response for the line pit.

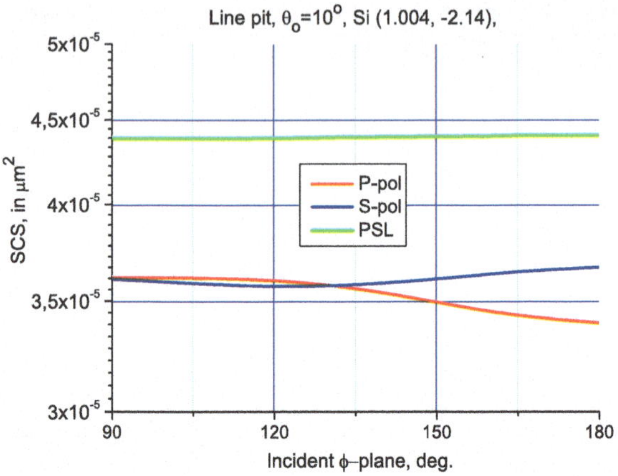

Fig. 4.6 Similar results: $\sigma^{P,S}(\theta_0, \varphi_0)$ versus φ_0 for the line pit

4.5 Conclusion

A new fictitious particle concept has been realized based on the Discrete Sources Method (DSM) to analyze polarized light scattering by line features on a plane substrate. We demonstrated that this extension enables to calculate light scattering by nano-dimensional line defects. This approach makes it possible to perform a comparative analysis of the scattering properties of a line bump, a line pit and standard PSL particles. The Total Scattering Cross-Section versus incident angle and incidence plane of exciting field have been examined and discussed. It was found that the total scattering intensity can be changed by an order of magnitude depending on the orientation of the linear defects with respect to the plane of external excitation.

Acknowledgements We gratefully acknowledge funding of this research by Deutsche Forschungsgemeinschaft (DFG) and Russian Foundation for Basic Research (RFBR).

References

1. N.G. Khlebtsov, L.A. Dykman, J. Quant. Spectrosc. Radiat. Transf. **111**, 1–35 (2010)
2. L. Novotny, N. van Hulst, Nat. Photonics **2**, 83–90 (2011)
3. H.-W. Chang, J. Lee, S. Hofmann, Y.H. Kim, L. Müller-Meskamp, B. Lüssem, C.-C. Wu, K. Leo, M.C. Gather, J. Appl. Phys. **113**, 204502 (2013)
4. E. Eremina, Yu. Eremin, T. Wriedt, J. Quant. Spectrosc. Radiat. Transf. **100**, 131–136 (2006)
5. A. Teulle, R. Marty, C. Girard, A. Arbouet, E. Dujardin, Opt. Commun. **291**, 412–415 (2013)
6. V. Jindal, Solid State Technol. **56**(3), 1–5 (2013)
7. A. Taflove, S. Hagness, *Computational Electrodynamics: The Finite-difference Time-domain Method* (Artech House, Norwood, 2005)
8. Ch. Hafner, Boundary methods for optical nano structures. J. Phys. Stat. Sol. B. **244**, 3435–3447 (2007)
9. I.R. Capoglu, G.S. Smith, IEEE Trans. Antennas Propag. **52**, 3805–3814 (2006)

10. J. Jin, *Finite Element Method in Electromagnetics* (Wiley, New York, 2002)
11. J. Alegret, P. Johansson, M. Kall, New J. Phys. **10**, 105004 (2008)
12. D. Colton, R. Kress, *Integral Equation Methods in Scattering Theory* (Wiley, New York, 1983)
13. Yu.A. Eremin, E.V. Zakharov, N.I. Nesmeyanova, The method of fundamental solutions in problems of diffraction of electromagnetic waves by bodies of revolution, *Seven Papers in Applied Mathematics*, vol. 125, American Mathematical Society Translations: Series 2 (1986), pp. 51–66
14. V.L.Y. Loke, M.P. Mengüç, T.A. Nieminen, J. Quant. Spectrosc. Radiat. Transf. **112**, 1711–1725 (2011)
15. M.A. Yurkin, Computational approaches for plasmonics, in *Handbook of Molecular Plasmonics*, ed. by S. D'Agostino (Pan Stanford Publishing, Singapore, 2013), pp. 83–135
16. T. Søndergaard, Phys. Status Solidi B Basic Res. **244**(10), 3448–3462 (2007)
17. Yu.A. Eremin, V.I. Ivakhnenko, Analysis of inhomogenities on substrates based on integral transformation method. Differ. Equ. **36**(9), 137–1380 (2000)
18. M.I. Mishchenko, N.T. Zakharova, N.G. Khlebtsov, T. Wriedt, G. Videen, J. Quant. Spectrosc. Radiat. Transf. **146**, 349–354 (2014)
19. A.M. Kern, O.J.F. Martin, J. Opt. Soc. Am. A **26**, 732–740 (2009)
20. Ch. Hafner, J. Smajic, M. Agio, Numerical methods for the electrodynamic analysis of nanostructures, in *Nanoclusters and Nanostructured Surfaces*, ed. by A.K. Ray (American Scientific Publishers, Stevenson Ranch, 2010), pp. 207–274
21. E. Eremina, Y. Eremin, T. Wriedt, J. Modern Opt. **58**(5–6), 384–399 (2011)
22. M. Kahnert, Numerical solutions of the macroscopic Maxwell equations for scattering by non-spherical particles: a tutorial review. J. Quant. Spectrosc. Radiat. Transf. **178**, 22–37 (2016)
23. B. Gallinet, J. Butet, O.J.F. Martin, Numerical methods for nanophotonics: standard problems and future challenges (Review). Laser Photonics Rev. **9**(6), 577–603 (2015)
24. Ch. Hafner, *Post-modern Electromagnetics. Using Intelligent MaXwell Solvers* (Wiley, New York, 1999)
25. Yu.A. Eremin, Differ. Equ. **45**(9), 1379–1385 (2009)
26. R. Schaback, H. Wendland, Acta Numer. **15**, 543–639 (2006)
27. Yu.A. Eremin, N.V. Orlov, A.G. Sveshnikov, in *Generalizes Multipole Techniques for Electromagnetic and Light Scattering*, ed. by T. Wriedt (Elsevier Science, Amsterdam, 1999), pp. 39–80
28. A. Doicu, Yu.A. Eremin, T. Wriedt, *Acoustic and Electromagnetic Scattering Analysis Using Discrete Sources* (Academic Press, New York, 2000)
29. H. Fujiwara, *Spectroscopic Ellipsometry. Principles and Applications* (Wiley, New York, 2007)
30. Yu.A. Eremin, A.G. Sveshnikov, Comput. Math. Math. Phys. **39**(12), 1967–1980 (1999)
31. W.C. Chew, *Waves and Fields in Inhomogeneous Media* (Wiley-IEEE Press, 1999)
32. N.S. Bahvalov, *Numerical Methods: Analysis, Algebra, Ordinary Differential Equations* (MIR, Moscow, 1977)
33. A. Ben-Israel, T.N.E. Greville, *Generalized Inverses: Theory and Applications* (Springer, New York, 2003)
34. V.A. Morozov, *Regularization Methods For Ill Posed Problems* (Springer, Berlin, 1984)
35. M. Mori, M. Sugihara, J. Comput. Appl. Math. **127**, 287–296 (2001)
36. E. Eremina, Y. Eremin, N. Grishina, T. Wriedt, J. Opt. **14**(1), 015001 (2012)
37. Yu.A. Eremin, J.C. Stover, N.V. Orlov, Opt. Eng. **38**(8), 1296–1304 (1999)
38. Yu. Eremin, T. Wriedt, Comput. Phys. Commun. **198**, 12–21 (2016)
39. http://refractiveindex.info/
40. J.C. Stover, *Optical Scattering: Measurement and Analysis* (SPIE Press, Bellingham, 2012)

Chapter 5
Convergent Fields Generated by Divergent Currents in the Method of Auxiliary Sources

George Fikioris and Nikolaos L. Tsitsas

Abstract A very simple scattering problem is considered, together with its solution obtained via the Method of Auxiliary Sources (MAS). We show that it is possible to concurrently have divergence of the auxiliary currents together with convergence of the scattered field generated by these divergent currents. The divergence manifests itself as rapid, unphysical oscillations in the auxiliary currents. It is stressed that the oscillations are not due to effects such as roundoff, matrix ill-conditioning, or to the well-studied phenomenon of internal resonances. We arrive at our conclusions using a number of means including asymptotic methods and a thorough discussion of the singularities of the analytic continuation of the scattered field. We also make a detailed comparison to corresponding discretizations of the Extended Integral Equation (EIE), in which similar phenomena do not occur. Analogies to superdirectivity and extensions to more complicated geometries are pointed out.

5.1 Introduction

Dealing theoretically with an electromagnetic or light-scattering problem is today often a routine task. With an abundance of numerical techniques to choose from, solving a problem frequently reduces simply to a choice between numerical methods, or between computer codes that implement the methods. As a consequence, it is usual to forget or ignore fundamental aspects of the numerical technique we use.

This chapter is an overview of the recent works [1–9]. It discusses a certain difficulty—which we consider fundamental and often overlooked or misunderstood—within certain types of Generalized Multipole Techniques (GMT). While intrinsic

G. Fikioris (✉)
School of Electrical and Computer Engineering, National Technical University of Athens, 15773 Athens, Greece
e-mail: gfiki@ece.ntua.gr

N. L. Tsitsas
Department of Informatics, Aristotle University of Thessaloniki, 54124 Thessaloniki, Greece
e-mail: ntsitsas@csd.auth.gr

T. Wriedt and Y. Eremin (eds.), *The Generalized Multipole Technique for Light Scattering*, Springer Series on Atomic, Optical, and Plasma Physics 99,
https://doi.org/10.1007/978-3-319-74890-0_5

to these techniques, the difficulty is nonetheless subtle because, even when it does occur, we can frequently obtain correct *final* results.

For definiteness, we discuss the difficulty within the Method of Auxiliary Sources (MAS). MAS is, in its simplest version, a particular implementation of GMT where we first seek electric currents that lie on a closed auxiliary surface within an impenetrable scatterer. These currents (we will call them "MAS currents") are such that the proper boundary conditions are satisfied. The MAS currents are intermediate results: after we find them (typically as the solution of an $N \times N$ linear algebraic system, where N is the number of MAS currents), we subsequently determine the field they generate. This field ("MAS field") is the MAS approximation to the true scattered field, and is usually the quantity we finally desire.

The difficulty to be discussed is easily visualized through example: Figs. 5.1 and 5.2 show the MAS currents and MAS field obtained for a certain two-dimensional scattering problem. The $N = 60$ MAS currents are numbered consecutively as we move around the auxiliary surface, and the abscissa in Fig. 5.1 is the index ℓ ($\ell = 0, 1, \ldots, N - 1$) of the MAS current I_ℓ (the real and imaginary part of this current appears as the ordinate of the figure). In Fig. 5.2, the abscissa is the azimuthal observation angle ϕ_{obs}. More details (a description of the particular scattering problem and of the specific way we implemented the MAS, problem dimensions, operating frequency, etc.) will be given later (in Sects. 5.2 and 5.3).

The MAS field shown in Fig. 5.2 is smooth and, one might suspect, free of serious difficulties. In fact, a comparison with the true scattered field shows excellent agreement: at least at the scale of the figure, the two fields coincide. Thus, the results in Fig. 5.2 are correct. In stark contrast, the MAS currents in Fig. 5.1 (more precisely, their real parts in the top Fig. 5.1) are certainly problematic, as they exhibit rapid and abnormal oscillations. Amplitudes vary rapidly across the figure; larger amplitudes occur at the figure's leftmost and rightmost parts; and there, adjacent currents alternate in sign.

How can such oscillating currents generate a field pattern as smooth as the one in Fig. 5.2? One might be reminded of antenna superdirectivity (whether in the microwave [10] or optical [11] frequency domain)—is this analogy useful? If yes, how do the well-known pitfalls of superdirectivity carry over to the MAS? More generally, under what conditions do MAS currents exhibit oscillations like those in Fig. 5.1? Larger values of N are generally associated with solution convergence[1]; the field values in Fig. 5.2 have already converged, but what happens *to the top* Fig. 5.1 if we increase N? Do we get less rapid oscillations? What happens if we use better software and hardware? In particular, what is the effect of using longer computer wordlength? The phenomenon of internal resonances has been well-studied in the literature (see, e.g., [12])—are the oscillations in Fig. 5.1 an effect due to an internal resonance? Are the oscillations related to the usual difficulties that arise when we discretize Fredholm integral equations of the first kind (see e.g. [13, 14] and Chap. 15 of [15])? Finally, why is the difficulty (i.e., the oscillations in Fig. 5.1) important in

[1]The value $N = 60$ in Figs. 5.1 and 5.2 roughly amounts to 46 points per wavelength on the auxiliary surface.

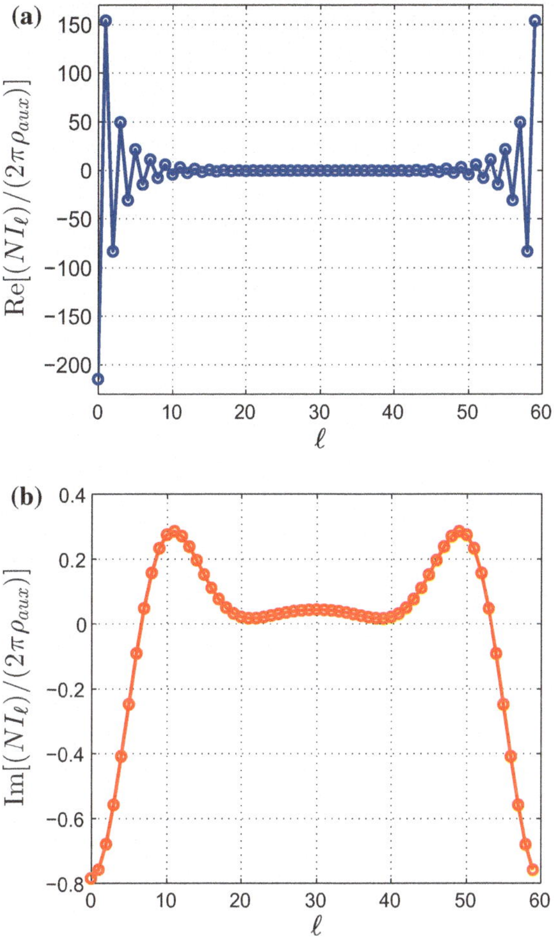

Fig. 5.1 Real (**a**) and imaginary (**b**) parts of the normalized MAS currents $NI_\ell/(2\pi\rho_{\mathrm{aux}})$ as functions of the index ℓ

the first place?—after all, the *final* results (i.e., the field values in Fig. 5.2) are correct *despite* the oscillations of their generating currents.

These are the main questions investigated in this chapter. For the most part, we address the questions using tools of asymptotic analysis. Specifically, we pick an exactly solvable scattering problem that is simple enough for us to develop an asymptotic formula ((5.33) below) for the oscillating currents. The mere existence of such a formula already answers one of the questions we posed: Since the oscillations can be predicted by asymptotic methods, they *cannot* be due to short computer wordlength nor, in particular, to roundoff error.

Our scattering problem's simplicity further enables us to *analytically* demonstrate that the large-N limit of the MAS field is the true scattered field. Therefore, the aforementioned correctness of the MAS field of Fig. 5.2 cannot be coincidental. And (at least for our simple scattering problem) the MAS field will remain correct even in

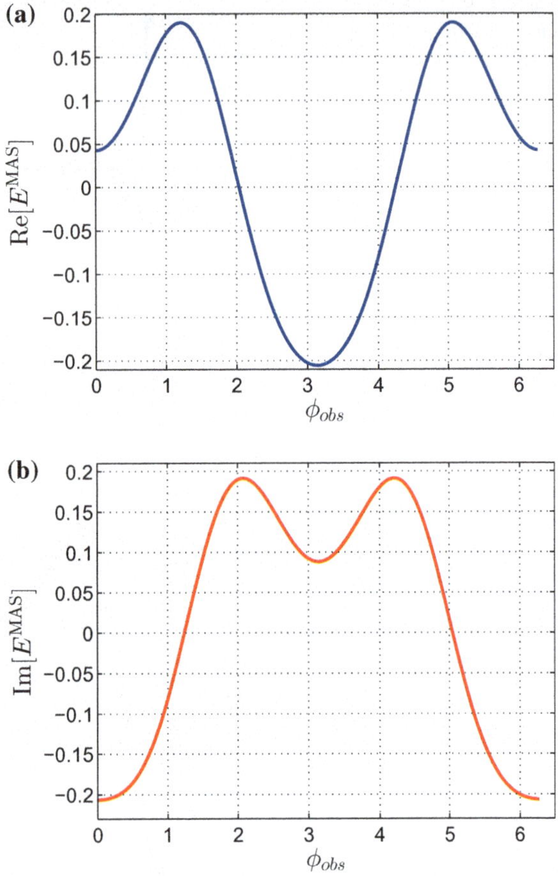

Fig. 5.2 Real (**a**) and imaginary (**b**) parts of the MAS field E^{MAS} as functions of ϕ_{obs}

cases where the oscillations are stronger and more rapid. As we will elaborate below (and as already suggested by the analogy with superdirectivity), obtaining MAS fields from yet-more-rapidly-oscillating MAS currents is a process that becomes increasingly susceptible to roundoff; this underscores the importance of showing convergence analytically.

The scattering problem we consider is simple, and this does present disadvantages; one might prefer to understand the oscillations and their consequences in the context of a real-life problem he or she is actually trying to solve. However, our main goal is to pinpoint and comprehend a difficulty. Choosing a simple problem as a means of doing so is advantageous for at least three reasons: First, a difficulty occurring in a simple problem—especially one that is not perplexed by shape-elongation effects—must also occur in more complicated ones. Second, a thorough study of a simple case can always help us understand more complicated ones—this is especially true here, where the difficulty we are interested in is clouded by the undesired effect of roundoff. Third, our simple problem enables an *analytical* comparison of MAS to certain discretizations of the Extended Integral Equation (EIE): as we will see, these discretizations *do not* present the aforementioned difficulties. This marked difference helps us better understand both MAS and the EIE.

A pioneering work, in Russian, that noted an analogy between MAS and superdirectivity appeared in 1981 [16]. Relations between [16] and our own work [6] have been recently discussed in [17]. Another work, also in Russian, that recently came to our attention is [18]; this important 1995 work discusses some relevant topics within the context of acoustics, with emphasis on the acoustic analogue of what we call the "continuous version of MAS" (Sect. 5.3.3). Apart from [16, 18], the first detailed discussions of the phenomena examined in this chapter are, to the best of our knowledge, the 2006 and 2007 works [1, 2], which are the first of the papers [1–9] overviewed in this chapter. Certain independent discussions can also be found in the 2008 work [19]. In this overview, we especially stress that the oscillations described herein are not related to roundoff, nor to internal resonances—and we do so using a number of different viewpoints, the comparison with the EIE being one. That is to say, oscillations, roundoff, and internal resonances are three separate effects, and it is easy to confuse them. Furthermore, it is also natural to try to *avoid* oscillations and to focus on cases where oscillations *do not* occur. These are, perhaps, the main reasons why such oscillations and their consequences are frequently not mentioned in the MAS-related literature.

5.2 Description of the Scattering Problem and Exact Solution

Our simple scattering problem is depicted in Fig. 5.3. An infinitely long (along the z-axis), perfect electric conducting (PEC) circular cylinder of radius ρ_{cyl} is illuminated by an external electric current filament I that lies on the x-axis. Let $(\rho_{\mathrm{cyl}}, \phi_{\mathrm{cyl}})$ and $(\rho_{\mathrm{obs}}, \phi_{\mathrm{obs}})$ denote the polar coordinates of a point on the PEC surface and of the observation point, respectively. The polar coordinates of the source are $(\rho_{\mathrm{fil}}, 0)$ with $\rho_{\mathrm{fil}} > \rho_{\mathrm{cyl}}$. The PEC cylinder is located in free-space, characterized by dielectric permittivity ε_0 and magnetic permeability μ_0.

In the sequel, we will use the notation $R_{A,B}$ to denote the distance from the point (ρ_A, ϕ_A) to the point (ρ_B, ϕ_B); for example

$$R_{\mathrm{fil,cyl}} = \sqrt{\rho_{\mathrm{fil}}^2 + \rho_{\mathrm{cyl}}^2 - 2\rho_{\mathrm{fil}}\rho_{\mathrm{cyl}} \cos\phi_{\mathrm{cyl}}}.$$

The electric field is z-directed, $\mathbf{E} = \hat{\mathbf{z}} E_z$, and the exact solution outside the PEC cylinder is (assuming an $e^{-i\omega t}$ time dependence, with ω being the angular frequency)

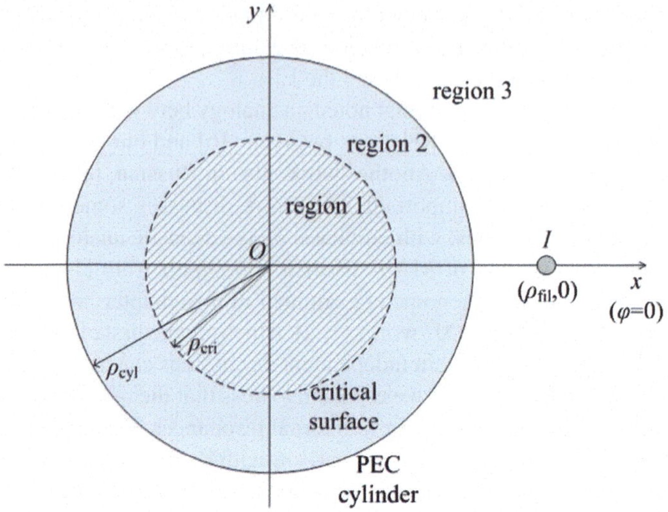

Fig. 5.3 Geometrical configuration of the three regions of the scattering problem

$$E_z = H_0(kR_{\rm obs,fil}) - \sum_{n=-\infty}^{\infty} \frac{J_n(k\rho_{\rm cyl})H_n(k\rho_{\rm fil})}{H_n(k\rho_{\rm cyl})} H_n(k\rho_{\rm obs})\, e^{in\phi_{\rm obs}}, \quad \rho_{\rm obs} > \rho_{\rm cyl}, \tag{5.1}$$

where J_n and $H_n \equiv H_n^{(1)}$ denote, respectively, the cylindrical Bessel and first kind Hankel functions of (integer) order n, while $k = \omega/c = \omega\sqrt{\varepsilon_0\mu_0}$ is the free-space wavenumber. In the right-hand side of (5.1), we have omitted the overall factor $-\frac{k^2 I}{4\omega\varepsilon_0}$, which is unimportant for our purposes; we do this to be consistent with [1]. In the RHS of (5.1), the first term is the incident field. The second term is the scattered field and is expressed as a Fourier series whose terms satisfy the wave equation (in the cylindrical coordinate system $(\rho_{\rm obs}, \phi_{\rm obs}, z)$), as well as the outgoing-wave condition.

By using the addition theorem

$$H_0\left(\sqrt{x_1^2 + x_2^2 - 2x_1x_2\cos\theta}\right) = \sum_{n=-\infty}^{\infty} J_n\left(\min\{x_1, x_2\}\right) H_n\left(\max\{x_1, x_2\}\right) e^{in\theta}, \tag{5.2}$$

(where $x_1 > 0$, $x_2 > 0$, θ real), (5.1) takes the form

$$E_z = \sum_{n=-\infty}^{\infty} H_n(k\rho_{\rm fil}) \frac{J_n(k\rho_{\rm obs})H_n(k\rho_{\rm cyl}) - J_n(k\rho_{\rm cyl})H_n(k\rho_{\rm obs})}{H_n(k\rho_{\rm cyl})} e^{in\phi_{\rm obs}}, \quad \rho_{\rm obs} < \rho_{\rm fil}, \tag{5.3}$$

from which it is readily verified that the total field E_z vanishes on the PEC surface, i.e. for $\rho_{\rm obs} = \rho_{\rm cyl}$.

The derivative of E_z with respect to ρ_{obs} is

$$\frac{\partial E_z}{\partial \rho_{\text{obs}}} = \frac{\partial}{\partial \rho_{\text{obs}}} H_0(k R_{\text{obs,fil}}) - \sum_{n=-\infty}^{\infty} \frac{J_n(k\rho_{\text{cyl}}) H_n(k\rho_{\text{fil}})}{H_n(k\rho_{\text{cyl}})} \frac{\partial}{\partial \rho_{\text{obs}}} H_n(k\rho_{\text{obs}})\, e^{in\phi_{\text{obs}}} \tag{5.4}$$

and, by Faraday's law, is proportional to H_ϕ, the ϕ-component of the magnetic field. By differentiating (5.3) and using the familiar Wronskian relation (see e.g. (10.5.2) of [20]), we get, for $(\rho_{\text{obs}}, \phi_{\text{obs}}) = (\rho_{\text{cyl}}, \phi_{\text{cyl}})$,

$$\frac{\partial E_z}{\partial \rho_{\text{obs}}} = -\frac{2i}{\pi \rho_{\text{cyl}}} \sum_{n=-\infty}^{\infty} \frac{H_n(k\rho_{\text{fil}})}{H_n(k\rho_{\text{cyl}})} e^{in\phi_{\text{cyl}}}, \quad (\rho_{\text{obs}}, \phi_{\text{obs}}) = (\rho_{\text{cyl}}, \phi_{\text{cyl}}), \tag{5.5}$$

which is proportional to the z-directed surface current density on the exterior surface of the PEC cylinder.

Convergence of the exact scattered field series in (5.1) can be examined by using the well-known, large-$|n|$ asymptotic approximations of J_n and H_n [20]

$$J_n(x) = (-1)^n J_{-n}(x) \sim \frac{1}{\sqrt{2\pi n}} \left(\frac{ex}{2n}\right)^n, \quad n \to +\infty, \tag{5.6}$$

$$H_n(x) = (-1)^n H_{-n}(x) \sim -i\sqrt{\frac{2}{\pi n}} \left(\frac{ex}{2n}\right)^{-n}, \quad n \to +\infty. \tag{5.7}$$

For large $|n|$, it follows that the nth term of this series behaves like

$$\frac{1}{n}\left(\frac{\rho_{\text{cri}}}{\rho_{\text{obs}}}\right)^{|n|} e^{i|n|\phi_{\text{obs}}},$$

with the distance ρ_{cri} defined by

$$\rho_{\text{cri}} = \frac{\rho_{\text{cyl}}^2}{\rho_{\text{fil}}}. \tag{5.8}$$

This *critical* radius ρ_{cri} is smaller than the radius ρ_{cyl} of the PEC cylinder.

The nth term of the series in the RHS of (5.1) thus behaves like the nth term of the Taylor series for $\ln(1+x)$ (see e.g. (4.6.1) of [20]), a series which converges for $|x| < 1$ and diverges for $|x| > 1$. Therefore, the series in (5.1) converges for all $\rho_{\text{obs}} > \rho_{\text{cri}}$, and diverges for all $\rho_{\text{obs}} < \rho_{\text{cri}}$; this is a well-known result found e.g. in [21–23].

By using (5.6) and (5.7), as well as the corresponding large-$|n|$ asymptotic approximations of the derivatives of the Bessel and Hankel functions, we conclude that the nth term of the scattered electric field's derivative series in (5.4) behaves like

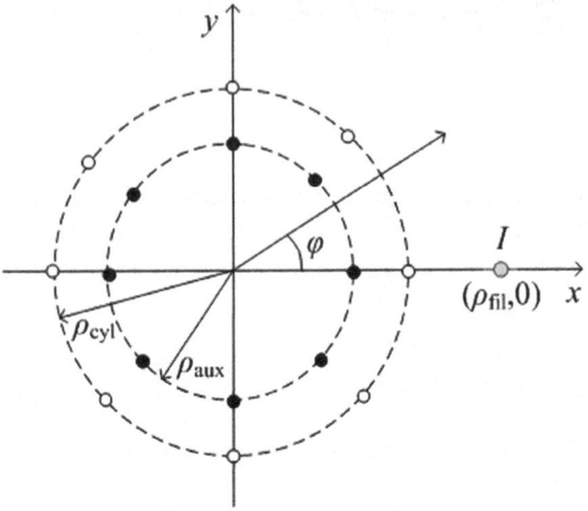

Fig. 5.4 Locations of the primary filamentary source (grey bullet), the auxiliary sources (black bullets) and the collocation points (white bullets) for $N = 8$

$$\left(\frac{\rho_{\rm cri}}{\rho_{\rm obs}}\right)^{|n|} e^{i|n|\phi_{\rm obs}},$$

i.e. like a geometric series. Therefore, this series also converges when $\rho_{\rm obs} > \rho_{\rm cri}$, and diverges when $\rho_{\rm obs} < \rho_{\rm cri}$.

The series solution (5.1), originally found for points $\rho_{\rm obs} > \rho_{\rm cyl}$ (Region 3 of Fig. 5.3), is thus also convergent and meaningful *inside* the PEC surface, until the critical radius ($\rho_{\rm obs} > \rho_{\rm cri}$). We have thus extended our solution (5.1) to Region 2 of Fig. 5.3. Since we found that the derivative (5.4) of (5.1) is also well-defined until the critical radius, the extended solution is in fact the analytic continuation (with respect to the single complex variable $\rho_{\rm obs}$) of the original solution.

5.3 Application of the MAS to the Scattering Problem

5.3.1 MAS Currents and Their Large-N Limit

In this section, we apply the MAS to the scattering problem of Sect. 5.2. Our N auxiliary sources are discrete electric current filaments that lie on a fictitious auxiliary surface of radius $\rho_{\rm aux}$ in the interior of the PEC cylinder (i.e. $0 < \rho_{\rm aux} < \rho_{\rm cyl}$). Filament #$\ell$ is located at $(\rho_{\rm aux}, \frac{2\pi\ell}{N})$; thus, the N sources are equispaced, see Fig. 5.4. The symbol I_ℓ will denote the current on filament #ℓ ($\ell = 0, \ldots, N-1$). The quantities $I_0, I_1, \ldots, I_{N-1}$ will be referred to as *MAS currents* and will be determined by approximately satisfying the PEC boundary condition as described below.

The distance between the ℓth MAS current and the observation point is denoted by

$$R_{\ell,\mathrm{obs}} = \sqrt{\rho_{\mathrm{aux}}^2 + \rho_{\mathrm{obs}}^2 - 2\rho_{\mathrm{aux}}\rho_{\mathrm{obs}} \cos\left(\phi_{\mathrm{obs}} - \frac{2\pi\ell}{N}\right)}, \quad \ell = 0, 1, \ldots, N-1, \tag{5.9}$$

while the total electric field (adopting the same normalization as in the previous section) is

$$E_z^N = H_0(kR_{\mathrm{fil,obs}}) + \sum_{\ell=0}^{N-1} I_\ell\, H_0(kR_{\ell,\mathrm{obs}}), \tag{5.10}$$

where the summation in the RHS is the scattered field that has been written as a sum over the N MAS currents. The superscript on E_z^N emphasizes the N-dependence of the quantity in (5.10), which is the *MAS field*.

Next, we take N equispaced collocation points on the PEC cylinder, with collocation point #p located at $(\rho_{\mathrm{cyl}}, \frac{2\pi p}{N})$; see Fig. 5.4. By enforcing the PEC boundary condition $E_z^N = 0$ for $(\rho_{\mathrm{obs}}, \phi_{\mathrm{obs}}) = (\rho_{\mathrm{cyl}}, \frac{2\pi p}{N})$, (5.10) gives

$$\sum_{\ell=0}^{N-1} H_0(kb_{p,\ell})I_\ell = -H_0(kd_p), \quad p = 0, 1, \ldots, N-1, \tag{5.11}$$

where

$$b_{p,\ell} = b_{\ell,p} = \sqrt{\rho_{\mathrm{aux}}^2 + \rho_{\mathrm{cyl}}^2 - 2\rho_{\mathrm{aux}}\rho_{\mathrm{cyl}} \cos\frac{2\pi(p-\ell)}{N}} \tag{5.12}$$

is the distance between auxiliary filament ℓ and collocation point p, and

$$d_p = \sqrt{\rho_{\mathrm{fil}}^2 + \rho_{\mathrm{cyl}}^2 - 2\rho_{\mathrm{fil}}\rho_{\mathrm{cyl}} \cos\frac{2\pi p}{N}} \tag{5.13}$$

is the distance between primary filament I and collocation point p.

Equation (5.11) constitutes a $N \times N$ system of linear algebraic equations with unknowns the MAS currents I_ℓ. These currents can be directly determined via standard linear-equation solvers. Still, we proceed to give a second method for the solution of (5.11); this will enable us to investigate MAS currents and fields analytically and, also, to have an independent method of solution that can help detect effects due to roundoff.

Specifically, the matrix in (5.11) is *circulant* due to $b_{p,\ell} = b_{0,\ell-p}$ and $b_{0,\ell+qN} = b_{0,\ell}$, $q \in \mathbb{Z}$ and hence the system (5.11) can be solved in closed form using the Discrete Fourier Transform (DFT). The relevant procedure and formulas are given in detail in Appendix B of [1] (see also [24–26]), where the solution is found to be

$$I_\ell = \sum_{m=0}^{N-1} I^{(m)} e^{i2\pi\ell m/N}, \quad \ell = 0, 1, \ldots, N-1, \tag{5.14}$$

where the DFT $I^{(m)}$ of I_ℓ is given by

$$I^{(m)} = \frac{1}{N}\frac{D^{(m)}}{B^{(N-m)}}, \quad m = 0, 1, \ldots, N-1, \tag{5.15}$$

with

$$B^{(m)} = \frac{1}{N}\sum_{\ell=0}^{N-1} H_0(kb_{0,\ell})e^{-i2\pi\ell m/N}, \tag{5.16}$$

and

$$D^{(m)} = -\frac{1}{N}\sum_{p=0}^{N-1} H_0(kd_p)e^{-i2\pi pm/N}, \tag{5.17}$$

being the DFT's of the first row of the system's matrix and the RHS vector, respectively.

By substituting (5.12) and (5.13) into (5.16) and (5.17), applying the addition theorem (5.2) to the Hankel functions, and interchanging the order of summation, we obtain

$$B^{(m)} = \sum_{q=-\infty}^{\infty} J_{qN+m}(k\rho_{\text{aux}})H_{qN+m}(k\rho_{\text{cyl}}), \tag{5.18}$$

$$D^{(m)} = -\sum_{q=-\infty}^{\infty} J_{qN+m}(k\rho_{\text{cyl}})H_{qN+m}(k\rho_{\text{fil}}). \tag{5.19}$$

It follows that $D^{(m)} = D^{(N-m)}$ and $B^{(m)} = B^{(N-m)}$, so that (5.15) implies

$$I^{(m)} = I^{(N-m)} = \frac{1}{N}\frac{D^{(m)}}{B^{(m)}}. \tag{5.20}$$

In the sequel we assume for simplicity that N is odd, so that (5.14) takes the form

$$I_\ell = I^{(0)} + 2\sum_{m=1}^{(N-1)/2} I^{(m)}\cos\left(\frac{2\pi\ell m}{N}\right), \quad \ell = 0, 1, \ldots, N-1. \tag{5.21}$$

Equations (5.14) and (5.21) are exact. Now, we determine the asymptotic behavior of I_ℓ, given by (5.21), in the limit $N \to \infty$; in this limit, the discrete MAS currents become a surface current density $J^s_{\text{limit}}(\phi_{\text{aux}})$. For $\phi_{\text{aux}} = 2\pi\ell/N$, $J^s_{\text{limit}}(\phi_{\text{aux}})$ equals the limit of the ratio of the MAS current I_ℓ to the arc-length distance between adjacent currents, i.e.

$$J^s_{\text{limit}}(\phi_{\text{aux}}) = \lim_{N\to\infty}\frac{NI_\ell}{2\pi\rho_{\text{aux}}}. \tag{5.22}$$

The quantities $\frac{NI_\ell}{2\pi\rho_{\text{aux}}}$ ($\ell = 0, 1, \ldots, N-1$) will be referred to as *normalized MAS currents*.

By using (5.6) and (5.7) and taking into account the fact that $0 \le m \le (N-1)/2$ in the RHS of (5.21), in the limit $N \to \infty$, only the $q = 0$ term in each sum (5.18) and (5.19) needs to be retained, and hence (5.20) gives the asymptotic relation

$$I^{(m)} \sim -\frac{1}{N}\frac{J_m(k\rho_{\text{cyl}})H_m(k\rho_{\text{fil}})}{J_m(k\rho_{\text{aux}})H_m(k\rho_{\text{cyl}})}, \quad m = 0, 1, \ldots, \frac{N-1}{2} \quad (N \to \infty). \tag{5.23}$$

By combining (5.21) and (5.23), we find that the limit of (5.22) is the following series,

$$J^s_{\text{limit}}(\phi_{\text{aux}}) = -\frac{1}{2\pi\rho_{\text{aux}}}\left[\frac{J_0(k\rho_{\text{cyl}})H_0(k\rho_{\text{fil}})}{J_0(k\rho_{\text{aux}})H_0(k\rho_{\text{cyl}})} + 2\sum_{m=1}^{\infty}\frac{J_m(k\rho_{\text{cyl}})H_m(k\rho_{\text{fil}})}{J_m(k\rho_{\text{aux}})H_m(k\rho_{\text{cyl}})}\cos(m\phi_{\text{aux}})\right]. \tag{5.24}$$

Using (5.6) and (5.7), it is easy to see that $J^s_{\text{limit}}(\phi_{\text{aux}})$

Case 1 exists when the discrete sources lie within Region 2 of Fig. 5.3 ($\rho_{\text{cri}} < \rho_{\text{aux}} < \rho_{\text{cyl}}$)

Case 2 diverges when the discrete sources lie within Region 1 ($\rho_{\text{aux}} < \rho_{\text{cri}}$).

The aforementioned convergence/divergence is consistent with Theorem 2.4 (p. 79) of [27], which holds for general geometries.

Besides being divergent in Case 2, the series in (5.24) is also meaningless in another case, when $J_m(k\rho_{\text{aux}}) = 0$ for some integer m. These discrete values of $k\rho_{\text{aux}}$ correspond to the internal resonances of the auxiliary surfaces, and we will return to them shortly.

5.3.2 Large-*N* Limit of MAS Field

We now determine the large-N behavior of the MAS field that we defined in (5.10). For this purpose, we use the DFT relation [1]

$$I^{(m)} = \frac{1}{N}\sum_{\ell=0}^{N-1} I_\ell e^{-i2\pi\ell m/N}, \tag{5.25}$$

which is the inverse of (5.21). Applying (5.2) to the second Hankel function in (5.10), interchanging the order of summation, and introducing $I^{(n)}$ through (5.25) we obtain

$$E_z^N = H_0(kR_{\text{fil,obs}}) + N\sum_{n=-\infty}^{\infty} I^{(n)}J_n(k\rho_{\text{aux}})H_n(k\rho_{\text{obs}})e^{in\phi_{\text{obs}}} \quad (\rho_{\text{obs}} > \rho_{\text{aux}}). \tag{5.26}$$

Equation (5.26) is exact. We now determine its asymptotic behavior as $N \to \infty$, irrespective of whether $\rho_{\rm aux} < \rho_{\rm cri}$ or not. For this purpose, we replace all the $I^{(n)}$ in (5.26) by their large-N forms in (5.23). Then, the factor $J_n(k\rho_{\rm aux})$ cancels out, and the result is

$$E_z^N \sim H_0(kR_{\rm fil,obs}) - \sum_{n=-\infty}^{\infty} \frac{J_n(k\rho_{\rm cyl})H_n(k\rho_{\rm fil})}{H_n(k\rho_{\rm cyl})} H_n(k\rho_{\rm obs})e^{in\phi_{\rm obs}} \quad (N \to \infty), \tag{5.27}$$

which is precisely (5.1).

As discussed in Sect. 5.2, the series in (5.27) converges when $\rho_{\rm obs} > \rho_{\rm cri}$, and diverges when $\rho_{\rm obs} < \rho_{\rm cri}$. Our findings in this section can be conveniently summarized as follows:

Case 1 ($\rho_{\rm cri} < \rho_{\rm aux} < \rho_{\rm cyl}$): We have found the expected result that the limit as $N \to \infty$ of the MAS field is the true field (5.1) outside the PEC cylinder ($\rho_{\rm obs} > \rho_{\rm cyl}$), and its analytic continuation between the auxiliary surface and the PEC cylinder ($\rho_{\rm aux} < \rho_{\rm obs} < \rho_{\rm cyl}$).

Case 2 ($\rho_{\rm aux} < \rho_{\rm cri}$): If the MAS currents are placed within Region 1 of Fig. 5.3 ($\rho_{\rm aux} < \rho_{\rm cri}$), then the limit of the MAS field

Case 2.1 exists and is the true field, in Region 3 of Fig. 5.3 ($\rho_{\rm obs} > \rho_{\rm cyl}$)

Case 2.2 exists and is the analytic continuation of the true field, in Region 2 of Fig. 5.3 ($\rho_{\rm cri} < \rho_{\rm obs} < \rho_{\rm cyl}$)

Case 2.3 does not exist, in the portion of Region 1 exterior to the auxiliary sources ($\rho_{\rm aux} < \rho_{\rm obs} < \rho_{\rm cri}$). (Only in this case does the series in (5.27) diverge.)

For Case 2, we found in Sect. 5.3.1 that the limit of the normalized MAS currents does not exist. Case 2.1 thus shows that it is possible, in the limit, to obtain the true electric field without having the normalized MAS currents converge to a surface current density. The explicit, analytical demonstration of this phenomenon is one of the central themes of this chapter. It explains why the numerically-obtained field values of Fig. 2 are the correct ones.

It is well known that fields behave smoothly away from their sources, so it may seem peculiar that the "field" in Case 2 behaves abruptly across the critical surface, i.e. during the transition from Case 2.2 to Case 2.3, as $\rho_{\rm obs}$ crosses the threshold at $\rho = \rho_{\rm cri}$. But this "field" is not a field produced by a current; it is merely a mathematical limit, as $N \to \infty$, of the field of N sources. There is no reason for this "field" to obey Maxwell's equations. For any *finite* N (however large), the field is a true field satisfying Maxwell's equations and does not behave abruptly.

5.3.3 Continuous Version of MAS

It is not hard to see that the system of linear equations (5.11) coincides with a certain discretization of the following Fredholm integral equation

$$\int_{-\pi}^{\pi} H_0(kR_{\text{aux,cyl}})J^s(\phi_{\text{aux}})\mathrm{d}\phi_{\text{aux}} = -\frac{1}{\rho_{\text{aux}}}H_0(kR_{\text{fil,cyl}}), \quad -\pi < \phi_{\text{cyl}} < \pi. \tag{5.28}$$

This discretization is, specifically, Nyström's method with the trapezoidal rule for quadrature (see e.g. Sect. 12.2 of [15]). In other words, (5.11) can be viewed as a result of applying the following procedure to the integral equation (5.28): (i) choose quadrature points ϕ_{aux} located at integer multiples of $2\pi/N$; (ii) approximate the integral on the left-hand side (LHS) of (5.28) by the composite rectangular rule [28] (which, because the integrand is periodic, coincides with the composite trapezoidal rule [28]); and then (iii) satisfy the resulting equation at the previously used quadrature points (i.e., at points $\phi_{\text{cyl}} = 2\pi p/N$, $p = 0, 1, \ldots, N-1$). A similar observation is made in p. 95 of [29].

The physical meaning of (5.28) is obvious: It is an integral equation whose unknown is an electric surface current density $J^s(\phi_{\text{aux}})$, located at $\rho = \rho_{\text{aux}}$, that is required to *exactly* satisfy the boundary condition at $\rho = \rho_{\text{cyl}}$. In this sense, (5.28) implements what we call the "continuous version of MAS," and will be referred to as the Continuous MAS Integral Equation (CMASIE).

It is shown in [1] (see also [18]) that we can formally solve (5.28) explicitly using Fourier series, and that the solution is

$$J^s(\phi_{\text{aux}}) = -\frac{1}{2\pi\rho_{\text{aux}}}\sum_{m=-\infty}^{\infty}\frac{J_m(k\rho_{\text{cyl}})H_m(k\rho_{\text{fil}})}{J_m(k\rho_{\text{aux}})H_m(k\rho_{\text{cyl}})}e^{im\phi_{\text{aux}}}. \tag{5.29}$$

This formal solution coincides with (5.24). It follows that, when $\rho_{\text{cri}} < \rho_{\text{aux}} < \rho_{\text{cyl}}$, the CMASIE is solvable. It also follows that, in Case 1 of Sect. 5.3.1, the large-N limit of the currents in the *discrete* version of MAS is the correct one. This correctness was to be expected, as it simply states that Nyström's method converges.

We can now further comprehend the exceptional case $J_m(k\rho_{\text{aux}}) = 0$: It is natural that $J_m(k\rho_{\text{aux}})$ appears in the denominators of (5.29) and (5.24), because the corresponding continuous surface current density $J^s(\phi_{\text{aux}}) = e^{im\phi_{\text{aux}}}$ is non-radiating [30–32]. In other words, the field external to the density $J^s(\phi_{\text{aux}}) = e^{im\phi_{\text{aux}}}$ is identically zero.

Viewing (5.11) as a discretization of the CMASIE (5.28) is useful for yet another reason: (5.28) is a first-kind integral equation, and such integral equations are ill-posed [13–15]. Therefore, any discretization of the CMASIE will be plagued by ill-conditioning and will be highly susceptible to effects such as roundoff. This is increasingly true as ρ_{aux} decreases, because the kernel of the CMASIE becomes smoother.

5.3.4 More on the Nonsolvable Case; Oscillations, Roundoff, and Internal Resonances

In the case of magnetostatics and Laplace's equation, it is well known that the scattering problem can be solved exactly by means of an image located at $(\rho, \phi) = (\rho_{\text{cri}}, 0)$, where ρ_{cri} is the critical distance given in (5.8) (for the corresponding electrostatic problem see [33]). Therefore, the position of the image is a singularity of the analytic continuation of the solution to the Laplace problem. It is also well-known [19, 34, 35] that the singularities for the Laplace case coincide with those of the Helmholtz case (but in the latter case, the singularity has no meaning as the position of an image).

The above assertions allow us to shed more light on Case 2 of Sects. 5.3.1 and 5.3.2, in which $\rho_{\text{aux}} < \rho_{\text{cri}}$. Exterior fields due to smooth surface current densities are analytic functions of the space variables, so that no smooth surface current density can produce a field that exhibits a singularity at a point exterior to this current density, something that also holds for smooth noncircular shapes. Consequently, a smooth current density located at $\rho = \rho_{\text{aux}} < \rho_{\text{cri}}$ cannot cause the singularity at $(\rho, \phi) = (\rho_{\text{cri}}, 0)$. It follows that, in Case 2, no smooth current can produce the desired field. This is why, in Case 2, there is no solution to the CMASIE (5.28); it is also why the discrete currents of Sect. 5.3.1 have no large-N limit.

Since the discrete currents cannot converge, they must behave in an unphysical and abrupt manner when N is large. This a priori reasoning is consistent with the oscillations shown in Fig. 5.1.

The correctness of the field, demonstrated in Sect. 5.3.2, already strongly indicates that the oscillations of Fig. 5.1 are not due to roundoff. As discussed in Sect. 5.3.3, however, roundoff error is generally a very important effect. Can the oscillations in Fig. 1 be due to roundoff and/or matrix ill-conditioning? We also solved the system using (5.14)–(5.17), which form a completely independent method. As it turned out, those results coincide (at least at the scale of the figure) with the ones in Fig. 1. This provides a very strong indication that the oscillations are *not* due to roundoff/matrix ill-conditioning: solving ill-conditioned systems with independent solvers generally produces different results.

Are the oscillations in Fig. 5.1 associated with internal resonances? Clearly, the answer is no: As is well-known, and as we saw in Sects. 5.3.1 and 5.3.2, the phenomenon of internal resonances occurs at discrete values of ρ_{aux} or, equivalently, at discrete frequencies. This is different from the phenomenon of oscillations, which occurs for all $\rho_{\text{aux}} < \rho_{\text{cri}}$. In terms of the formal Fourier-series solution in (5.29), internal resonances occur when a single term in the series becomes infinite. By contrast, oscillations occur when the series for the continuous auxiliary source diverges because of the rapid growth of its terms (for large values of the summation index m).

5.3.5 Asymptotic Formula for the Oscillating MAS Currents

We now provide yet further—and, in our opinion, conclusive—evidence that the oscillations are not due to roundoff, by providing a large-N asymptotic formula for the oscillating values. In addition to

$$N >> 1, \tag{5.30}$$

and the nonsolvability condition

$$\rho_{\text{aux}} < \rho_{\text{cri}}, \tag{5.31}$$

we further assume

$$k\rho_{\text{cyl}} = \mathscr{O}(1), \quad k\rho_{\text{fil}} = \mathscr{O}(1), \quad k\rho_{\text{aux}} = \mathscr{O}(1). \tag{5.32}$$

Subject to the above conditions, in Appendix D of [2], the following asymptotic formula is derived (for even N)

$$I_\ell \sim \frac{I}{N}(-1)^{\ell+1} t^{N/2} \frac{t+1}{t-1} D_\ell \tag{5.33}$$

where

$$t = \frac{\rho_{\text{cri}}}{\rho_{\text{aux}}} \quad (t > 1), \tag{5.34}$$

while

$$D_\ell = \frac{(t-1)^2}{t^2 - 2t\cos(\frac{2\pi\ell}{N}) + 1} \tag{5.35}$$

is a *correction factor* in the sense that it equals 1 when $\ell = 0$ and decreases monotonically away from 1 when ℓ grows. For the validity of (5.33) it is required [2] that t is not very close to 1, and that ℓ is not too large. A similar equation to (5.33) can also be derived for odd N [2]. In the asymptotic formula (5.33), it is worth pointing out that the various radii do not appear separately, but only through the ratio t.

Equation (5.33) reveals a number of properties of the oscillating values: The MAS current I_0 (which is the MAS current closest to the original filament I) is real and, because of the factor $t^{N/2}$, exponentially large in N. Because of the factor $(-1)^{\ell+1}$, adjacent currents have different signs. The correction factor D_ℓ, which is real, is the envelope of the oscillating values. There is less tapering when t grows, so that the oscillations spread around as one moves the auxiliary source closer to the origin and away from the critical radius ρ_{cri} (for the related details see Fig. 7 of [2]).

Good agreement is observed when the asymptotic formula (5.33) is compared to results obtained by directly applying the MAS, that is by solving the $N \times N$ system (5.11). For example, for the parameter values $N = 70$, $k\rho_{\text{cyl}} = 2.1$, $k\rho_{\text{fil}} = 3$,

$k\rho_{\mathrm{aux}} = 1.3$ (which give $k\rho_{\mathrm{cri}} = 1.47$ and $t \simeq 1.131$) the error between $\mathrm{Re}\{I_\ell/I\}$ and (5.33) is 3.2% when $\ell = 0$, and 19% when $\ell = 7$. Also, errors generally decrease with increasing N, e.g. the error for $\ell = 7$ decreases to 15% when N is increased from 70 to 100.

Such good agreement occurs in the absence of roundoff errors. A separate asymptotic formula, that we will not give here, has been derived for the matrix condition number [36]; that formula shows that condition numbers also grow exponentially with N. Therefore, roundoff error will quickly become dominant (for example, if one increases N, leaving the other parameters fixed), and will mask the true behavior of the MAS currents. The true behavior, by contrast, is well-predicted by the formula (5.33). The distinction is important because effects due to roundoff are hardware and software dependent, and can possibly be overcome by using more powerful computers and algorithms, e.g., longer computer wordlengths and better system-solving routines. By contrast, the discussed oscillations can never be avoided in this manner.

5.3.6 *Analogies with Superdirectivity*

In the context of antenna arrays, superdirectivity is typically accomplished by small inter-element spacings and large oscillating currents [10]. Thus, there are obvious parallels between the oscillations discussed herein and the more familiar phenomenon of array superdirectivity. Reference [6] (see also [17]) goes beyond this, performing an asymptotic study that focuses on similarities between the corresponding near fields. In both cases, this field decays rapidly within a small distance from the oscillating sources. Furthermore, the great importance of roundoff to MAS is analogous to the high sensitivity (e.g., to mechanical tolerances) of superdirective arrays.

When we apply MAS, we are attempting to obtain the field scattered by the perfect conductor by using a current-carrying region (the auxiliary surface) that is smaller in size than the actual scatterer. This, also, is true of superdirective antennas, where one attempts to obtain a field produced by a "normal" antenna by using an antenna that is smaller in size.

5.4 Comparisons with the Extended Integral Equation (EIE)

Our previous discussions concern nonsolvability of the CMASIE (5.28) and the ensuing oscillations in solutions of the system (5.11), which is a discretized version of the CMASIE. In the present section, we perform a detailed comparison to the so-called Extended Integral Equation (EIE) and its corresponding discretized version. In the EIE, the unknown is the actual surface current *on* the PEC; this unknown is required to cancel the incident field on an auxiliary surface *inside* the PEC. By contrast, the unknown of the CMASIE is an auxiliary surface current *inside* the PEC, required to cancel the incident field *on* the PEC. In other words, both the EIE and the CMASIE use two surfaces (namely, the PEC surface and an auxiliary surface), but interchange the roles of the two.

Because of this similarity, one might surmise that the EIE gives rise to oscillations like those in Fig. 5.1. In this section, for the simple scattering problem of Sect. 5.2, we demonstrate analytically that this is not true. In particular, we show *irrespective of the auxiliary surface's position* that the EIE is a solvable equation and that, when normalized, the solution to the discrete version of the EIE converges, as $N \to \infty$, to the correct solution. Therefore, any oscillations that do occur when discretizing the EIE are likely to be caused by matrix ill-conditioning, an effect magnifying errors that would normally be negligible.

The EIE implements the well-known *Extended Boundary Condition* (EBC): For external illumination of closed and smooth PEC scatterers, the familiar boundary condition requires a total tangential electric field that is zero on the PEC surface. The EBC requires cancellation in this surface's *interior*. The EIE (also called null-field integral equation) amounts to satisfying the EBC in a proper closed interior auxiliary surface [37–40]. The EIE is the first step of many techniques including (i) the Null-Field Method (NFM) [37–42], (also known as the T-matrix method [43] or the Extended Boundary Condition Method [44, 45]; see also the review papers [46, 47]), (ii) the Dual-Surface Integral Equations Method (DSIEM) [48], (iii) the Combined Helmholtz Integral Equation Formulation (CHIEF) [49], and (iv) the Null-Field Method with Discrete Sources (NFM-DS) [29, 50].

5.4.1 Solvability of the EIE

We consider the same scattering problem with Sect. 5.2. The exact surface current $\mathbf{J}_s = J_{sz}\hat{\mathbf{z}}$ on the PEC surface $\rho = \rho_{\text{cyl}}$ is given by the convergent series

$$J_{sz}(\phi_{\text{cyl}}) = -\frac{I}{2\pi\rho_{\text{cyl}}} \sum_{n=-\infty}^{+\infty} \frac{H_n(k\rho_{\text{fil}})}{H_n(k\rho_{\text{cyl}})} e^{in\phi_{\text{cyl}}}. \tag{5.36}$$

In order to be consistent with the notations and conventions of [4] (and, also [3]), here and throughout Sect. 5.4, we have restored the multiplicative factor omitted in (5.5).

At an observation point $(\rho_{\text{obs}}, \phi_{\text{obs}})$, the total field is written as the incident plus the scattered fields according to

$$E_z(\rho_{\text{obs}}, \phi_{\text{obs}}) = -\frac{k^2}{4\omega\varepsilon_0}\left[I H_0(kR_{\text{fil,obs}}) + \rho_{\text{cyl}} \int_{-\pi}^{\pi} H_0(kR_{\text{cyl,obs}}) J_{sz}(\phi_{\text{cyl}}) d\phi_{\text{cyl}} \right] \tag{5.37}$$

The EIE results heuristically from (5.37) by demanding that $E_z = 0$ for $(\rho_{\text{obs}}, \phi_{\text{obs}}) = (\rho_{\text{aux}}, \phi_{\text{aux}})$, where $\rho_{\text{aux}} < \rho_{\text{cyl}}$. It is

$$\int_{-\pi}^{\pi} H_0(kR_{\text{cyl,aux}}) J_{sz}(\phi_{\text{cyl}}) d\phi_{\text{cyl}} = -\frac{I}{\rho_{\text{cyl}}} H_0(kR_{\text{fil,aux}}), \qquad -\pi < \phi_{\text{aux}} < \pi. \quad (5.38)$$

For the special case of our simple scattering problem, (5.38) is the 2-D analogue of the integral equation due to Waterman [37], who seems to first have used the term "extended integral equation" (see also the discussion in [39]). A more rigorous statement and derivation of the EIE can be found in [39, 41]. Note that the EIE (5.38) is very similar to the CMASIE (5.28); in particular, the two equations have the same kernel $H_0(kR_{\text{cyl,aux}})$. (Recall that the missing factor I in the RHS of (5.28) is due to the different notation employed in Sect. 5.3.) Since $R_{\text{cyl,aux}} \neq 0$, the kernel $H_0(kR_{\text{cyl,aux}})$ is analytic and defines a compact integral operator [15].

We can solve the EIE explicitly just as before (see Appendix A of [1]): By the addition theorem, the Fourier series of the right-hand side (RHS) and the kernel are, respectively,

$$-\frac{I}{\rho_{\text{cyl}}} H_0(kR_{\text{fil,aux}}) = -\frac{I}{\rho_{\text{cyl}}} \sum_{n=-\infty}^{+\infty} J_n(k\rho_{\text{aux}}) H_n(k\rho_{\text{fil}}) e^{in\phi_{\text{aux}}}, \quad (5.39)$$

$$H_0(kR_{\text{cyl,aux}}) = \sum_{n=-\infty}^{+\infty} J_n(k\rho_{\text{aux}}) H_n(k\rho_{\text{cyl}}) e^{in(\phi_{\text{cyl}}-\phi_{\text{aux}})}. \quad (5.40)$$

The Fourier coefficients of the solution to (5.38) are proportional to the ratio of the Fourier coefficients in (5.39) and (5.40). Thus, the factors $J_n(k\rho_{\text{aux}})$ cancel and the solution is

$$J_{sz}(\phi_{\text{cyl}}) = -\frac{I}{2\pi\rho_{\text{cyl}}} \sum_{n=-\infty}^{+\infty} \frac{H_n(k\rho_{\text{fil}})}{H_n(k\rho_{\text{cyl}})} e^{in\phi_{\text{cyl}}}. \quad (5.41)$$

Hence, the EIE (5.38) is solvable *irrespective of the position of the auxiliary surface*, and its solution (5.41) equals the true surface current (5.36). Using $\rho_{\text{cyl}} < \rho_{\text{fil}}$ and (5.7), it is easy to show that $J_{sz}(\phi_{\text{cyl}})$ belongs to $L^2[-\pi, \pi]$. Thus, Picard's criterion [15]—which is a necessary and sufficient condition for L^2-solvability—must be satisfied for the EIE (5.38), and this can indeed be verified directly (see Sect. IV of [3]).

By contrast, the CMASIE (5.28) is nonsolvable when ρ_{aux} is smaller than the critical value $\rho_{\text{cri}} = \rho_{\text{cyl}}^2/\rho_{\text{fil}}$ because the series (5.29), corresponding to (5.41), diverges when $\rho_{\text{aux}} < \rho_{\text{cri}}$ (but converges otherwise). To be more precise, when $\rho_{\text{aux}} < \rho_{\text{cri}}$ Picard's criterion is *not* satisfied (see Sect. IV of [3]), meaning that there is no L^2-solution.

Another important difference between the solution (5.29) of the CMASIE and (5.41) is that (5.41) is well defined for all values of $k\rho_{\text{aux}}$ (as opposed to (5.41),

(5.29), as discussed previously, is not well-defined at the eigenvalues of the interior problem). This indicates that (5.41) is the unique solution of (5.38). A rigorous discussion of unique solvability of the null-field equations—which stem from the EIE—can be found in [39, 41].

5.4.2 Discretization of the EIE

We discretize the EIE (5.38) just as we discretized the CMASIE (Sect. 5.3.3), by applying Nyström's method with quadrature points $(\rho_{\text{aux}}, 2\pi p/N)$, and with the composite trapezoidal rule for quadrature. In this way, we obtain the linear system

$$\sum_{\ell=0}^{N-1} I_\ell H_0(kb_{p,\ell}) = -I H_0(kd_p), \quad p = 0, \ldots, N-1, \tag{5.42}$$

where $b_{p,\ell}$ is the distance between $(\rho_{\text{cyl}}, 2\pi\ell/N)$ and $(\rho_{\text{aux}}, 2\pi p/N)$, d_p is the distance between $(\rho_{\text{fil}}, 0)$ and $(\rho_{\text{aux}}, 2\pi p/N)$, and $I_\ell = 2\pi\rho_{\text{cyl}} J_{sz}(2\pi\ell/N)/N$. Equation (5.42) is identical to (5.11) except for the definition of d_p. We can also arrive at (5.42) independently of the EIE (5.38), by defining the unknowns I_ℓ as N discrete currents on the PEC surface that must exactly cancel the incident field at the N auxiliary points.

In view of the solvability of (5.38), it is natural to expect Nyström's method to converge. We can verify this by exploiting the fact that the matrix in (5.42) is circulant and, as in Sect. 5.3.1, solving (5.42) using the DFT. We get (for odd N)

$$\frac{I_\ell}{I} = I^{(0)} + 2 \sum_{m=1}^{(N-1)/2} I^{(m)} \cos\left(\frac{2\pi\ell m}{N}\right), \quad \ell = 0, \ldots, N-1, \tag{5.43}$$

where

$$I^{(m)} = \frac{1}{N}\frac{D^{(m)}}{B^{(m)}} \tag{5.44}$$

$$B^{(m)} = J_m(k\rho_{\text{aux}}) H_m(k\rho_{\text{cyl}}) + \left(\sum_{n=-\infty}^{-1} + \sum_{n=1}^{\infty}\right) J_{nN+m}(k\rho_{\text{aux}}) H_{nN+m}(k\rho_{\text{cyl}}) \tag{5.45}$$

$$D^{(m)} = -J_m(k\rho_{\text{aux}}) H_m(k\rho_{\text{fil}}) - \left(\sum_{n=-\infty}^{-1} + \sum_{n=1}^{\infty}\right) J_{nN+m}(k\rho_{\text{aux}}) H_{nN+m}(k\rho_{\text{fil}}) \tag{5.46}$$

The terms for $n = 0$ are independent of N and have been written separately. Using (5.6) and (5.7) we see that the nth term in the second series in (5.45) is of the order of $(\rho_{\text{aux}}/\rho_{\text{cyl}})^{nN}/(nN)$ for large n. Since $\rho_{\text{aux}} < \rho_{\text{cyl}}$, this behaves like the nth term of the uniformly convergent Taylor series for $\ln(1 + x)$ ($|x| < 1$). We can thus find the limit of the series as $N \to \infty$ term-by-term, and the limit is zero. Similarly, the limit of the other three series in (5.45) and (5.46) is also zero. Consequently, in the large-N limit, we keep only the $n = 0$ (separate) terms in (5.45) and (5.46), and thus (5.44) reduces to

$$I^{(m)} \sim -\frac{1}{N}\frac{H_m(k\rho_{\text{fil}})}{H_m(k\rho_{\text{cyl}})} \quad (N \to \infty), \tag{5.47}$$

which is independent of ρ_{aux}. By (5.43) and (5.47), the limit

$$\lim_{N\to\infty} \frac{NI_\ell}{2\pi\rho_{\text{cyl}}} \quad \left(\phi_{\text{cyl}} = \frac{2\pi\ell}{N}\right) \tag{5.48}$$

exists *irrespective of the position of the auxiliary surface* and equals the $J_{sz}(\phi_{\text{cyl}})$ in (5.41). Hence, the discretization method converges. This implies that—as opposed to MAS—there are no oscillations.

5.4.3 Analytic Continuation of the Scattered Field

For the CMASIE to be solvable and for the MAS solution to be oscillation-free, the auxiliary surface must enclose the singularity of the analytic continuation of the scattered field (i.e., $(\rho, \phi) = (\rho_{\text{cri}}, 0)$ in the cylindrical problem). A similar restriction must also hold for more complicated scattering problems. By contrast, our discussions of the EIE brought out no restrictions of this type. Thus, the singularity of the analytic continuation, while important for MAS, is not relevant to the EIE. This is logical, as the analytic continuation of E_z (E_z is the total exterior field) *does not* vanish in the interior–if it did, it would vanish everywhere [39].

Nonetheless, issues of this sort have raised controversy in recent publications [51–56]. In particular, A.G. Kyurkchan writes, "the surface where null field condition is set has to include the set of singularities of analytical continuation of the wave field into non-physical area" [53]. The opposite point of view can be found in works by Yu.A. Eremin, A.G. Sveshnikov, and S.P. Skobelev, who in [55] (see also [54]) write, "The equivalence of the NFM to the original diffraction problem has been unquestionable for a long time," and "an internal surface can be chosen arbitrarily, provided that it is nonresonant." Here, NFM means null-field method. In this section, we provide a detailed explanation [4] (see also [8]) to why the analytic continuation is irrelevant to the EIE for the cylindrical problem. Our explanation focuses on differences between (i) the analytic continuation of the interior field due to $J_{sz}(\phi_{\text{cyl}})$; and (ii) the exterior field due to $J_{sz}(\phi_{\text{cyl}})$.

We specifically show that the incident field (which is well-defined in the interior) *is not* cancelled by the analytic continuation of E_z^s, i.e., the analytic continuation of

$$B = -\frac{k^2 \rho_{\mathrm{cyl}}}{4\omega\varepsilon_0} \int_{-\pi}^{\pi} H_0(kR_{\mathrm{cyl,obs}}) J_{sz}(\phi_{\mathrm{cyl}}) \mathrm{d}\phi_{\mathrm{cyl}} \quad (\rho_{\mathrm{obs}} > \rho_{\mathrm{cyl}}) \tag{5.49}$$

to points $(\rho_{\mathrm{obs}}, \phi_{\mathrm{obs}})$ with $\rho_{\mathrm{obs}} < \rho_{\mathrm{cyl}}$. Rather, it is cancelled by the interior field due to $J_{sz}(\phi_{\mathrm{cyl}})$, which is

$$A = -\frac{k^2 \rho_{\mathrm{cyl}}}{4\omega\varepsilon_0} \int_{-\pi}^{\pi} H_0(kR_{\mathrm{cyl,obs}}) J_{sz}(\phi_{\mathrm{cyl}}) \mathrm{d}\phi_{\mathrm{cyl}} \quad (\rho_{\mathrm{cyl}} > \rho_{\mathrm{obs}}). \tag{5.50}$$

We accomplish this by explicitly calculating A and B (whose definitions appear very similar) and showing that they are indeed different: To find B, we use the addition theorem

$$H_0(kR_{\mathrm{cyl,obs}}) = \sum_{n=-\infty}^{+\infty} J_n(k\rho_{\mathrm{cyl}}) H_n(k\rho_{\mathrm{obs}}) e^{in(\phi_{\mathrm{obs}} - \phi_{\mathrm{cyl}})} \quad (\rho_{\mathrm{obs}} > \rho_{\mathrm{cyl}}), \tag{5.51}$$

in (5.49), interchange the orders of integration and summation, and introduce the Fourier-series coefficients of $J_{sz}(\phi_{\mathrm{cyl}})$ from (5.41) to get, for $\rho_{\mathrm{obs}} > \rho_{\mathrm{cyl}}$,

$$B = \frac{k^2 I}{4\omega\varepsilon_0} \sum_{n=-\infty}^{+\infty} J_n(k\rho_{\mathrm{cyl}}) H_n(k\rho_{\mathrm{obs}}) \frac{H_n(k\rho_{\mathrm{fil}})}{H_n(k\rho_{\mathrm{cyl}})} e^{in\phi_{\mathrm{obs}}}. \tag{5.52}$$

The latter series is convergent, not only when $\rho_{\mathrm{obs}} > \rho_{\mathrm{cyl}}$, but in the extended region $\rho_{\mathrm{obs}} > \rho_{\mathrm{cri}}$. Moreover, the series for the derivative $\partial B / \partial \rho_{\mathrm{obs}}$ also converges in $\rho_{\mathrm{obs}} > \rho_{\mathrm{cri}}$. Hence, the series in (5.52) is analytic and can be used to define the analytic continuation of E_z^s to the interior, at least when $\rho_{\mathrm{cri}} < \rho_{\mathrm{obs}} < \rho_{\mathrm{cyl}}$.

On the other hand, to find A we use the addition theorem

$$H_0(kR_{\mathrm{cyl,obs}}) = \sum_{n=-\infty}^{+\infty} J_n(k\rho_{\mathrm{obs}}) H_n(k\rho_{\mathrm{cyl}}) e^{in(\phi_{\mathrm{obs}} - \phi_{\mathrm{cyl}})} \quad (\rho_{\mathrm{cyl}} > \rho_{\mathrm{obs}}) \tag{5.53}$$

in order to finally obtain

$$A = \frac{k^2 I}{4\omega\varepsilon_0} \sum_{n=-\infty}^{+\infty} J_n(k\rho_{\mathrm{obs}}) H_n(k\rho_{\mathrm{fil}}) e^{in\phi_{\mathrm{obs}}}. \tag{5.54}$$

Clearly, A differs from B.[2] Furthermore, use of the addition theorem and comparison with the first term in the RHS of (5.37) shows that A is minus the incident field, just as expected.

We have thus explained that, from the point of view of solvability and oscillations, the analytic continuation of the scattered field is not relevant to the EIE. This is true, at least, for our simple scattering problem.

5.4.4 More on the Effects of Roundoff; Relevance to More Complicated Problems

As we have seen, oscillations such as those arising in the MAS do not occur in EIE (at least for the "sufficiently simple" problem we have examined). Hence, any oscillations appearing in EIE solutions are probably due to matrix ill-conditioning, which can greatly amplify errors due to both software and hardware. For EIE solutions, the cause of such errors is finite computer wordlength and imperfect special-function computation. The fact that effects such as roundoff can result in oscillations that resemble the MAS-type ones has been pointed out in the context of wire-antenna integral equations [3, 57], where matrix condition numbers increase exponentially with matrix size, as they do in our problem (see Sect. 5.3.5). Further instances of first-kind Fredholm integral equations that are *solvable* and that have oscillating discretizations can be found in [58–61]. For Hermitian kernels (the kernel of CMASIE/EIE is non-Hermitian), the solution can be written as a series involving eigenvalues and eigenfunctions, and this leads to the following observation [61]: "Changes in the eigenvalues due to rounding and truncation errors can radically alter the coefficients of the higher eigenfunctions and cause characteristic spurious oscillations in numerical results." With the aid of Picard's theorem [15], the above observation can be extended to other compact operators. As it turns out, for the simple cylindrical problem, the high-order singular functions are highly oscillatory (see also [3]).

Since oscillations appearing in the solutions of the EIE can be similar in their form to the MAS-type ones, it is important to distinguish the origin of the oscillations and not to confuse the two different types. To this end, we performed careful numerical experiments (for details, see [4]) for an elliptical 2-D PEC scatterer illuminated by a plane wave. It was found that the real and imaginary parts of the numerically-obtained normalized currents I_ℓ/I (solved via a standard Matlab routine that explicitly computes matrix inverses) indeed present oscillations with respect to source index ℓ (see Fig. 2 of [4]). These oscillations resemble the MAS-type ones of Fig. 5.1a. However, we verified that these EIE oscillations are of a *very different nature*. We did this by using a better system solver, namely the Matlab Gaussian-elimination routine. For this routine the curves of I_ℓ/I barely present oscillations at

[2] They are equal, however, in the limit $\rho_{\text{obs}} \to \rho_{\text{cyl}}$. This is apparent from (5.52) and (5.54), and explained by the continuity of the electric field across the surface current density $J_{sz}(\phi_{\text{cyl}})$ at $\rho = \rho_{\text{cyl}}$.

the scale of the aforementioned figure. Hence, at least in this case, the oscillations are due to the original solver, which magnifies finite-computer-wordlength errors and imperfections in special-function calculations. In other words, in conjunction with the original solver the discretization method satisfactorily converges, and gives correct results, before the occurrence of noticeable oscillations. However, extensive experiments with larger values of N resulted in larger condition numbers and more serious oscillations; such problems eventually appear even when the original solver is used.

In cases more complicated than the simple one discussed herein, it may be hard to correct the problematic results. Still, understanding why they occur is of importance: it may eventually help one correct, or make one certain that using better hardware/software cannot lead to correction (as is the case with the MAS-type oscillations). Besides computing condition numbers, one should check whether the oscillations depend on the particular hardware and software. For example, if changing the system-solving and quadrature routines leads to the same results, then one can deduce that the oscillations are probably similar in nature to those in MAS. But if the results *differ* (without necessarily becoming better), then EIE-type effects are surely present.

5.5 Selected Conclusions; Additional Remarks

In this concluding section, we do not attempt to summarize all our findings on MAS-type oscillations. Rather, we selectively emphasize certain results that are expected to carry over to more complicated problems. We then explain, in detail, why MAS-type oscillations should generally be avoided.

1. It is expected that MAS-type oscillations will occur in more complicated problems, e.g., noncircular or non-PEC scatterers. This has already been demonstrated for a dielectric circular cylindrical scatterer in [5] and a PEC 2-D scatterer of elliptical shape in [2].
2. For sufficiently smooth 2-D PEC scatterers, oscillations in the MAS currents will occur if the auxiliary surface does not enclose all singularities of the scattered field's analytic continuation.
3. When oscillations do occur, they increase sharply in magnitude when the number N of auxiliary sources increases.
4. In the oscillating case, and for sufficiently smooth geometries, the true field is still obtained despite the auxiliary-current oscillations.

Why should oscillations, in general, be avoided? If we were not to assume 4, this would be a trivial question. But even in problems in which 4 is true (so that the currents are intermediate results, not useful for applications), it is misguided—at least for large N—to believe that oscillations do not matter. As stressed by Higham [62], the design of stable algorithms involves: (i) avoiding subtractions of quantities contaminated by error; (ii) minimizing the size (in comparison to the finally-desired solution) of intermediary results; and (iii) looking for formulations that are mathematically but not numerically equivalent. Higham also suggests looking at the numbers generated by the computer (a common practice during the first generation of electronic computing).

Since the scattered electric field is expected to be small while (at least some of) the intermediate quantities I_ℓ are very large, subtractive cancellation is bound to occur. And because the I_ℓ have arisen by solving a highly ill-conditioned system, they are sensitive to roundoff and highly likely to be corrupted by error. Increasing N will cause a great increase in both I_ℓ and the matrix condition number. As a result, the computed electric field will deteriorate rapidly as N increases.

It is thus desirable to follow Higham's third guideline and to enclose all singularities. In complicated practical problems, however, the singularities are difficult to determine beforehand (but for some work toward this end, see [12, 21, 63]). When this is the case, looking at the generated numbers is probably the best we can do. In particular, we should scrutinize the I_ℓ—experimenting, when possible, with N and the shape of the auxiliary surface.

Since we analyzed a difficulty that is generally to be avoided, this chapter has an outlook that is, to some extent, negative. Although negative scientific results in general are often useful [64], one might be led to think that *all* results in this chapter are negative. This, however, is not true. In fact, our result that is most worthy of further exploration undoubtedly has a positive twist. This result is the MAS field's *convergence* (in spite of the MAS currents' divergence and oscillations). Here, we are specifically referring to the convergence of the field obtained via the solutions I_ℓ of the system (5.11), which implements what we called the discrete version of MAS and which has a straightforward and widely used extension to more complicated geometries.

The aforementioned MAS-field convergence occurs, at least, in the circular geometries discussed herein. Does it occur in more general ones? If yes, for what class of noncircular geometries? Is it possible to analytically demonstrate MAS-field convergence (regardless of the divergence or convergence of the auxiliary I_ℓ) for some practically interesting class of scatterers? This would be a very interesting mathematical and theoretical finding and would have positive consequences for applications. Here, we are not only referring to "one-shot" applications of the MAS (where one does not or cannot experiment with the shape and size of the auxiliary surface); more generally, such a finding would allow us to view oscillations in MAS currents as a drawback which does not automatically invalidate a MAS solution. Rather, we would balance this drawback against possible benefits associated with the particular MAS solution. It is even plausible to find MAS solutions, with not-too-rapidly-oscillating MAS currents, in which some instability due to oscillations is outweighed by advantages such as a rapidity of convergence. The potential benefit of such solutions can be viewed as an analogue to *practical* superdirective antennas (whose difficulties are mild enough to be outweighed by advantages such as increased directivity).

References

1. G. Fikioris, On two types of convergence in the method of auxiliary sources. IEEE Trans. Antennas Propag. **54**, 2022–2033 (2006)
2. G. Fikioris, I. Psarros, On the phenomenon of oscillations in the method of auxiliary sources. IEEE Trans. Antennas Propag. **55**, 1293–1304 (2007)
3. P.J. Papakanellos, G. Fikioris, A. Michalopoulou, On the oscillations appearing in numerical solutions of solvable and nonsolvable integral equations for thin-wire antennas. IEEE Trans. Antennas Propag. **58**, 1635–1644 (2010)
4. G. Fikioris, N.L. Tsitsas, I. Psarros, On the nature of oscillations in discretizations of the extended integral equation. IEEE Trans. Antennas Propag. **59**, 1415–1419 (2011)
5. C.A. Valagiannopoulos, N.L. Tsitsas, G. Fikioris, Convergence analysis and oscillations in the method of fictitious sources applied to dielectric scattering problems. J. Opt. Soc. Am. A. **29**, 1–10 (2012)
6. P. Andrianesis, G. Fikioris, Superdirective-type near fields in the method of auxiliary sources (MAS). IEEE Trans. Antennas Propag. **60**, 3056–3060 (2012)
7. G. Fikioris, N.L. Tsitsas, G.K. Charitos, Spurious oscillations in a combined method-of-auxiliary-sources/extended-integral-equation solution to a simple scattering problem. J. Quant. Spectrosc. Radiat. Transf. (Special Issue, Peter C. Waterman and his Scientific Legacy) **123**, 30–40 (2013)
8. G. Fikioris, O.N. Bakas, Study of convergence, divergence, and oscillations in method-of-auxiliary-sources (MAS) and extended-integral-equation (EIE) solutions to a simple cavity problem. IEEE Trans. Microw. Theory Tech. **61**, 2773–2782 (2013)
9. G. Fikioris, N.L. Tsitsas, On convergence and inherent oscillations within computational methods employing fictitious sources. Comput. Math. Appl. **69**, 636–649 (2015)
10. C.A. Balanis, *Antenna Theory, Analysis and Design*, 4th edn. (Wiley, New York, 2016) (sec. 6.9)
11. A.E. Krasnok, P.A. Belov, A.E. Miroshnichenko, A.I. Kuznetsov, B.S. Luk'yanchuk, Yu.S. Kivshar, All-dielectric optical nanoantennas, in *Progress in Compact Antennas*, ed. by L. Huitema (2014), http://www.intechopen.com/books/progress-in-compactantennas/all-dielectric-optical-nanoantennas
12. N.L. Tsitsas, E.G. Alivizatos, H.T. Anastassiu, D.I. Kaklamani, Optimization of the method of auxiliary sources (MAS) for scattering by an infinite cylinder under oblique incidence. Electromagnetics **25**, 39–54 (2005)
13. C.W. Groetsch, *Inverse Problems in the Mathematical Sciences* (Vieweg, Braunschweig, 1993), p. 92
14. C.W. Groetsch, Integral equations of the first kind, inverse problems and regularization: a crash course. Proc. J. Phys.: Conf. Ser. **73**, 1–32 (2007)
15. R. Kress, *Linear Integral Equations*, 3rd edn. (Springer, New York, 2014)
16. N.P. Malakshinov, V.G. Yerikhov, On one numerical method for solving diffraction problems. Antennas **25**, 53–65 (1977) (in Russian)
17. S.P. Skobelev, Comments on "superdirective-type near fields in the method of auxiliary sources" (with authors reply). IEEE Trans. Antennas Propag. **61**, 2360 (2013)
18. Y.I. Bobrovniskii, T.M. Tomilina, General properties and fundamental errors of the method of equivalent sources. Acoust. Phys. **41**, 737–750 (1995) (in Russian)
19. A.H. Barnett, T. Betcke, Stability and convergence of the method of fundamental solutions for Helmholtz problems on analytic domains. J. Comput. Phys. **227**, 7003–7026 (2008)
20. Digital Library of Mathematical Functions. National Institute of Standards and Technology, http://dlmf.nist.gov/
21. Y. Leviatan, Analytic continuation considerations when using generalized formulations for scattering problems. IEEE Trans. Antennas Propag. **38**, 1259–1263 (1990)
22. M.L. Burrows, Example of the generalised-function validity of the Rayleigh hypothesis. Electron. Lett. **5**, 694–695 (1969)

23. Y. Leviatan, A. Boag, A. Boag, Generalized formulations for electromagnetic scattering from perfectly conducting and homogeneous material bodies—theory and numerical solution. IEEE Trans. Antennas Propag. **36**, 1722–1734 (1988)
24. R.M. Gray, Toeplitz and circulant matrices: a review (2002), http://ee.stanford.edu/~gray/toeplitz.pdf
25. R. Vescovo, Inversion of block-circulant matrices and circular array approach. IEEE Trans. Antennas Propag. **45**, 1565–1567 (1997)
26. R.W.P. King, G. Fikioris, R.B. Mack, *Cylindrical Antennas and Arrays* (Cambridge University Press, Cambridge, 2002) (secs. 4.5 and 10.3)
27. A. Doicu, Y. Eremin, T. Wriedt, *Acoustic and Electromagnetic Scattering Using Discrete Sources* (Academic Press, London, 2000)
28. R. Kress, *Numerical Analysis* (Springer, New York, 1998) (sec. 9.4)
29. T. Wriedt (ed.), Generalized multipole techniques for electromagnetic and light scattering, *Computational Methods in Mechanics* (Elsevier, Amsterdam, 1999)
30. B.Z. Katsenelenbaum, Nonapproximable diagrams and nonradiating currents. J. Commun. Technol. Electron. **38**, 112–118 (1993) (originally published in Radiotekhnika i elektronika **6**, 998–1005 (1993))
31. B.Z. Katsenelenbaum, *Electromagnetic Fields: Restrictions and Approximations* (Wiley-VCH, Weinheim, 2003)
32. D. Margetis, G. Fikioris, J.M. Myers, T.T. Wu, Highly directive current distributions: general theory. Phys. Rev. E **58**, 2531–2547 (1998)
33. S. Ramo, J.R. Whinnery, T. Van Duzer, *Fields and Waves in Communication Electronics* (Wiley, New York, 1984), p. 51
34. R.F. Millar, The analytic continuation of solutions to elliptic boundary value problems in two independent variables. J. Math. Anal. Appl. **76**, 498–515 (1980)
35. R.F. Millar, Singularities and the Rayleigh hypothesis for solutions to the Helmholtz equation. IMA J. Appl. Math. **37**, 155–171 (1986)
36. H.T. Anastassiu, D.G. Lymperopoulos, D.I. Kaklamani, Accuracy analysis and optimization of the method of auxiliary sources (MAS) for scattering by a circular cylinder. IEEE Trans. Antennas Propag. **52**, 1541–1547 (2004)
37. P.C. Waterman, Matrix formulation for electromagnetic scattering. Proc. IEEE **53**, 805–812 (1965)
38. P.C. Waterman, New formulation of acoustic scattering. J. Acoust. Soc. Am. **45**, 1417–1429 (1968)
39. P.A. Martin, On the null-field equations for the exterior problems of acoustics. Q. J. Mech. Appl. Math. **33**, 385–396 (1980)
40. M.I. Mishchenko, L.D. Travis, A.A. Lacis, *Scattering, Absorption, and Emission of Light by Small Particles* (Cambridge University Press, Cambridge, 2002)
41. D. Colton, R. Kress, The unique solvability of the null field equations of acoustics. Q. J. Mech. Appl. Math. **36**, 87–95 (1983)
42. A.G. Dallas, On the convergence and numerical stability of the second Waterman scheme for approximation of the acoustic field scattered by a hard object. Technical report, Department of Mathematical Sciences, University of Delaware, No. 2000-7:1-35 (2000)
43. B. Peterson, S. Ström, T matrix formulation of electromagnetic scattering from multilayered scatterers. Phys. Rev. D **10**, 2670–2684 (1974)
44. T. Wriedt, A. Doicu, Comparison between various formulations of the extended boundary condition method. Opt. Commun. **142**, 91–98 (1997)
45. W.C. Chew, *Waves and Fields in Inhomogeneous Media* (IEEE Press, New York, 1995), pp. 453–460
46. F.M. Kahnert, Numerical methods in electromagnetic scattering theory. J. Quant. Spectrosc. Radiat. Transf. **79–80**, 775–824 (2003)
47. M.I. Mishchenko, G. Videen, V.A. Babenko, N.I. Khlebtsov, T. Wriedt, T-matrix theory of electromagnetic scattering by particles and its applications: a comprehensive reference database. J. Quant. Spectrosc. Radiat. Transf. **88**, 357–406 (2004)

48. R.A. Shore, A.D. Yaghjian, Dual-surface integral equations in electromagnetic scattering. IEEE Trans. Antennas Propag. **53**, 1706–1709 (2005)
49. H.A. Schenck, Improved integral formulation for acoustic radiation problems. J. Acoust. Soc. Am. **44**, 41–58 (1967)
50. T. Wriedt, Review of the null-field method with discrete sources. J. Quant. Spectrosc. Radiat. Transf. **106**, 535–545 (2007)
51. A.G. Kyurkchan, N.I. Smirnova, Generalization of the method of extended boundary conditions. J. Commun. Technol. Electron. **53**, 767–774 (2008)
52. A.G. Kyurkchan, N.I. Smirnova, Solution of wave diffraction problems by the null-field method. Acoust. Phys. 55, 691–687 (2009)
53. A.G. Kyurkchan, Using apriory information about analytic properties of the solution on mathematical modeling in wave diffraction and propagation theory, in *Proceedings of the 13th International Conference on Mathematics Methods in Electromagnetic Theory, Kyiv, Ukraine* (2010), pp. 57–60
54. S.P. Skobelev, Some features of the null field method and method of auxiliary sources, in *Proceedings of the 13th International Conference on Mathematics Methods in Electromagnetic Theory, Kyiv, Ukraine* (2010), pp. 350–353
55. Yu.A. Eremin, A.G. Sveshnikov, S.P. Skobelev, Null field method in wave diffraction problems. Comput. Math. Math. Phys. **51**, 1391–1394 (2011)
56. Yu.A. Eremin, S.P. Skobelev, On a generalization of the extended boundary condition method. J. Commun. Technol. Electron. **56**, 1305–1310 (2011)
57. G. Fikioris, J. Lionas, C.G. Lioutas, The use of the frill generator in thin-wire integral equations. IEEE Trans. Antennas Propag. **51**, 1847–1854 (2003)
58. A.N. Tikhonov, V.Y. Arsenin, Solutions of ill-posed problems (translated from the Russian by F. John) (Wiley, New York, 1977), p. 82
59. V.B. Glasko, *Inverse Problems of Mathematical Physics* (American Institute of Physics, New York, 1988), p. 9
60. A. Kirsch, *An Introduction to the Mathematical Theory of Inverse Problems* (Springer, New York, 1996), p. 13
61. B. Noble, The numerical solution of integral equations, in *The State of the Art in Numerical Analysis*, ed. by D. Jacobs (Academic Press, New York, 1976), p. 939
62. N.J. Higham, *Accuracy and Stability of Numerical Algorithms*, 2nd edn. (SIAM Publications, Philadelphia, 2002) (sec. 1.18)
63. N.L. Tsitsas, E.G. Alivizatos, H.T. Anastassiu, D.I. Kaklamani, Optimization of the method of auxiliary sources (MAS) for oblique incidence scattering by an infinite dielectric cylinder. Electr. Eng. **89**, 353–361 (2007)
64. G. Anderson, H. Sprott, B.R. Olsen, Opinion: publish negative results; non-confirmatory or negative results are not worthless. The Scientist, 15 Jan 2013, http://www.the-scientist.com/?articles.view/articleNo/33968/title/Opinion--Publish-Negative-Results/

Chapter 6
MMP Simulation of Plasmonic Particles on Substrate Under E-Beam Illumination

Ueli Koch, Jens Niegemann, Christian Hafner and Juerg Leuthold

Abstract A novel numerical approach to investigate the resonance behavior of plasmonic particles on a substrate under e-beam illumination is presented. The method is based on the Multiple Multipole Program (MMP), a generalized point matching technique, and is augmented by the ability to compute layered media and electron energy loss spectroscopy (EELS) measurements. Furthermore, the whole framework is complemented by a mesh-based method that automatically places the multipole expansions and matching points for arbitrary three-dimensional geometries. The performance of our technique is analyzed by a series of numerical experiments. The EELS responses of a plasmonic split-ring resonator in free space and a plasmonic disk dimer on a membrane, as well as the resonant modes, are simulated. Finally, our implementation is compared to the established discontinuous Galerkin time domain (DGTD) method with respect to its computational efficiency. We show significantly improved performance especially for the computation of EELS resonance maps.

U. Koch (✉) · C. Hafner · J. Leuthold
Department of Information Technology and Electrical Engineering, Institute of Electromagnetic Fields, ETH Zurich, Gloriastrasse 35, 8092 Zurich, Switzerland
e-mail: uelikoch@ethz.ch

C. Hafner
e-mail: hafner@ethz.ch

J. Leuthold
e-mail: leuthold@ethz.ch

J. Niegemann
Lumerical Computational Solutions Inc., 1700-1095, West Pender Street, Vancouver, BC, Canada
e-mail: jniegemann@lumerical.com

T. Wriedt and Y. Eremin (eds.), *The Generalized Multipole Technique for Light Scattering*, Springer Series on Atomic, Optical, and Plasma Physics 99,
https://doi.org/10.1007/978-3-319-74890-0_6

6.1 Introduction

With the progress in nanofabrication over the past decades, metallic nanostructures with feature sizes well below 100 nm can today be produced routinely and reliably. This ability to structure metals on the nanoscale was instrumental in enabling the field of plasmonics, where the interaction between light and such tailored metallic nanostructures is studied and exploited for a variety of applications (e.g., see [1] and references therein). Despite the tremendous progress in the field, plasmonic nanostructures still present a significant challenge for both experimental and numerical characterization. Since the size of the particles is often well below the diffraction limit, one typically has to rely on electron microscopy to image the structures under investigation. A particularly interesting technique in this regard is to employ scanning transmission electron microscopy (STEM) in conjunction with electron energy loss spectroscopy (EELS) [2]. In this approach, fast electrons are passing nearby or through the plasmonic structure. Due to their local electric field, the electrons can excite plasmonic modes of the system and therefore lose a corresponding amount of kinetic energy. By performing spectroscopy on the transmitted electrons, one can therefore locally measure the excitation spectrum of the plasmonic structure. By raster-scanning over the sample, one can obtain EELS maps, showing such a spectrum at every point in a plane. One advantage of EELS over optical spectroscopy lies in its much better spatial resolution which is not limited by the diffraction of light. In addition, EELS allows to excite resonances that are optically dark due to their vanishing electric dipole moments.

In order to better understand experimental measurements and to properly interpret the EELS results, reliable EELS simulations play an important role. Traditionally, either the Finite Element Method (FEM) [3] or the Boundary Element Method (BEM) [4] were used to model EELS experiments. More recently, other numerical methods such as the Discrete Dipole Approximation (DDA) [5] and the Discontinuous Galerkin Time-Domain (DGTD) [6] approach were also shown to be suitable methods.

Since the computation of EELS maps requires a large number of calculations with identical geometry but different excitations (electron beams at different positions), numerical methods with multiple right-hand-sides features (multiple excitations) are favorable. The Multiple Multipole Program (MMP) [7, 8] supported multiple right-hand-sides already in early implementations. The latest implementation of MMP contained in the OpenMaXwell package [9] contains many additional advanced features for modeling and efficiently solving 2D and 3D electrodynamic problems, including eigenvalue problems (computation of resonators and of guided waves) and scattering problems in presence of strongly dispersive materials such as metals at optical frequencies. MMP provides built-in error estimates and a fast convergence if the interfaces of all objects are sufficiently smooth. As a consequence, MMP is very attractive for plasmonics and for EELS simulations in particular.

In the following, we will describe how to perform realistic EELS simulations with MMP. To this end, we first propose a novel method to automatically place the multipole expansions and matching points for arbitrary three-dimensional geometries. Since the dielectric properties of the substrate strongly influence the measurements, we also give details of how to properly handle substrates and stratified media in the MMP context. Finally, we provide the explicit expressions to include a relativistic electron beam as a source of excitation. To demonstrate the efficiency of our novel MMP approach, we perform a series of numerical experiments and compare the performance of our implementation to the established DGTD method.

6.2 Generic MMP Simulation

Although MMP is a flexible and powerful method for electromagnetic simulations, it also has some disadvantages. Probably the biggest inconveniences are the manual selection and placement of expansions and matching points. Even for an experienced user, this step can be time consuming when modeling complex three-dimensional geometries. To mitigate this problem, we have developed a novel generic approach which places the expansions and matching points automatically. Similarly to the classical boundary element method, it is based on a surface mesh. In the following, we will focus on triangular meshes, but our method is readily extended to any kind of surface discretization.

For a given mesh, we first compute the center $\mathbf{c}_E$ and the outward pointing normal vector $\mathbf{n}_E$ of each surface element E. Then, we place our expansions e at a distance d both inwards and outwards from the center, so the positions are

$$\mathbf{r}_{e,\text{in}} = \mathbf{c}_E + d\mathbf{n}_E \ , \tag{6.1a}$$

$$\mathbf{r}_{e,\text{out}} = \mathbf{c}_E - d\mathbf{n}_E \ . \tag{6.1b}$$

The distance d still remains to be defined. A first idea would be to choose d based on the area of the respective element. However, we found that this approach can lead to poor conditioning and therefore reduces the robustness of our method. Instead, we employ an even simpler choice for d, namely a constant for the entire mesh. As we will demonstrate in Sect. 6.5.2.4, a distance of approximately 90% of the minimal curvature radius of the geometry usually produces good results while still remaining numerically stable.

In a second step, the matching points and their weights need to generated. For this, an idea from the finite element method is adapted. Since the error is distributed over the entire surface, it makes sense to use a two-dimensional quadrature rule on the surface elements. Generally, any order of quadrature rule could be chosen, but we found that only a second order rule provides a good balance between accuracy and computational effort (see Sect. 6.5.2.2). The locations of the matching points are exactly on the quadrature nodes. The corresponding weights for all boundary conditions of a single matching point (m) are then given by $w_m = \sqrt{A_E w_{\text{quad}}}$, where A_E and w_{quad} are the element area and the quadrature weight, respectively.

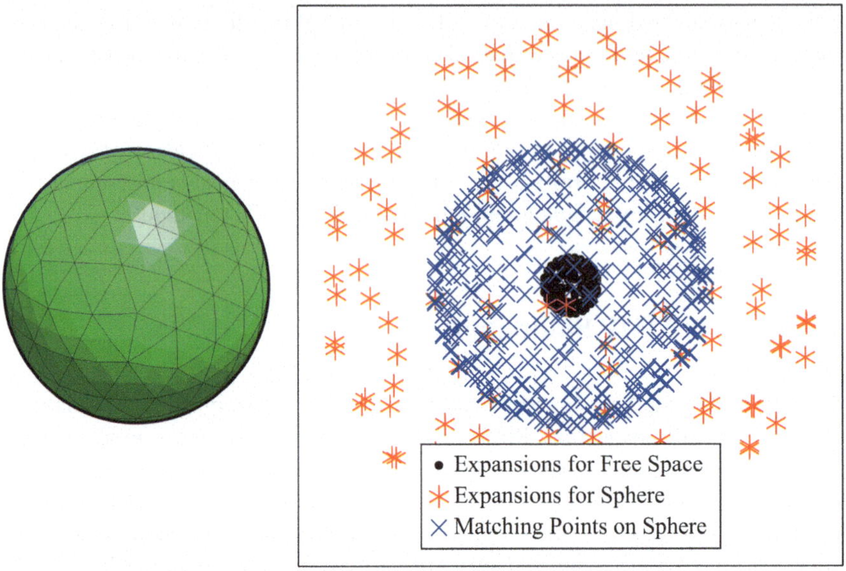

Fig. 6.1 Surface mesh with curvilinear elements generated by Netgen [10] and the generated distribution of the expansions and matching points for a floating sphere

With this simple procedure, the user only needs to provide a surface mesh and a distance parameter d instead of spending a lot of time on the three-dimensional MMP modeling. To illustrate this generic process, we model a basic sphere. The surface mesh and the resulting distribution of expansions and matching points are shown in Fig. 6.1. We should mention that a sphere could probably be modeled more efficiently by manually placing higher order spherical expansions in the center. However, for more complex geometries the placement is usually not as obvious and a manual placement becomes time-consuming and inefficient.

6.3 Dipoles in Layered Media

The generic approach, presented in the previous section, allows us to efficiently model complex three-dimensional geometries with MMP. However, it does not make sense to also use this method to discretize infinitely extended structures such as substrates or membranes. We therefore need an alternative way to incorporate the effect of a substrate into our calculations. At this point, the flexibility of MMP with respect to the expansions comes into play. Instead of using standard dipoles (or higher-order multipoles), we can also employ layered dipoles which directly incorporate the effect of the substrate into the basis functions. The electromagnetic fields caused by dipoles in stratified media are a well-studied subject [11–13]. We took advantage of this knowledge and implemented the layered dipole expansion in MMP [14] which saves us from explicitly discretizing the boundaries of layers and therefore reduces the memory requirements and the computational cost significantly. Since the computation of such a layered dipole expansion is fairly involved, we now present the most important steps in more detail.

6.3.1 Layered Media

In general, we follow the procedure and notation used in [11, Chap. 2] with some minor deviations. The basic idea consists of writing the field as an infinite superposition of plane waves,

$$\mathbf{E} = \mathbf{E}_0 \mathrm{e}^{\mathrm{i}\mathbf{k}\cdot\mathbf{r}} , \tag{6.2a}$$

$$\mathbf{H} = \mathbf{H}_0 \mathrm{e}^{\mathrm{i}\mathbf{k}\cdot\mathbf{r}} , \tag{6.2b}$$

where $\mathbf{k} = \left(k_x, k_y, k_z\right)^{\mathrm{T}}$ is the wave vector. Due to the continuity of the tangential component, $k_\rho = \sqrt{k_x^2 + k_y^2}$ is identical in all layers and it suffices to derive the reflections and transmissions for all k_z-values using the continuity conditions at the material interfaces.

The reader should be aware of the different ways of numbering the layers. Here, we count the layers from bottom to top as sketched in Fig. 6.2. This is opposite to the numbering used in [11].

6.3.1.1 Reflection Coefficients

Starting with the simple case of two half-spaces, reflections and transmissions are occurring at a single layer interface. This directly implies that no multiple reflections need to be considered leading to the Fresnel reflection coefficients R. When splitting

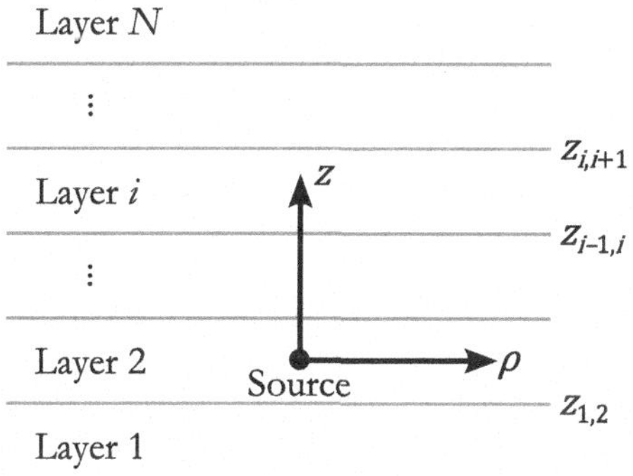

Fig. 6.2 Planar structure of a layered medium

the field into transverse electric (TE) and transverse magnetic (TM) modes, one obtains for the two half-spaces i and j

$$R_{i,j}^{\mathrm{TE}} = \frac{\mu_j k_{z,i} - \mu_i k_{z,j}}{\mu_j k_{z,i} + \mu_i k_{z,j}}, \tag{6.3a}$$

$$R_{i,j}^{\mathrm{TM}} = \frac{\varepsilon_j k_{z,i} - \varepsilon_i k_{z,j}}{\varepsilon_j k_{z,i} + \varepsilon_i k_{z,j}}, \tag{6.3b}$$

where ε, μ and k_z are the permittivity, permeability and the normal component of the wave vector, respectively. For better readability the superscripts TE and TM will be omitted, if the equation holds for both cases. From the definitions (6.3), the antisymmetric property $R_{j,i} = -R_{i,j}$ of Fresnel reflection coefficients can be easily extracted.

Once we introduce finite layers, multiple reflections start occurring and the reflection coefficients need to be generalized. For the interface between layers i and j we then use

$$\tilde{R}_{i,j} = \frac{R_{i,j} + \tilde{R}_{j,k} \mathrm{e}^{2\mathrm{i}k_{z,j}d_j}}{1 + R_{i,j}\tilde{R}_{j,k} \mathrm{e}^{2\mathrm{i}k_{z,j}d_j}}, \tag{6.4}$$

where d_j is the thickness of layer j and the index k represents the other layer next to layer j (opposite to layer i). In (6.4), the two uppermost and the two lowermost layers are special cases. The topmost layer N is a half-space and therefore has no reflections from above, so $\tilde{R}_{N,N+1} = 0$. The second uppermost layer $N-1$ has no transmitted waves from top layer reflections (i.e. $\tilde{R}_{N,N+1}\mathrm{e}^{2\mathrm{i}k_{z,N}d_N} = 0$ although $d_N \to \infty$). This implies that the generalized reflection coefficient is equal to the Fresnel coefficient, $\tilde{R}_{N-1,N} = R_{N-1,N}$. For all other layers, the relation (6.4) can be applied from top to bottom to get all reflection coefficients of the form $\tilde{R}_{i,i+1}$. An equivalent procedure from bottom to top holds for the coefficients of the form $\tilde{R}_{i,i-1}$ where the two lowermost layers are special cases.

Equation (6.4) only holds in the case of a sourceless layer. Dipole sources in layered media are covered in Sect. 6.3.2, but nonetheless an additional term for the source layer s has to be defined first. It reads

$$\tilde{M}_s = \left[1 - \tilde{R}_{s,s+1}\tilde{R}_{s,s-1}\mathrm{e}^{2\mathrm{i}k_{z,s}d_s}\right]^{-1} \tag{6.5}$$

and describes the magnitude enhancement or attenuation by reflection and transmission. There is no common name for this term, but in this work we will call it generalized source magnitude factor.

6.3.1.2 Continuity at Layer Interfaces

This last part about layered structures covers the propagation of waves through layers (cf. [11]). Again, they should be separated in TE- and TM-waves, but most equations are valid for both modes. For a complete coverage, four different cases must be considered due to upward (↑) and downward (↓) propagation and positive (+) and negative (−) superposition with reflected waves. First, we discuss a positive superposition and start with upward propagation. The wave in layer i is then given by

$$A_{i,\uparrow}^{+}\left[\mathrm{e}^{\mathrm{i}k_{z,i}z}+\tilde{R}_{i,i+1}\mathrm{e}^{-\mathrm{i}k_{z,i}z}\mathrm{e}^{\mathrm{i}k_{z,i}2z_{i,i+1}}\right], \tag{6.6}$$

where $A_{i,\uparrow}^{+}$ is the unknown amplitude and $z_{i,i+1}$ is the z-position of the boundary between layers i and $i+1$. Assuming we know the amplitude $A_{i-1,\uparrow}^{+}$, we can enforce equality at the layer boundary $z_{i-1,i}$ to determine the amplitude in layer i. This results in the recursion

$$A_{i,\uparrow}^{+}=A_{i-1,\uparrow}^{+}\frac{1+R_{i-1,i}}{1+R_{i-1,i}\tilde{R}_{i,i+1}\mathrm{e}^{\mathrm{i}k_{z,i}2d_i}}\mathrm{e}^{\mathrm{i}\left(k_{z,i-1}-k_{z,i}\right)z_{i-1,i}}. \tag{6.7}$$

For downward propagating waves with field

$$A_{i,\downarrow}^{+}\left[\mathrm{e}^{-\mathrm{i}k_{z,i}z}+\tilde{R}_{i,i-1}\mathrm{e}^{\mathrm{i}k_{z,i}z}\mathrm{e}^{-\mathrm{i}k_{z,i}2z_{i-1,i}}\right], \tag{6.8}$$

the derivation is very similar and only the direction of propagation as well as the signs in the subscripts need to be flipped. After exploiting the antisymmetry of the Fresnel coefficients, the amplitude transfer factor reads

$$A_{i,\downarrow}^{+}=A_{i+1,\downarrow}^{+}\frac{1-R_{i,i+1}}{1-R_{i,i+1}\tilde{R}_{i,i-1}\mathrm{e}^{\mathrm{i}k_{z,i}2d_i}}\mathrm{e}^{-\mathrm{i}\left(k_{z,i+1}-k_{z,i}\right)z_{i,i+1}}. \tag{6.9}$$

The same can also be done for waves with negative reflection superposition. The fields now have the form

$$A_{i,\uparrow}^{-}\left[\mathrm{e}^{\mathrm{i}k_{z,i}z}-\tilde{R}_{i,i+1}\mathrm{e}^{-\mathrm{i}k_{z,i}z}\mathrm{e}^{\mathrm{i}k_{z,i}2z_{i,i+1}}\right] \tag{6.10}$$

for upward propagation and

$$A_{i,\downarrow}^{-}\left[\mathrm{e}^{-\mathrm{i}k_{z,i}z}-\tilde{R}_{i,i-1}\mathrm{e}^{\mathrm{i}k_{z,i}z}\mathrm{e}^{-\mathrm{i}k_{z,i}2z_{i-1,i}}\right] \tag{6.11}$$

for downward propagation. Assuming that the neighboring amplitude is known, the current one can be computed in a similar manner as above. Finally, this leads to the amplitude transfer factors

$$A_{i,\uparrow}^{-}=A_{i-1,\uparrow}^{-}\frac{1-R_{i-1,i}}{1+R_{i-1,i}\tilde{R}_{i,i+1}\mathrm{e}^{\mathrm{i}k_{z,i}2d_i}}\mathrm{e}^{\mathrm{i}\left(k_{z,i-1}-k_{z,i}\right)z_{i-1,i}}, \tag{6.12}$$

$$A^{-}_{i,\downarrow} = A^{-}_{i+1,\downarrow} \frac{1 + R_{i,i+1}}{1 - R_{i,i+1}\tilde{R}_{i,i-1}\mathrm{e}^{\mathrm{i}k_{z,i}2d_i}} \mathrm{e}^{-\mathrm{i}\left(k_{z,i+1}-k_{z,i}\right)z_{i,i+1}} , \tag{6.13}$$

for upward and downward propagation, respectively.

Similarly to (6.5), starting equations in the source layer can be defined. They are then used for the above iterative amplitude computation for all other layers. For TE-waves and a horizontally (h) oriented source at distance h from the lower layer boundary, they read as:

$$A^{\mathrm{TE}}_{s,\uparrow,h} = \left(1 + \tilde{R}^{\mathrm{TE}}_{s,s-1}\mathrm{e}^{\mathrm{i}k_{z,s}2h}\right)\tilde{M}^{\mathrm{TE}}_{s} , \tag{6.14a}$$

$$A^{\mathrm{TE}}_{s,\downarrow,h} = \left(1 + \tilde{R}^{\mathrm{TE}}_{s,s+1}\mathrm{e}^{\mathrm{i}k_{z,s}2(d_s-h)}\right)\tilde{M}^{\mathrm{TE}}_{s} , \tag{6.14b}$$

whereas for TM-waves, they are given by

$$A^{\mathrm{TM}}_{s,\uparrow,h} = \left(-1 + \tilde{R}^{\mathrm{TM}}_{s,s-1}\mathrm{e}^{\mathrm{i}k_{z,s}2h}\right)\tilde{M}^{\mathrm{TM}}_{s} , \tag{6.15a}$$

$$A^{\mathrm{TM}}_{s,\downarrow,h} = \left(-1 + \tilde{R}^{\mathrm{TM}}_{s,s+1}\mathrm{e}^{\mathrm{i}k_{z,s}2(d_s-h)}\right)\tilde{M}^{\mathrm{TM}}_{s} . \tag{6.15b}$$

In the case of a vertically (v) oriented source, only TM-waves need to be considered and the starting amplitudes can be written as

$$A^{\mathrm{TM}}_{s,\uparrow,v} = \left(1 + \tilde{R}^{\mathrm{TM}}_{s,s-1}\mathrm{e}^{\mathrm{i}k_{z,s}2h}\right)\tilde{M}^{\mathrm{TM}}_{s} , \tag{6.16a}$$

$$A^{\mathrm{TM}}_{s,\downarrow,v} = \left(1 + \tilde{R}^{\mathrm{TM}}_{s,s+1}\mathrm{e}^{\mathrm{i}k_{z,s}2(d_s-h)}\right)\tilde{M}^{\mathrm{TM}}_{s} . \tag{6.16b}$$

6.3.2 Layered Dipole

With the reflection coefficients at hand, we can now proceed to compute the emission of a dipole in a stratified medium. As before, we refrain from giving a full derivation and only present the main steps. To obtain the physical fields, all propagating ($k_z^2 > 0$) and evanescent ($k_z^2 < 0$) contributions have to be properly summed up. One therefore has to evaluate Sommerfeld integrals of the form [11–13]

$$F_{ij} = \frac{1}{4\pi}\int_0^{\infty} G^F_{ij} J_0\left(k_\rho \rho\right) k_\rho \,\mathrm{d}k_\rho . \tag{6.17}$$

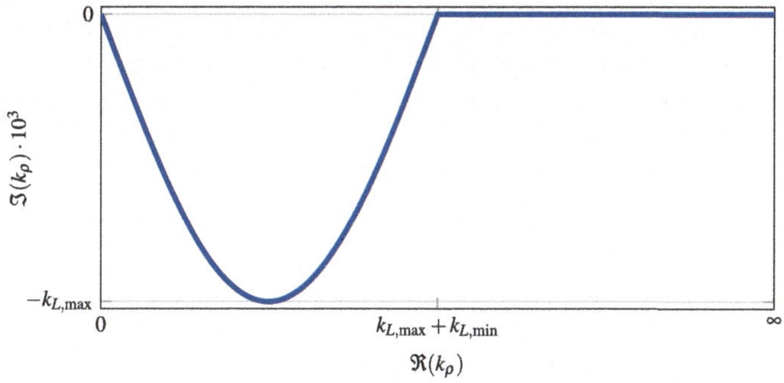

Fig. 6.3 Schematic of the Sommerfeld integration path to avoid singularities on the real axis

Here, $J_\nu(\cdot)$ is the Bessel function of first kind of order ν and G_{ij}^F is the Green's function operator for a field component F in coordinate direction i of a dipole with axis in direction j. An additional factor $1/2$ was already pulled out for a simplification of the Green's functions later in this section. The integrand contains singularities along the real axis, so one typically moves the integration path into the complex plane. In our case, we employ a sinusoidal path as sketched in Fig. 6.3, where $k_{L,\max}$ and $k_{L,\min}$ denote the largest and smallest wavenumber in the layer materials, respectively. As an additional complication, the integration along the tail can be highly oscillatory. This is why the integration is often converging slowly when the observation point is close to the source. There are several methods proposed in literature (e.g. see [15]) to accelerate convergence in such cases. Here, for the numerical integration, we simply subtract the direct term, which can be computed analytically. The analytic result is then added afterwards.

The fields $\mathbf{F} = (E_x, E_y, E_z, H_x, H_y, H_z)^{\mathrm{T}}$ of the layered dipole expansion e can be written as

$$\mathbf{F} = \mathsf{A}_{e,r}\mathbf{x}_e \,, \tag{6.18}$$

where $\mathbf{x}_e = (x_{e,x}, x_{e,y}, x_{e,z})^{\mathrm{T}}$ is the expansion strength and $\mathsf{A}_{e,r}$ is the layered dipole matrix for observation location r with radial distance ρ. The full layered dipole matrix reads

$$\mathsf{A}_{e,r} = \frac{\mathsf{W}_{bc}}{4\pi}\int_0^\infty \begin{bmatrix} G_{xx}^E & G_{xy}^E & G_{xz}^E \\ G_{yx}^E & G_{yy}^E & G_{yz}^E \\ G_{zx}^E & G_{zy}^E & G_{zz}^E \\ G_{xx}^H & G_{xy}^H & G_{xz}^H \\ G_{yx}^H & G_{yy}^H & G_{yz}^H \\ G_{zx}^H & G_{zy}^H & G_{zz}^H \end{bmatrix}_{e,r} J_0\left(k_\rho \rho\right) k_\rho \,\mathrm{d}k_\rho \,, \tag{6.19}$$

where the row-weights $\mathsf{W}_{bc} = \mathrm{diag}\left(1, 1, \varepsilon_s/\varepsilon_o, 1, 1, \mu_s/\mu_o\right)$ for source layer s and observation layer o were introduced to ensure that the continuity conditions are fulfilled at the material interfaces. The integration in (6.19) is meant to be computed element-wise and the Green's functions act as operators in space.

To compute the Green's functions in (6.19), we first derive them from Green's functions for the vector potential **A** and the scalar potential ϕ [16] as

$$G^E_{x_i x_j} = \mathrm{i}\omega G^A_{x_i x_j} - \frac{1}{\mathrm{i}\omega}\frac{\partial^2}{\partial x_i \partial x_j} G^\phi_{x_j} , \tag{6.20a}$$

$$G^H_{x_i x_j} = \frac{1}{\mu}\left(\frac{\partial}{\partial x_{i_+}} G^A_{x_{i_-} x_j} - \frac{\partial}{\partial x_{i_-}} G^A_{x_{i_+} x_j}\right) . \tag{6.20b}$$

Here, $G^H_{x_i x_j}$ results from a curl operation, so x_{i_-} and x_{i_+} represent the previous and next coordinate direction relative to x_i, respectively. Since the layers are stacked in vertical direction, the Green's functions are only dependent on z, while the horizontal dependence comes from the Bessel function $J_0\left(k_\rho \rho\right)$. Therefore, all derivatives with respect to x and y do not affect the Green's functions. Furthermore, some of the vector potential Green's functions are zero, namely $G^A_{xy} = G^A_{yx} = G^A_{xz} = G^A_{yz} = 0$. Others are identical due to symmetry, such as $G^A_{xx} = G^A_{yy}$, $G^\phi_x = G^\phi_y$. To derive the final expressions, three cases must be distinguished:

- observation in source layer,
- observation above source layer,
- observation below source layer.

6.3.2.1 Observation in Source Layer

For the case when the source and the observation point lie in the same layer, only the indirect terms originating from reflections are integrated. To do so, we first specify the amplitude coefficients [13] for horizontal dipoles

$$A_h = \mathrm{e}^{\mathrm{i}k_{z,s}(d_s-h)} \tilde{R}^{\mathrm{TE}}_{s,s+1} \left[\mathrm{e}^{\mathrm{i}k_{z,s}(d_s-h)} + \tilde{R}^{\mathrm{TE}}_{s,s-1} \mathrm{e}^{\mathrm{i}k_{z,s}(d_s+h)}\right] \tilde{M}^{\mathrm{TE}}_s , \tag{6.21a}$$

$$B_h = \mathrm{e}^{\mathrm{i}k_{z,s}(d_s-h)} \tilde{R}^{\mathrm{TM}}_{s,s+1} \left[\mathrm{e}^{\mathrm{i}k_{z,s}(d_s-h)} - \tilde{R}^{\mathrm{TM}}_{s,s-1} \mathrm{e}^{\mathrm{i}k_{z,s}(d_s+h)}\right] \tilde{M}^{\mathrm{TM}}_s , \tag{6.21b}$$

$$C_h = \mathrm{e}^{\mathrm{i}k_{z,s}h} \tilde{R}^{\mathrm{TE}}_{s,s-1} \left[\mathrm{e}^{\mathrm{i}k_{z,s}h} + \tilde{R}^{\mathrm{TE}}_{s,s+1} \mathrm{e}^{\mathrm{i}k_{z,s}(2d_s-h)}\right] \tilde{M}^{\mathrm{TE}}_s , \tag{6.21c}$$

$$D_h = \mathrm{e}^{\mathrm{i}k_{z,s}h} \tilde{R}^{\mathrm{TM}}_{s,s-1} \left[\mathrm{e}^{\mathrm{i}k_{z,s}h} - \tilde{R}^{\mathrm{TM}}_{s,s+1} \mathrm{e}^{\mathrm{i}k_{z,s}(2d_s-h)}\right] \tilde{M}^{\mathrm{TM}}_s . \tag{6.21d}$$

Similarly, for vertical dipoles

$$A_v = \mathrm{e}^{\mathrm{i}k_{z,s}h} \tilde{R}^{\mathrm{TM}}_{s,s-1} \left[\mathrm{e}^{\mathrm{i}k_{z,s}h} + \tilde{R}^{\mathrm{TM}}_{s,s+1} \mathrm{e}^{\mathrm{i}k_{z,s}(2d_s-h)}\right] \tilde{M}^{\mathrm{TM}}_s , \tag{6.22a}$$

$$B_v = \mathrm{e}^{\mathrm{i}k_{z,s}(d_s-h)} \tilde{R}^{\mathrm{TM}}_{s,s+1} \left[\mathrm{e}^{\mathrm{i}k_{z,s}(d_s-h)} + \tilde{R}^{\mathrm{TM}}_{s,s-1} \mathrm{e}^{\mathrm{i}k_{z,s}(d_s+h)}\right] \tilde{M}^{\mathrm{TM}}_s , \tag{6.22b}$$

where the naming conventions from Sect. 6.3.1.1 were adopted. With these coefficients at hand, we can explicitly state the Green's functions as

$$G^A_{xx} = G^A_{yy} = -\frac{\mu_s}{\mathrm{i}k_{z,s}}\left[A_h \mathrm{e}^{-\mathrm{i}k_{z,s}z} + C_h \mathrm{e}^{\mathrm{i}k_{z,s}z}\right], \tag{6.23a}$$

$$G^A_{zz} = -\frac{\mu_s}{\mathrm{i}k_{z,s}}\left[A_v \mathrm{e}^{\mathrm{i}k_{z,s}z} + B_v \mathrm{e}^{-\mathrm{i}k_{z,s}z}\right], \tag{6.23b}$$

$$G^A_{zx} = -\frac{\mu_s}{k_\rho^2}\left[(A_h + B_h)\,\mathrm{e}^{-\mathrm{i}k_{z,s}z} - (C_h + D_h)\,\mathrm{e}^{\mathrm{i}k_{z,s}z}\right]\frac{\partial}{\partial x}, \tag{6.23c}$$

$$G^A_{zy} = -\frac{\mu_s}{k_\rho^2}\left[(A_h + B_h)\,\mathrm{e}^{-\mathrm{i}k_{z,s}z} - (C_h + D_h)\,\mathrm{e}^{\mathrm{i}k_{z,s}z}\right]\frac{\partial}{\partial y}, \tag{6.23d}$$

$$G^\phi_x = G^\phi_y = -\frac{1}{\varepsilon_s \mathrm{i}k_{z,s}}\left[\frac{k_s^2 A_h + k_{z,s}^2 B_h}{k_\rho^2}\mathrm{e}^{-\mathrm{i}k_{z,s}z} + \frac{k_s^2 C_h + k_{z,s}^2 D_h}{k_\rho^2}\mathrm{e}^{\mathrm{i}k_{z,s}z}\right], \tag{6.23e}$$

$$G^\phi_z = -\frac{1}{\varepsilon_s \mathrm{i}k_{z,s}}\left[A_v \mathrm{e}^{\mathrm{i}k_{z,s}z} + B_v \mathrm{e}^{-\mathrm{i}k_{z,s}z}\right]. \tag{6.23f}$$

Inserting these expressions into (6.20), evaluating the derivatives and integrating the result yields the full indirect electromagnetic field matrix. Afterwards, the contribution of an identical dipole in a homogeneous medium needs to be added to obtain the final result.

6.3.2.2 Observation Above Source Layer

When the observation point is in layer o above the source, the amplitude transitions from Sect. 6.3.1.2 are required. To find the Green's functions in this case, multiple steps need to be performed. First, the Green's functions for the source layer (6.23) are evaluated at the upper layer boundary of the source layer. They can be reformulated such that their amplitudes have the form of the factors $A_{s,\uparrow}$ in (6.14a) and (6.15a) or (6.16a), depending on orientation and wave type. The propagation through the layers can then be computed by the expressions given in Sect. 6.3.1.2. To simplify the computation, it is helpful to split the amplitude into a source term and a propagation factor. The missing propagation factor is retrieved by starting with a unit amplitude factor in the source layer and iteratively applying (6.7) or (6.12) until the observation layer has been reached. With the amplitudes given, the Green's functions for the potentials can be stated as

$$G^A_{xx} = G^A_{yy} = -\frac{\mu_s}{\mathrm{i}k_{z,s}}\left[A^{\mathrm{TE}}_{s,\uparrow,h} A^{+,\mathrm{TE}}_{o,\uparrow}\left(\mathrm{e}^{\mathrm{i}k_{z,o}z} + \tilde{R}^{\mathrm{TE}}_{o,o+1}\mathrm{e}^{\mathrm{i}k_{z,o}(2z_{o,o+1}-z)}\right)\right], \tag{6.24a}$$

$$G^A_{zz} = -\frac{\mu_s}{\mathrm{i}k_{z,s}}\left[A^{\mathrm{TM}}_{s,\uparrow,v} A^{+,\mathrm{TM}}_{o,\uparrow}\left(\mathrm{e}^{\mathrm{i}k_{z,o}z} + \tilde{R}^{\mathrm{TM}}_{o,o+1}\mathrm{e}^{\mathrm{i}k_{z,o}(2z_{o,o+1}-z)}\right)\right], \tag{6.24b}$$

$$G^A_{zx} = -\frac{\mu_s}{k_\rho^2}\Big[-A^{\mathrm{TE}}_{s,\uparrow,h} A^{-,\mathrm{TE}}_{o,\uparrow}\left(\mathrm{e}^{\mathrm{i}k_{z,o}z} - \tilde{R}^{\mathrm{TE}}_{o,o+1}\mathrm{e}^{\mathrm{i}k_{z,o}(2z_{o,o+1}-z)}\right) -$$

$$- A^{\mathrm{TM}}_{s,\uparrow,h} A^{+,\mathrm{TM}}_{o,\uparrow} \left(\mathrm{e}^{\mathrm{i}k_{z,o}z} + \tilde{R}^{\mathrm{TM}}_{o,o+1} \mathrm{e}^{\mathrm{i}k_{z,o}(2z_{o,o+1}-z)} \right) \Big] \frac{\partial}{\partial x}, \tag{6.24c}$$

$$G^A_{zy} = -\frac{\mu_s}{k_\rho^2} \Big[- A^{\mathrm{TE}}_{s,\uparrow,h} A^{-,\mathrm{TE}}_{o,\uparrow} \left(\mathrm{e}^{\mathrm{i}k_{z,o}z} - \tilde{R}^{\mathrm{TE}}_{o,o+1} \mathrm{e}^{\mathrm{i}k_{z,o}(2z_{o,o+1}-z)} \right) -$$

$$- A^{\mathrm{TM}}_{s,\uparrow,h} A^{+,\mathrm{TM}}_{o,\uparrow} \left(\mathrm{e}^{\mathrm{i}k_{z,o}z} + \tilde{R}^{\mathrm{TM}}_{o,o+1} \mathrm{e}^{\mathrm{i}k_{z,o}(2z_{o,o+1}-z)} \right) \Big] \frac{\partial}{\partial y}, \tag{6.24d}$$

$$G^\phi_x = G^\phi_y = -\frac{1}{\varepsilon_s \mathrm{i}k_{z,s}} \Big[\frac{k_s^2}{k_\rho^2} A^{\mathrm{TE}}_{s,\uparrow,h} A^{+,\mathrm{TE}}_{o,\uparrow} \left(\mathrm{e}^{\mathrm{i}k_{z,o}z} + \tilde{R}^{\mathrm{TE}}_{o,o+1} \mathrm{e}^{\mathrm{i}k_{z,o}(2z_{o,o+1}-z)} \right) +$$

$$+ \frac{k_{z,s}^2}{k_\rho^2} A^{\mathrm{TM}}_{s,\uparrow,h} A^{-,\mathrm{TM}}_{o,\uparrow} \left(\mathrm{e}^{\mathrm{i}k_{z,o}z} - \tilde{R}^{\mathrm{TM}}_{o,o+1} \mathrm{e}^{\mathrm{i}k_{z,o}(2z_{o,o+1}-z)} \right) \Big], \tag{6.24e}$$

$$G^\phi_z = -\frac{1}{\varepsilon_s \mathrm{i}k_{z,s}} \left[A^{\mathrm{TM}}_{s,\uparrow,v} A^{+,\mathrm{TM}}_{o,\uparrow} \left(\mathrm{e}^{\mathrm{i}k_{z,o}z} + \tilde{R}^{\mathrm{TM}}_{o,o+1} \mathrm{e}^{\mathrm{i}k_{z,o}(2z_{o,o+1}-z)} \right) \right]. \tag{6.24f}$$

From here, the same steps as discussed in the previous section must be carried out to obtain the final matrix. In this case, no direct term needs to be added.

6.3.2.3 Observation Below Source Layer

The case when the observation layer o is located below the source layer is very similar to the one above, but requires a change in the propagation direction. Hence, the final formulas read

$$G^A_{xx} = G^A_{yy} = -\frac{\mu_s}{\mathrm{i}k_{z,s}} \left[A^{\mathrm{TE}}_{s,\downarrow,h} A^{+,\mathrm{TE}}_{o,\downarrow} \left(\mathrm{e}^{-\mathrm{i}k_{z,o}z} + \tilde{R}^{\mathrm{TE}}_{o,o-1} \mathrm{e}^{-\mathrm{i}k_{z,o}(2z_{o-1,o}-z)} \right) \right] \tag{6.25a}$$

$$G^A_{zz} = -\frac{\mu_s}{\mathrm{i}k_{z,s}} \left[A^{\mathrm{TM}}_{s,\downarrow,v} A^{+,\mathrm{TM}}_{o,\downarrow} \left(\mathrm{e}^{-\mathrm{i}k_{z,o}z} + \tilde{R}^{\mathrm{TM}}_{o,o-1} \mathrm{e}^{-\mathrm{i}k_{z,o}(2z_{o-1,o}-z)} \right) \right], \tag{6.25b}$$

$$G^A_{zx} = -\frac{\mu_s}{k_\rho^2} \Big[A^{\mathrm{TE}}_{s,\downarrow,h} A^{-,\mathrm{TE}}_{o,\downarrow} \left(\mathrm{e}^{-\mathrm{i}k_{z,o}z} - \tilde{R}^{\mathrm{TE}}_{o,o-1} \mathrm{e}^{-\mathrm{i}k_{z,o}(2z_{o-1,o}-z)} \right) +$$

$$+ A^{\mathrm{TM}}_{s,\downarrow,h} A^{+,\mathrm{TM}}_{o,\downarrow} \left(\mathrm{e}^{-\mathrm{i}k_{z,o}z} + \tilde{R}^{\mathrm{TM}}_{o,o-1} \mathrm{e}^{-\mathrm{i}k_{z,o}(2z_{o-1,o}-z)} \right) \Big] \frac{\partial}{\partial x}, \tag{6.25c}$$

$$G^A_{zy} = -\frac{\mu_s}{k_\rho^2} \Big[A^{\mathrm{TE}}_{s,\downarrow,h} A^{-,\mathrm{TE}}_{o,\downarrow} \left(\mathrm{e}^{-\mathrm{i}k_{z,o}z} - \tilde{R}^{\mathrm{TE}}_{o,o-1} \mathrm{e}^{-\mathrm{i}k_{z,o}(2z_{o-1,o}-z)} \right) +$$

$$+ A^{\mathrm{TM}}_{s,\downarrow,h} A^{+,\mathrm{TM}}_{o,\downarrow} \left(\mathrm{e}^{-\mathrm{i}k_{z,o}z} + \tilde{R}^{\mathrm{TM}}_{o,o-1} \mathrm{e}^{-\mathrm{i}k_{z,o}(2z_{o-1,o}-z)} \right) \Big] \frac{\partial}{\partial y}, \tag{6.25d}$$

$$G^\phi_x = G^\phi_y = -\frac{1}{\varepsilon_s \mathrm{i}k_{z,s}} \Big[\frac{k_s^2}{k_\rho^2} A^{\mathrm{TE}}_{s,\downarrow,h} A^{+,\mathrm{TE}}_{o,\downarrow} \left(\mathrm{e}^{-\mathrm{i}k_{z,o}z} + \tilde{R}^{\mathrm{TE}}_{o,o-1} \mathrm{e}^{-\mathrm{i}k_{z,o}(2z_{o-1,o}-z)} \right) +$$

$$+\frac{k_{z,s}^2}{k_\rho^2} A_{s,\downarrow,h}^{\text{TM}} A_{o,\downarrow}^{-,\text{TM}} \left(\mathrm{e}^{-ik_{z,o}z} - \tilde{R}_{o,o-1}^{\text{TM}} \mathrm{e}^{-ik_{z,o}(2z_{o-1,o}-z)} \right) \Bigg] , \tag{6.25e}$$

$$G_z^\phi = -\frac{1}{\varepsilon_s \mathrm{i}k_{z,s}} \left[A_{s,\downarrow,v}^{\text{TM}} A_{o,\downarrow}^{+,\text{TM}} \left(\mathrm{e}^{-ik_{z,o}z} + \tilde{R}_{o,o-1}^{\text{TM}} \mathrm{e}^{-ik_{z,o}(2z_{o-1,o}-z)} \right) \right] . \tag{6.25f}$$

Again, the final matrix is found by transformation and integration and no direct term needs to be added.

6.4 Electron Energy Loss Spectroscopy

As discussed in the introduction, electron energy loss spectroscopy (EELS) is a technique to experimentally determine optical excitation spectra using a scanning transmission electron microscope (STEM). To simulate the physical processes in an EELS measurement requires a few extensions to MMP which will be discussed in the following.

6.4.1 Electron Beam Expansion

First, the electromagnetic fields caused by a relativistic electron beam must be computed. Here, the only free parameters are the position and the velocity v of the electrons. Assuming a propagation along the z-axis, the electric and magnetic field can be written as in [2]

$$E_x = -\frac{2e\omega}{v^2\gamma\varepsilon} \mathrm{e}^{\mathrm{i}\omega\frac{z}{v}} K_1\left(\frac{\omega\rho}{v\gamma}\right)\frac{x}{\rho} , \tag{6.26a}$$

$$E_y = -\frac{2e\omega}{v^2\gamma\varepsilon} \mathrm{e}^{\mathrm{i}\omega\frac{z}{v}} K_1\left(\frac{\omega\rho}{v\gamma}\right)\frac{y}{\rho} , \tag{6.26b}$$

$$E_z = \frac{2e\omega}{v^2\gamma\varepsilon} \mathrm{e}^{\mathrm{i}\omega\frac{z}{v}} K_0\left(\frac{\omega\rho}{v\gamma}\right)\frac{\mathrm{i}}{\gamma} , \tag{6.26c}$$

$$H_x = \frac{2e\omega}{v\gamma} \mathrm{e}^{\mathrm{i}\omega\frac{z}{v}} K_1\left(\frac{\omega\rho}{v\gamma}\right)\frac{y}{\rho} , \tag{6.26d}$$

$$H_y = -\frac{2e\omega}{v\gamma} \mathrm{e}^{\mathrm{i}\omega\frac{z}{v}} K_1\left(\frac{\omega\rho}{v\gamma}\right)\frac{x}{\rho} , \tag{6.26e}$$

$$H_z = 0 . \tag{6.26f}$$

Here, we introduced

- the elementary charge e,
- the Lorentz contraction factor $\gamma = \left(\sqrt{1 - v^2/c^2}\right)^{-1}$,
- the speed of light in dielectric material c,
- the modified Bessel function of second kind $K_\nu(\cdot)$ of order ν,
- the distance to the beam $\rho = \sqrt{(x - x_0)^2 + (y - y_0)^2}$.

The electric and magnetic field caused by an electron beam are decaying exponentially from the center and have a mainly radial and azimuthal orientation, respectively. They are therefore similar to the fields generated by a line current but include a retardation effect.

6.4.2 *Electron Energy Loss Computation*

With the source terms at hand, we use MMP to compute the scattering caused by a structure close to the electron beam. To then obtain an actual loss spectrum, some further post-processing is required. The frequency-dependent total energy loss $P(\omega)$ can be computed by integration of the energy loss along the electron path [2] as

$$P(\omega) = \frac{1}{\pi \hbar \omega} \int_{-\infty}^{\infty} \Re\left\{ e^{-i\omega t} v E_z^{\text{scat}}(z(t), \omega) \right\} \mathrm{d}t\ , \tag{6.27}$$

where the electron beam direction is again chosen to be in local z-direction without loss of generality. By inserting the transformation $t = \frac{z(t)-z_0}{v}$ and using setting $z_0 = 0$, the integration can be performed over space instead of time. Thus, we need to evaluate

$$P(\omega) = \frac{1}{\pi \hbar \omega} \int_{-\infty}^{\infty} \Re\left\{ e^{-i\omega \frac{z}{v}} E_z^{\text{scat}}(z, \omega) \right\} \mathrm{d}z\ , \tag{6.28}$$

which can be done using an appropriate numerical quadrature rule.

6.5 Numerical Experiments

In the following, we present a series of numerical experiments to demonstrate the convergence behavior and performance of our MMP implementation. For the numerical quadrature, we implemented a routine similar to the one proposed in [17]. For the evaluation of Bessel functions with complex arguments, the Fortran library of D.E. Amos [18] was ported to C++, giving highly accurate values at the cost of somewhat limited performance. As the main parts of MMP contain linear algebra, the versatile and powerful library *Eigen* [19] was used for all vector or matrix algebra functions. To accelerate the computation, a basic parallelization was implemented using Intels *Math Kernel Library (MKL)* [20] and *Threading Building Blocks (TBB)* [21] library.

6.5.1 Plane Wave Excitation of a Dielectric Sphere

The first example serves solely to test the convergence of our mesh-based MMP approach. To this end, we model a dielectric sphere with relative permittivity $\varepsilon_r = 4$. We excite it from the top by a plane wave with a wavelength equal to the diameter of the sphere. To compute the electromagnetic fields, we first need to fix the expansions and matching points. We start with the mesh and the corresponding distribution shown in Fig. 6.1.

Then, the excited fields can be computed following the standard MMP procedure. The real part of the scattered electric field on the polarization axis (in-plane) and the real part of the scattered magnetic field (out-of-plane) are depicted in Fig. 6.4 in a vertical cut plane through the center of the sphere.

6.5.2 Properties of Mesh-Based MMP

To determine optimal parameters for an efficient and accurate numerical simulation with the mesh-based MMP approach, a series of numerical convergence studies was performed.

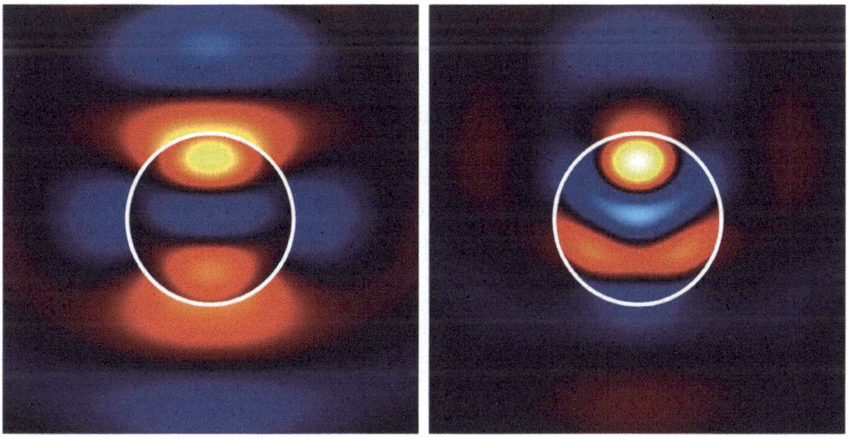

Fig. 6.4 Real part of the scattered electric and the scattered magnetic field of a dielectric sphere excited by a plane wave

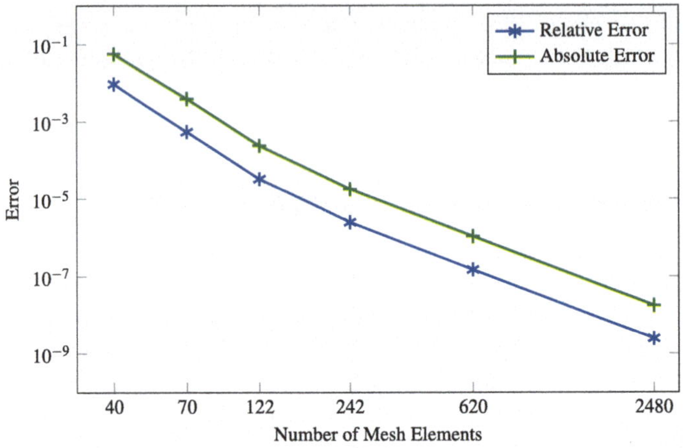

Fig. 6.5 Mesh refinement to observe convergence of the absolute and relative error using the example of a sphere excited by a plane wave

6.5.2.1 Mesh-Refinement

First, we test how many mesh elements are required to model the fields accurately. For this purpose, the mesh was successively refined. To compute error estimates, the field mismatch is integrated over the entire surface using a quadrature rule of higher order than the one used to generate the matching points. The graph in Fig. 6.5 shows the behavior of relative and absolute error over the number of mesh elements. We observe an exponential decay with a slightly flattened slope after a kink at 122 elements. Since a relative error of around 10^{-5} is sufficient for most practical purposes, we will use the mesh with 122 elements for all further tests.

6.5.2.2 Quadrature-Refinement

The next important parameter is the order of the quadrature rule applied on the mesh elements for the matching point placement. The quadrature order directly influences the factor of overdetermination of the MMP system. Naturally, this has a large impact on the memory consumption and on the computational time. Table 6.1 shows the overdetermination depending on the quadrature order.

Table 6.1 Overdetermination caused by given quadrature order

Quadrature order	Overdetermination
1	1
2	3
3	6
4	6

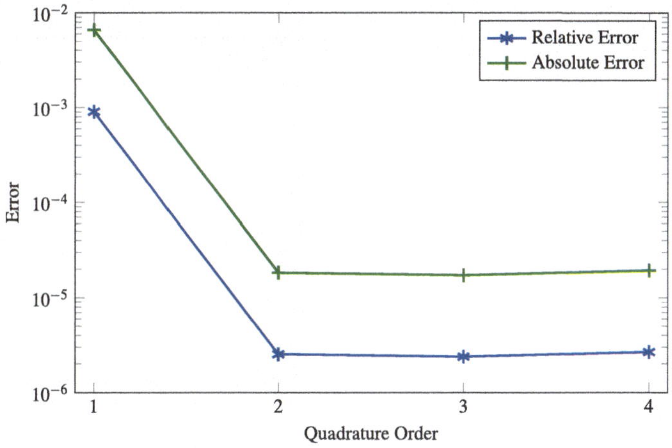

Fig. 6.6 Absolute and relative error for a sphere excited by a plane wave as a function of the quadrature order

Since the rule of 1st order only has one node in the center, it leads to a square MMP matrix. Unfortunately, it is well known that a square system can result in oscillatory or unphysical solutions. This behavior can also be observed in the error plot depicted in Fig. 6.6. On the other hand, triangular quadrature rules of 3rd and 4th order have the same number of nodes, so we are left with the choice between a 2nd order and a 4th order rule. From the data in Fig. 6.6 we clearly see that a 4th order rule does not provide any extra accuracy, so we employ 2nd order rules for the distribution of matching points in all further MMP calculations.

6.5.2.3 Mesh Element Order

Finally, we briefly discuss the use of curvilinear elements versus using straight elements. However, since the geometry only affects the preprocessing, namely the generation of expansions and matching points, it has no significant influence of the computational time. It therefore makes sense to use the most accurate geometry representation, i.e., curvilinear elements, if available. To support this statement, Table 6.2 shows the errors for both kind of elements for the spherical test system.

As expected, we find that the error reduces significantly when using curvilinear elements.

Table 6.2 Absolute and relative errors for planar or curvilinear mesh elements using the example of a sphere excited by a plane wave

Error	Linear	Curved
Relative	$2.0397 \cdot 10^{-5}$	$2.5381 \cdot 10^{-6}$
Absolute	$1.4712 \cdot 10^{-4}$	$1.8393 \cdot 10^{-5}$

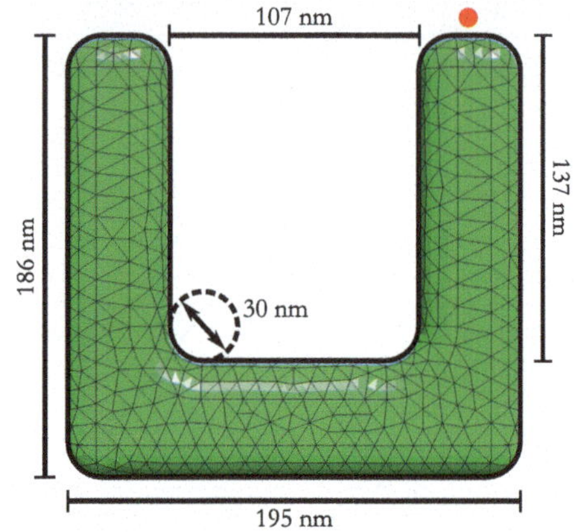

Fig. 6.7 Split-ring resonator geometry and mesh to show the experimental setup. The red dot indicates the location of the electron beam excitation

6.5.2.4 Expansion Distance to Particle Surface

The final test of our MMP implementation concerns the influence of the expansion distance. For this particular case, we switch to a more challenging test geometry. Specifically, we will use a split-ring resonator similar to the one studied in [22]. The mesh is shown in Fig. 6.7 and the structure is excited by an electron beam at the position indicated by the red dot. Then, an electron energy loss spectrum is computed for different distances of the expansions to the split-ring surface, starting at half the minimal curvature and going up to 90% thereof. The resulting spectra are depicted in Fig. 6.8. These results clearly demonstrate the importance and difficulty of placing the expansions optimally. For short distances, the signal is strongly

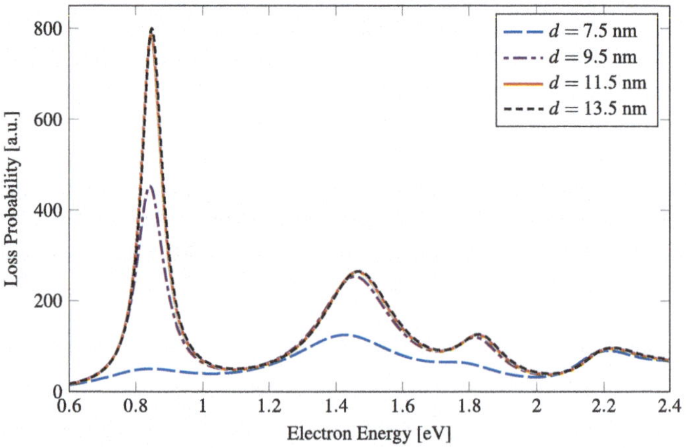

Fig. 6.8 Electron-energy loss spectra for different distances of the expansions to the surface. Increasing the distance improves the accuracy of the method considerably

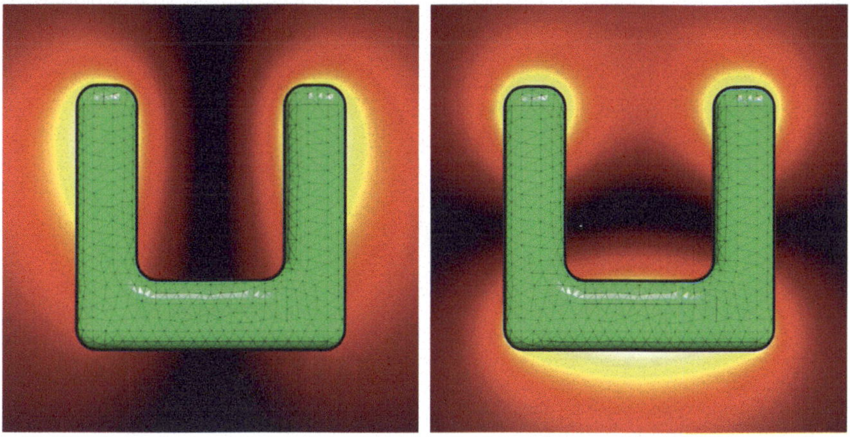

(a) $\lambda = 1460\,\text{nm}$ (b) $\lambda = 845\,\text{nm}$

Fig. 6.9 Normalized EELS maps for the first two modes of a plasmonic split-ring resonator

suppressed. When moving the expansions further away, the spectrum converges towards its correct solution. For our example, the result does not change for distances above $d = 13.5\,\text{nm}$. However, when the expansions are placed even further away from the surface, their locations become too close to each other. As a result, the condition number of the MMP matrix deteriorates, which can lead to significant instabilities. We find that using a distance of around 90% of the smallest curvature radius in the geometry works well in most cases.

6.5.3 Electron Energy Loss Spectroscopy of a Plasmonic Split-Ring Resonator in Free Space

After the convergence studies in the previous sections, we now demonstrate the advantages of MMP when computing EELS maps. Since the MMP matrix is independent of the excitation, we can setup and factorize the matrix once and then efficiently solve the system for a large number of different sources. To compute the EELS maps shown in Fig. 6.9, we solved the system for 201 by 201 different beam positions. We find that both the resonance frequencies and the EELS maps are qualitatively similar to the results published in [22], but the resonance are slightly blue-shifted. Those deviations are readily explained by the omission of the silicon nitride membrane in our MMP simulations.

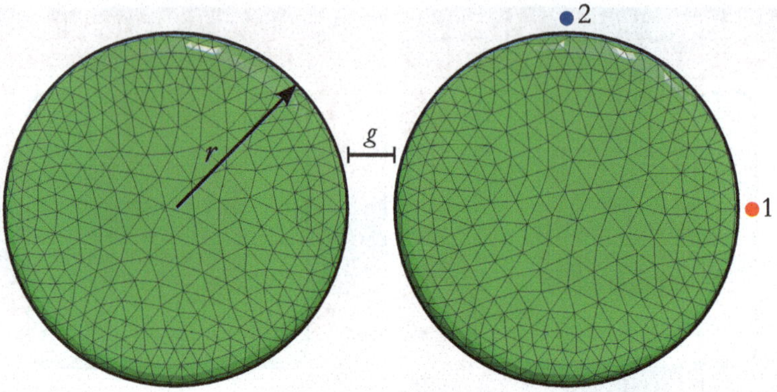

Fig. 6.10 Geometry and mesh for the disk-dimer experiment. The colored dots again indicate the positions of the electron beam excitation

6.5.4 Electron Energy Loss Spectroscopy of a Plasmonic Disk-Dimer on a Membrane

As discussed above, to obtain results comparable to experimental measurements, the inclusion of the substrate or membrane is essential. Here, we demonstrate that MMP can handle this by using the layered dipole expansions presented in Sect. 6.3.2. To this end, we model a plasmonic disk dimer made out of gold and deposited on a 30 nm thick silicon nitride membrane ($\varepsilon_r = 4$). The disk radius is $r = 75$ nm and its height is $t = 30$ nm. The distance between the disks is taken to be $g = 20$ nm. The geometry and the corresponding mesh are sketched in Fig. 6.10, where the dots indicate the positions of the electron beam excitation for EELS spectra computation. The resulting spectra are shown in Fig. 6.11.

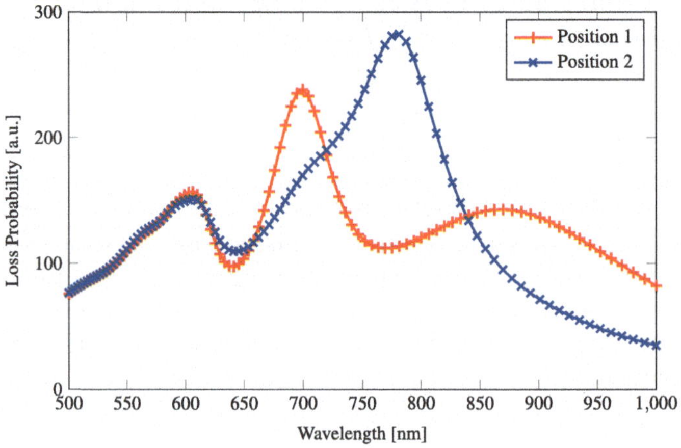

Fig. 6.11 Electron energy loss spectrum of a plasmonic disk-dimer calculated at two different electron beam positions

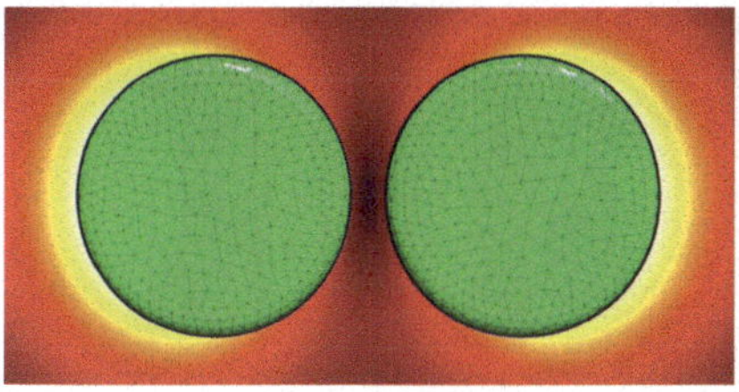

(a) $\lambda = 870\,\text{nm}$

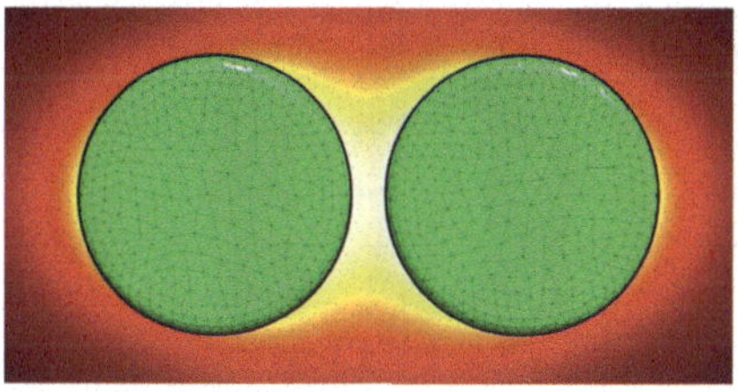

(b) $\lambda = 700\,\text{nm}$

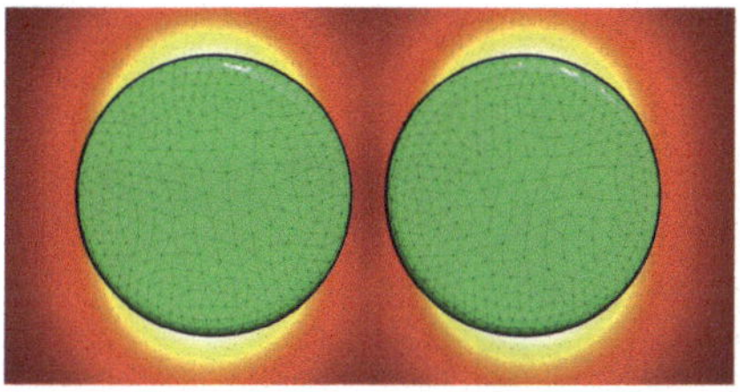

(c) $\lambda = 773\,\text{nm}$

Fig. 6.12 Normalized EELS maps for the first three modes of the plasmonic disk-dimer

We can clearly observe how the two different excitation positions lead to distinct resonance peaks. This can be explained by looking at the corresponding EELS maps depicted in Fig. 6.12.

Here, we see that the mode with a resonance at 773 nm shows very little signal at position #1. In contrast, the modes at 870 nm and 700 nm are comparatively weak at position #2. To gain further insight into the nature of the different modes, we also use MMP to compute the distribution of the electric field in a plane slightly above the dimer. The results for the z-component (perpendicular to the cut plane) can be found in Fig. 6.13. These plots contain information about the relative phase, which allows us to identify the mode at 870 nm as the symmetric and the mode at 700 nm as the anti-symmetric combination of the fundamental mode of an individual disk (cf. [23]). Here, the higher-energetic mode at 773 nm corresponds to a dark mode and would not be visible when using classical optical spectroscopy.

6.5.5 Comparison of MMP and DGTD for Electron Energy Loss Spectroscopy Calculations

Finally, we compare the performance of our novel MMP implementation to an established Discontinuous Galerkin time-domain (DGTD) solver [23–25]. It is important to keep in mind that MMP is a frequency-domain method while DGTD works in the time-domain. For this comparison, we return to the EELS experiment with the golden split-ring resonator of Sect. 6.5.3 and compare the computation times of the

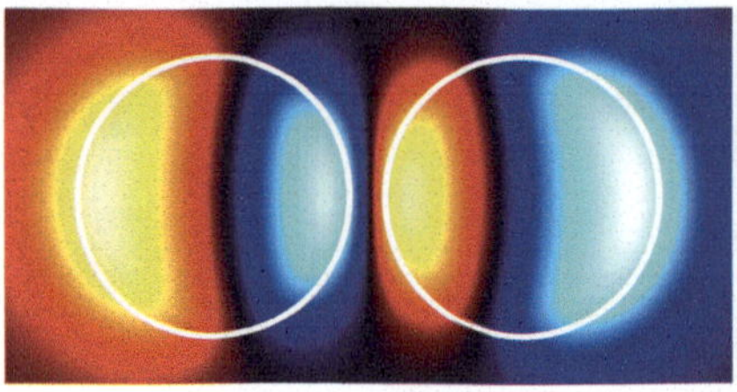

(a) $\lambda = 870\,\text{nm}$

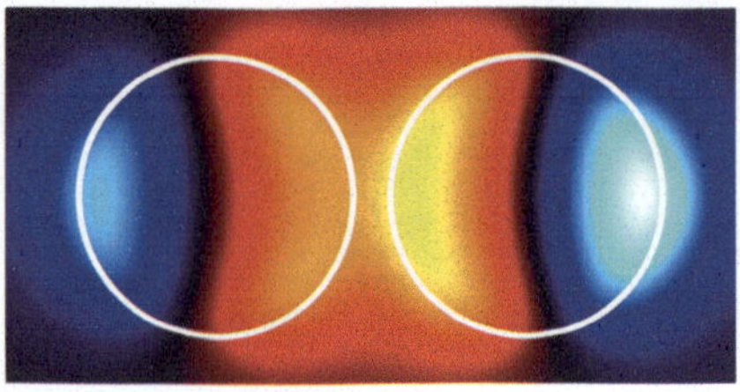

(b) $\lambda = 700\,\text{nm}$

(c) $\lambda = 773\,\text{nm}$

Fig. 6.13 Normalized electric field in z-direction for the first three modes of a plasmonic disk-dimer

two solvers for both EELS spectrum and EELS map computations. The simulations were run on a desktop computer with Intel Xeon W3680 processor with 6 cores at 3.33 GHz and 12 GB RAM. We restricted both solvers to run on a single core to get a fair comparison independent of the implementation of parallelization. To guarantee a similar accuracy of both methods, the same surface mesh was used for the split-ring resonator. For DGTD method, the given surface mesh was extended to a tetrahedral volume mesh using *Netgen* [10].

First, we computed the EELS spectrum and measured the time consumption of both methods. The timings of the individual steps performed in the MMP solver and the total time for DGTD are listed in Table 6.3.

As a time-domain solver, DGTD computes the entire spectrum in a single run, while MMP requires a separate simulation per frequency point. Therefore, the time

Table 6.3 Computation times for EELS spectrum simulations with MMP or DGTD. In DGTD, the full spectrum is received in one simulation while in MMP it scales with the number of frequency points n_f

	MMP	DGTD
Assembly (min)	$0.5 \cdot n_f$	–
QR decomposition (min)	$50.0 \cdot n_f$	–
Solving equation (min)	$<0.1 \cdot n_f$	–
Loss evaluation (min)	$<0.1 \cdot n_f$	–
Total time (min)	$50.6 \cdot n_f$	2505

Table 6.4 Computation times for EELS map simulations with MMP or DGTD. In DGTD, a full simulation is required for each point while in MMP the QR decomposition in the setup can be reused for all beam positions. Here, n_r is the number of raster points

	MMP	DGTD
Assembly (min)	0.5	–
QR decomposition (min)	50.0	–
Solving and Evaluation (min)	$0.06 \cdot n_r$	–
Total time (min)	$50.0 + 0.06 \cdot n_r$	$2505 \cdot n_r$

per frequency point in MMP needs to be multiplied by the number of points in a spectrum. Hence, for a good spectral resolution it is beneficial to use a time-domain solver while frequency-domain methods are only appropriate for a small number of frequencies. In our example, DGTD performs more efficiently once the number of frequency points exceeds 50.

Second, the rasterization time for an EELS pattern was measured. Since MMP is a boundary discretization method in the frequency-domain, it can reuse the matrix decomposition for the different beam positions as they are just right-hand-sides. In contrast, DGTD requires a full calculation per point. While this does also deliver a full spectrum, maps are usually only required for very few selected frequencies. Hence, MMP easily outperforms DGTD for the computation of EELS maps. The corresponding timings can be found in Table 6.4.

From this data, we find that both frequency and time-domain methods have their respective advantages. For the computationally more demanding EELS map computation, MMP is clearly preferable in our case. However, if we had used an example with substrate, the timings would have been more favorable for DGTD, since the evaluation of the Sommerfeld integrals is computationally expensive.

Finally, it should be noted that the absolute times of both methods can be improved significantly by parallelization. Simply using a parallel implementation of the matrix factorization and then distributing the different positions for an EELS map on multiple cores leads to an almost perfect scaling. Similarly, since each position corresponds to a separate computation, DGTD can also be expected to scale very well with the number of cores and processors. Additionally, as was shown in [23], the EELS computation with the DGTD method can be massively accelerated by using the computational power of GPUs.

6.6 Summary and Outlook

We presented a number of extensions to the well established multiple multipole program (MMP), which enhance the flexibility and usability of this technique. Our novel procedure to automatically generate a distribution of both expansions and matching points from a given surface mesh leads to a significant reduction in preparation time and allows to apply MMP to arbitrary geometries. In addition, the introduction of layered dipoles as expansions allows a more efficient simulation of realistic experiments where the nanostructures are usually deposited on substrates or membranes. Finally, with the introduction of an electron beam source and a suitable adaptive integration routine, MMP can also be used to compute electron energy loss spectra and EELS maps.

Overall, we showed that MMP is a well-tested, robust and fast simulation technique which is particularly well suited for the analysis of plasmonic nanostructures. Especially for the computation of EELS maps, the MMP method has strong advantages over alternative simulation techniques such as classical FEM or DGTD. Due to the small size of the system matrix, MMP allows to precompute and store a matrix decomposition. Then, even very large maps can be computed with relatively little computational effort. If, on the other hand, only the spectrum at a single point is required, MMP has some disadvantages over time-domain simulations.

As an outlook, we expect that a more sophisticated placement of the expansions based on the curvature of the surface could boost the efficiency enormously. Unfortunately, constructing an algorithm that reliably works for arbitrary geometries without sacrificing stability has proved difficult and further research is needed to achieve this goal.

References

1. S.A. Maier, H.A. Atwater, J. Appl. Phys. **98**(1) (2005)
2. F.J. García de Abajo, Rev. Mod. Phys. **82**(1), 209 (2010)
3. B.W. Reed, J.M. Chen, N.C. MacDonald, J. Silcox, G.F. Bertsch, Phys. Rev. B **60**, 5641 (1999)
4. F.J. García de Abajo, A. Howie, Phys. Rev. B **65**, 115418 (2002)
5. N. Geuquet, L. Henrard, Ultramicroscopy **110**(8), 1075 (2010)
6. C. Matyssek, J. Niegemann, W. Hergert, K. Busch, Photonics Nan. Fundam. Appl. **9**(4), 367 (2011)
7. C. Hafner, *Post-modern Electromagnetics Using Intelligent MaXwell Solvers* (Wiley, Chichester, 1999)
8. C. Hafner, Numerische Berechnung elektromagnetischer Felder - Grundlagen, Methoden, Anwendungen. PhD Thesis, ETH Zurich, Zurich (1987)
9. C. Hafner, OpenMaXwell, http://openmax.ethz.ch/
10. J. Schöberl, Netgen mesh generator, http://sourceforge.net/projects/netgen-mesher/
11. W.C. Chew, *Waves and Fields in Inhomogeneous Media*, IEEE Press series on Electromagnetic Waves (IEEE Press, New York, 1995)
12. G. Dural, M.I. Aksun, IEEE Trans. Microw. Theory and Tech. **43**(7), 1545 (1995)
13. A. Alparslan, *Layered Media Green's Functions: Derivation and Approximation Techniques for All Ranges and Materials* (LAMBERT Academic Publishing, Saarbrücken, 2011)
14. A. Alparslan, Numerical analysis of photonic nano structures in layered geometries. PhD Thesis, ETH Zurich, Zurich (2013)
15. R. Golubovic, A.G. Polimeridis, J.R. Mosig, IEEE Trans. Antennas Propag. **60**(5), 2409 (2012)
16. M. Dogan, M.I. Aksun, A.K. Swan, B.B. Goldberg, M.S. Unlu, J. Opt. Soc. Am. A Opt. Image Sci. Vis. **26**(6), 1458 (2009)
17. L.F. Shampine, J. Comput. Appl. Math. **211**(2), 131 (2008)
18. D.E. Amos, Package for bessel functions of a complex argument and nonnegative order, http://www.netlib.org/amos/index.html

19. Tuxfamily, Eigen 3.1.2, http://eigen.tuxfamily.org/
20. Intel, Intel math kernel library 11.0, http://software.intel.com/en-us/intel-mkl
21. Intel, Intel threading building blocks 4.1, http://software.intel.com/en-us/intel-tbb
22. F. von Cube, S. Irsen, J. Niegemann, C. Matyssek, W. Hergert, K. Busch, S. Linden, Opt. Mater. Express **1**(5), 1009 (2011)
23. F. von Cube, S. Irsen, R. Diehl, J. Niegemann, K. Busch, S. Linden, Nano Lett. **13**(2), 703 (2013)
24. K. Busch, M. König, J. Niegemann, Laser Photonics Rev. **5**(6), 773 (2011)
25. F. von Cube, J. Niegemann, S. Irsen, D.C. Bell, S. Linden, Phys. Rev. B **89**(11), 115434 (2014)

Chapter 7
The Generalized Multipole Technique for the Simulation of Low-Loss Electron Energy Loss Spectroscopy

Lars Kiewidt and Mirza Karamehmedović

Abstract In this study, we demonstrate the use of a Generalized Multipole Technique (GMT) to simulate low-loss Electron Energy Loss Spectroscopy (EELS) spectra of isolated spheriodal nanoparticles. The GMT provides certain properties, such as semi-analytical description of the electromagnetic fields, efficient solution of the underlying electromagnetic model, accurate description of the near field, and flexibility regarding the position and direction of the incident electron beam, that are advantageous for computation of EELS spectra. Within the chapter, we provide a derivation of the electromagnetic model and its connection to EELS spectra, and comprehensive validation of the implemented GMT regarding electromagnetic scattering and EELS.

7.1 Introduction to Generalized Multipole Techniques and Their Use in the Simulation of EELS

Generalized Multipole Techniques (GMT) were developed independently by different researchers in the early eighties to solve electromagnetic scattering problems [13, 16, 22, 23, 30]. Later Ludwig [24] suggested the generic term *Generalized Multipole Technique* to summarize the above mentioned methods. Major developments of the GMT were done, among others, by Hafner [14], Hafner and Bomholt [15],

L. Kiewidt (✉)
Center for Environmental Research and Sustainable Technology,
University of Bremen, Leobener Straße 6, 28359 Bremen, Germany
e-mail: kiewidt@uni-bremen.de; lars.kiewidt@wur.nl

L. Kiewidt
Biobased Chemistry and Technology (BCT), Wageningen University, P.O. Box 17,
6700 AA Wageningen, The Netherlands

M. Karamehmedović
Department of Applied Mathematics and Computer Science
and Department of Physics, Technical University of Denmark,
Matematiktorvet 303B, 2800 Kgs. Lyngby, Denmark
e-mail: mika@dtu.dk

T. Wriedt and Y. Eremin (eds.), *The Generalized Multipole Technique for Light Scattering*, Springer Series on Atomic, Optical, and Plasma Physics 99,
https://doi.org/10.1007/978-3-319-74890-0_7

Leidenberger et al. [21], Piller and Martin [31], and S. Mohsen Raeis Zadeh Bajestani [27]. A comprehensive review of different GMTs is given by Wriedt [37]. In this chapter, however, the term GMT explicitly refers to our implementation of a numerical field solver.

In general, the GMT is well-suited for the computation of EEL spectra because of several advantageous properties that will be explained later in this chapter. The current method based on the GMT was developed to provide a robust, accurate, and fast model for the numerical inversion of EEL spectra of nanoparticles. A detailed description and validation of the method are already given in [20]. Nevertheless, the fundamentals of the underlying classical electromagnetic model, descriptions of the implemented method, and numerical results are presented in this chapter again to give a comprehensive and self-contained overview of the simulation of EELS in this edited book.

The fundamentals of the classical electromagnetic model are explained in Sect. 7.2, followed by a detailed description of the implementation of the current method in Sect. 7.3. Validation of the implemented method is provided by comparisons to available analytical solutions for classical electromagnetic scattering and EELS in Sect. 7.4. Finally, in Sect. 7.5, the results are summarized and conclusions regarding extensions, improvements, and future applications of the method are drawn.

7.2 The Classical Electromagnetic Model and the Computation of the Electron Energy Loss Probability

The fundamentals of the electromagnetic EELS model and the calculation of the electron energy loss probability $P(\omega)$ were already outlined by Matyssek et al. [25] and presented in detail by Kiewidt et al. [20]. Nevertheless, due to the great significance of the electron energy loss probability to EELS, its derivation in light of the classical electromagnetic model is repeated here.

Within experimental EELS measurements a large number of electrons interact with the specimen under investigation. Due to the quantum mechanical *probabilistic* nature of the interactions the energy loss of the electrons is distributed over some range. The *mean expected energy loss* is then given by (see [10, p. 214])

$$\Delta E_{\text{prob}} = \int_0^\infty \hbar\omega P(\omega)\mathrm{d}\omega, \tag{7.1}$$

where $P(\omega)$ is the probability of an energy loss of $\hbar\omega$ per unit of ω. The quantities $\hbar \approx 1.055 \times 10^{-34}$ Js and ω are the reduced Planck constant and the angular frequency associated with the energy loss, respectively.

In classical *deterministic* electromagnetic theory, however, the energy loss of a single electron moving in an electromagnetic field is calculated by integrating the

Lorentz force along the path of the electron (see again [10, p. 214]),

$$\Delta E_{\text{det}} = e \int_{-\infty}^{\infty} \left[\mathbf{E}^{\text{sca}}(\mathbf{r}_e(t), t)\right] \cdot \mathbf{v}\, \mathrm{d}t. \tag{7.2}$$

Here, $\mathbf{E}^{\text{sca}}(\mathbf{r}_e(t), t)$ is the scattered electric field intensity along the path of the electron caused by the oscillations of the electrons in the specimen, and $-e \approx 1.602 \times 10^{-19}$ C, $\mathbf{v}$, and $\mathbf{r}_e(t)$ are the elementary charge, the velocity vector of the electron, and the time-dependent path of the electron, respectively. Equation (7.2) already contains two simplifications: First, the contribution of the scattered magnetic field to the Lorentz force is not included because the vector $\mathbf{v}$ is orthogonal to $\mathbf{v} \times \mathbf{B}^{\text{sca}}(\mathbf{r}_e(t), t)$. Second, the negative signs of the Lorentz force and the elementary charge cancel.

As a further simplification, we assume the velocity and the trajectory of the electron to be constant in the following derivation because the energy loss of the incident electrons (approx. 0.5–50 eV) is small compared to their initial kinetic energy (approx. 50–300 keV). This simplification, known as the *no-recoil approximation*, is frequently used in the simulation of EELS [25, 26].

By equating (7.1) and (7.2) a *sufficient* condition on the electron energy loss probability $P(\omega)$ is derived for

$$\Delta E_{\text{det}} = \Delta E_{\text{prob}} \tag{7.3}$$

to hold true.

Substituting (7.1) and (7.2) into (7.3) yields

$$e \int_{-\infty}^{\infty} \mathbf{v} \cdot \mathbf{E}^{\text{sca}}(\mathbf{r}_e(t), t)\mathrm{d}t = \int_{0}^{\infty} \hbar\omega P(\omega)\mathrm{d}\omega. \tag{7.4}$$

As the right-hand side of (7.4) is formulated in the frequency domain we replace the scattered electric field intensity in the time domain on the left-hand side by its inverse Fourier transform, defined by

$$\left(\mathscr{F}^{-1}\hat{f}\right)(t) = f(t) = \frac{1}{2\pi} \int_{-\infty}^{\infty} \hat{f}(\omega)\mathrm{e}^{-\mathrm{i}\omega t}\mathrm{d}\omega, \quad t \in \mathbb{R}, \tag{7.5}$$

to get

$$\frac{e}{2\pi} \int_{-\infty}^{\infty}\int_{-\infty}^{\infty} \mathbf{v} \cdot \mathbf{E}^{\text{sca}}(\mathbf{r}_e(t), \omega)\mathrm{e}^{-\mathrm{i}\omega t}\mathrm{d}\omega\mathrm{d}t = \int_{0}^{\infty} \hbar\omega P(\omega)\mathrm{d}\omega. \tag{7.6}$$

Note that we use the time-dependence factor $\mathrm{e}^{-\mathrm{i}\omega t}$ throughout this chapter. Using the identity that (see [25])

$$\mathbf{E}^{\mathrm{sca}}(\mathbf{r}_e(t), -\omega) = \mathbf{E}^{\mathrm{sca}}(\mathbf{r}_e(t), \omega)^*, \tag{7.7}$$

where $\mathbf{E}^*$ is the complex conjugate of $\mathbf{E}$, (7.6) can be rewritten

$$\frac{e}{\pi}\int_{-\infty}^{\infty}\int_{0}^{\infty} \mathrm{Re}\left\{\mathbf{v}\cdot\mathbf{E}^{\mathrm{sca}}(\mathbf{r}_e(t), \omega)\mathrm{e}^{-\mathrm{i}\omega t}\right\} \mathrm{d}\omega \mathrm{d}t = \int_{0}^{\infty} \hbar\omega P(\omega)\mathrm{d}\omega. \tag{7.8}$$

After changing the order of integration on the left-hand side, and after some rearrangement, (7.8) yields

$$\int_{0}^{\infty}\left(\frac{e}{\pi}\int_{-\infty}^{\infty} \mathrm{Re}\left\{\mathbf{v}\cdot\mathbf{E}^{\mathrm{sca}}(\mathbf{r}_e(t), \omega)\ \mathrm{e}^{-\mathrm{i}\omega t}\right\} \mathrm{d}t - \hbar\omega P(\omega)\right)\mathrm{d}\omega = 0. \tag{7.9}$$

A *sufficient* condition for (7.9) to hold is that the outer integrand vanishes, that is,

$$\frac{e}{\pi}\int_{-\infty}^{\infty} \mathrm{Re}\left\{\mathbf{v}\cdot\mathbf{E}^{\mathrm{sca}}(\mathbf{r}_e(t), \omega)\mathrm{e}^{-\mathrm{i}\omega t}\right\} \mathrm{d}t - \hbar\omega P(\omega) = 0. \tag{7.10}$$

Finally, solving (7.10) for the electron energy loss probability $P(\omega)$ results in

$$P(\omega) = \frac{e}{\pi\hbar\omega}\int_{-\infty}^{\infty} \mathrm{Re}\left\{\mathbf{v}\cdot\mathbf{E}^{\mathrm{sca}}(\mathbf{r}_e(t), \omega)\mathrm{e}^{-\mathrm{i}\omega t}\right\} \mathrm{d}t. \tag{7.11}$$

It is seen from (7.11) that the energy loss probability $P(\omega)$ is expressed in terms of the scattered electric field intensity $\mathbf{E}^{\mathrm{sca}}(\mathbf{r}_e(t), \omega)$ in the frequency domain along the path of the electron, and that the computation of the electron energy loss probability reduces to solving a sequence of classical electromagnetic scattering problems with subsequent integration of the electric scattered field intensity along the path of the electron. Hence, (7.11) is the basis of the electromagnetic EELS model, also called classical dielectric formalism [10, Sect. II.B], and thus enables the use of established numerical methods, like the GMT, to compute EEL spectra.

7.3 Implementation of the GMT to Compute Low-Loss EELS

7.3.1 *The Electromagnetic Model*

Figure 7.1 illustrates the non-penetrable EELS configuration considered in the current model. The nanoparticle under investigation is modeled by a spheroid of semidiameters a and c. The boundary of the spheroid $\partial\Omega^-$, that separates the interior domain Ω^- from the exterior domain Ω^+ of constitutive parameters $\varepsilon_0 \approx 8.854 \times 10^{-12}$As V^{-1}m^{-1} and $\mu_0 = 4\pi \times 10^{-7}$ NA^{-1}, is parametrized by

$$\left.\begin{aligned} x &= a \sin\theta \cos\varphi \\ y &= a \sin\theta \sin\varphi \\ z &= c \cos\theta \end{aligned}\right\} \quad 0 \le \theta \le \pi,\ 0 \le \varphi < 2\pi. \tag{7.12}$$

A cartesian coordinate system is placed with the origin at the center of the spheroid, and with the z-axis aligned with the axis of rotational symmetry of the spheroid. According the definition of the spherical coordinates θ is the polar angle measured from the positive z-axis, and φ is the azimuthal angle in the xy-plane measured from the positive x-axis towards the positive y-axis.

The parametrization (7.12) allows modeling a variety of different particle shapes ranging from extremely oblate nanodiscs ($c \ll a$), over perfect spheres ($a = c$), to extremely elongated shapes such as nanotubes ($c \gg a$). Also, other analytical parametrizations and numerical representations of the surface of the particle can readily be included into the current model.

The incident electrons pass the nanoparticle with velocity vector $\mathbf{v}$ and impact parameter $\mathbf{b}$. Note that the impact parameter is measured as the orthogonal distance from the origin of the coordinate system to the path of the electron $\mathbf{r}_e(t)$.

The particle itself is assumed to be homogeneous and the material is modeled by its permittivity ε and permeability μ. Here, we only consider non-magnetic materials so that $\mu_r = 1$ and consequently $\mu = \mu_0$. The relative frequency-dependent permittivity $\varepsilon_r(\omega)$ is either obtained from experimental data or modeled by the well-known Drude model [4, 5] named after Paul Drude,

$$\varepsilon_r(\omega) = 1 - \frac{\omega_p^2}{\omega^2 + i\eta\omega} \quad \text{with} \quad \omega_p = \sqrt{\frac{n_e e^2}{\varepsilon_0 m_e}}, \tag{7.13}$$

where ω_p is the plasma frequency, and η is the collision rate of the free electrons in the material with the immobile cores. Further, m_e, $-e$, n_e are the electron mass, electron charge, and electron density, respectively. A modern discussion of the Drude model is given by Dressel and Grüner [3, Sects. 5 and 6].

A more advanced model for the dielectric function is the Drude–Lorentz model (see [3, Sect. 6]), that also accounts for interband transitions by adding several higher

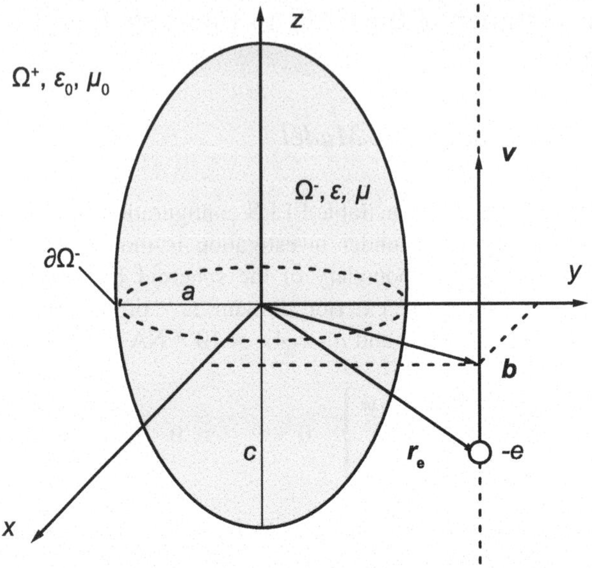

Fig. 7.1 Mathematical model of the EELS configuration considered in this chapter

order terms,

$$\varepsilon_r(\omega) = 1 - \frac{f_0\omega_p^2}{\omega^2 + i\eta_0\omega} - \sum_{k=1}^{N} \frac{f_k\omega_p^2}{\omega^2 - \omega_{e,k}^2 + i\eta_k\omega}. \tag{7.14}$$

Here, f_k is the so-called oscillator strength, and $\omega_{e,k}$ is the eigenfrequency of the kth oscillator.

Figure 7.2 shows a comparison of the simple Drude model, (7.13), and the advanced Drude–Lorentz model, (7.14), for aluminum against experimental data of Rakić [34]. The corresponding parameters in the Drude model are $\hbar\omega_p = 15.8\,\text{eV}$ and $\hbar\eta = 0.6\,\text{eV}$ [10]. For the Drude–Lorentz model the parameter set of Rakić et al. [35] with $N = 4$ Lorentzian terms is used. It is seen that the Drude model describes the real and imaginary component of the dielectric function quite well in the interval $\lambda_0 \in [100\,\text{nm}, 550\,\text{nm}]$ but deviates for larger wavelengths. Furthermore, the Drude model does not capture the local extrema caused by strong interband transitions around $\lambda_0 \approx 800\,\text{nm}$ (approx. 1.5 eV).

The Lorentz–Drude model, in contrast, matches well with the experimental data in the full range under consideration for both the real and imaginary component of the dielectric function. Also, it describes the strong interband transitions around $\lambda_0 \approx 800\,\text{nm}$ well.

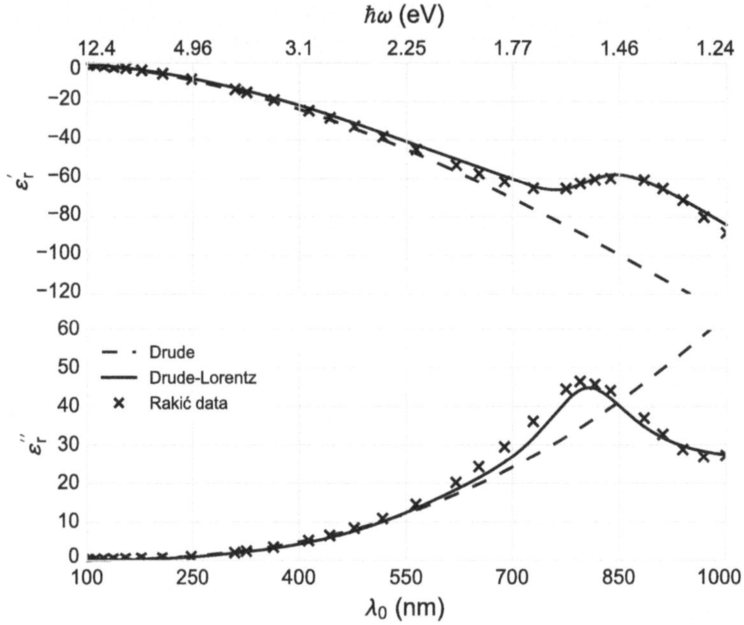

Fig. 7.2 Comparison of the Drude model (7.13) and the Drude–Lorentz model (7.14) against experimental data for aluminum [34]

7.3.2 Computation of the Electromagnetic Fields Using the GMT

To solve the scattering problem, the exterior electric and magnetic field intensities are decomposed in contributions from the incident fields, caused by the passing electron, and the scattered field,

$$\begin{cases} \mathbf{E}^{\text{ext}}(\mathbf{r}, \omega) = \mathbf{E}^{\text{inc}}(\mathbf{r}, \omega) + \mathbf{E}^{\text{sca}}(\mathbf{r}, \omega), \\ \mathbf{H}^{\text{ext}}(\mathbf{r}, \omega) = \mathbf{H}^{\text{inc}}(\mathbf{r}, \omega) + \mathbf{H}^{\text{sca}}(\mathbf{r}, \omega), \end{cases} \tag{7.15}$$

where $\mathbf{r} \in \Omega^+$ is the position vector of any point in the outer domain.

Within the electromagnetic approach the incident electron is modeled by the electromagnetic fields caused by a point charge in uniform straight motion. Analytic expressions for the electromagnetic fields caused by a point charge in uniform straight motion along the z-axis are given in [6, Sects. 5.4d and 5.4e]. Considering the translation of the trajectory of the electron in Fig. 7.1, the components of the incident electric and magnetic field intensities read

$$\begin{cases} \mathbf{E}_x^{\mathrm{inc}}(\mathbf{r},\omega) = \dfrac{x-b_x}{\rho}\dfrac{ev\alpha\hat{\kappa}}{i4\omega\varepsilon_0}\mathrm{e}^{\mathrm{i}\alpha z}H_1^{(1)}(\hat{\kappa}\rho), \\ \mathbf{E}_y^{\mathrm{inc}}(\mathbf{r},\omega) = \dfrac{y-b_y}{\rho}\dfrac{ev\alpha\hat{\kappa}}{\mathrm{i}4\omega\varepsilon_0}\mathrm{e}^{\mathrm{i}\alpha z}H_1^{(1)}(\hat{\kappa}\rho), \\ \mathbf{E}_z^{\mathrm{inc}}(\mathbf{r},\omega) = \dfrac{ev\hat{\kappa}^2}{4\omega\varepsilon_0}\mathrm{e}^{\mathrm{i}\alpha z}H_0^{(1)}(\hat{\kappa}\rho), \\ \mathbf{H}_x^{\mathrm{inc}}(\mathbf{r},\omega) = \dfrac{y-b_y}{\rho}\dfrac{\mathrm{i}ev\hat{\kappa}}{4}\mathrm{e}^{\mathrm{i}\alpha z}H_1^{(1)}(\hat{\kappa}\rho), \\ \mathbf{H}_y^{\mathrm{inc}}(\mathbf{r},\omega) = -\dfrac{x-b_x}{\rho}\dfrac{\mathrm{i}ev\hat{\kappa}}{4}\mathrm{e}^{\mathrm{i}\alpha z}H_1^{(1)}(\hat{\kappa}\rho), \\ \mathbf{H}_z^{\mathrm{inc}}(\mathbf{r},\omega) = 0, \end{cases} \tag{7.16}$$

with

$$\alpha = \frac{\omega}{\mathrm{v}}, \quad k_0 = \frac{2\pi}{\lambda_0}, \quad \hat{\kappa}^2 = k_0^2 - \alpha^2, \tag{7.17}$$

and where

$$\rho = \sqrt{(x-b_x)^2 + (y-b_y)^2} \tag{7.18}$$

is the orthogonal distance of the observation point to the path of the electron. Further, the quantities λ_0 and k_0 are the free-space wavelength and the free-space wavenumber, respectively, and v is the magnitude of the velocity vector $\mathbf{v}$. The function $H_\nu^{(k)}$ is the cylindrical Hankel function of νth order and kth kind.

As already pointed out in Sect. 7.2, the computation of low-loss EEL spectra using the electromagnetic model requires the solution of several independent classical scattering problems. Hence, the electric scattered field intensity along the path of the integration path is calculated using the GMT. Figure 7.3 shows the general concept of the GMT to compute the scattered and interior electromagnetic field intensities. Several exterior and interior multipole origins are distributed in space to approximate the scattered and interior electromagnetic field intensities. In the case of spheroidal nanoparticles the multipole origins are typically distributed along the axis of rotational symmetry of the particle. The scattered and interior electric field intensities are then computed as a superposition of all multipoles,

$$\mathbf{E}^{\mathrm{sca}}(\mathbf{r},\omega) \approx \sum_{j=1}^{N_{\mathrm{sca}}}\sum_{l=1}^{M_{\mathrm{sca}}^j}\sum_{m=-l}^{l}\left(a_{lm}^{\mathrm{sca},j}\mathbf{M}_{lm}^{(3),j} + b_{lm}^{\mathrm{sca},j}\mathbf{N}_{lm}^{(3),j}\right), \ \mathbf{r} \in \Omega^+ \tag{7.19}$$

$$\mathbf{E}^{\mathrm{int}}(\mathbf{r},\omega) \approx \sum_{j=1}^{N_{\mathrm{int}}}\sum_{l=1}^{M_{\mathrm{int}}^j}\sum_{m=-l}^{l}\left(a_{lm}^{\mathrm{int},j}\mathbf{M}_{lm}^{(1),j} + b_{lm}^{\mathrm{int},j}\mathbf{N}_{lm}^{(1),j}\right), \ \mathbf{r} \in \Omega^-. \tag{7.20}$$

Here, $\mathbf{E}^{\mathrm{int}}$ and $\mathbf{E}^{\mathrm{sca}}$ are the total interior and scattered exterior electric field intensities, respectively. The quantities $a_{lm}^{\mathrm{int},j}$ and $b_{lm}^{\mathrm{int},j}$, and $a_{lm}^{\mathrm{sca},j}$ and $b_{lm}^{\mathrm{sca},j}$ are unknown expansions coefficients of the total interior and scattered exterior electric field intensities of the jth multipole that have to be determined numerically. The functions

$\mathbf{M}_{lm}^{(1),j}$ and $\mathbf{N}_{lm}^{(1),j}$, and $\mathbf{M}_{lm}^{(3),j}$ and $\mathbf{N}_{lm}^{(3),j}$ are the regular and radiating *spherical vector wave functions (SVWF)* [2], respectively. The latter satisfy the Silver–Müller radiation condition,

$$\left| \varepsilon^{1/2}\mathbf{E}^{\text{sca}}(\mathbf{r}, \omega) - \mu^{1/2}\mathbf{H}^{\text{sca}}(\mathbf{r}, \omega) \times \frac{\mathbf{r}}{|\mathbf{r}|} \right| = \mathcal{O}\left(\frac{1}{|\mathbf{r}|}\right). \tag{7.21}$$

Of course similar expansions exist for the scattered and interior magnetic field intensities. For simplicity of notation, however, they are omitted in this chapter.

Explicit expressions for the SVWF are given by Nédélec [29, Sect. 5.3, p. 185 ff.]. A summary of all the equations implemented in the current model can be found in [19, Appendices A and B].

The SVWF arise from the solution of the vectorial wave equation using a separation of variables approach. A derivation of the vectorial wave equation from the time-harmonic Maxwell equations is given, among others, in [19]. Consequently, the SVWF are also exact solutions of the Maxwell equations. Furthermore, the SVWF form a basis in the space $L^2_{\text{tan}}(\partial\Omega^-)$ of vectors with square-integrable tangential components, for any bounded C^2 domain Ω^- [2, Theorem 1.9 on p. 159], [29, Theorems 2.4.7 and 2.4.8 on pp. 36–40].

The unknown expansion coefficients in (7.19) and (7.20), sometimes called amplitudes, are computed numerically using a *Generalized Point Matching (GPM)* method [14, Sects. 5.5.3 and 7.3.1], [18, Sect. 3.5] that enforces the transmission conditions,

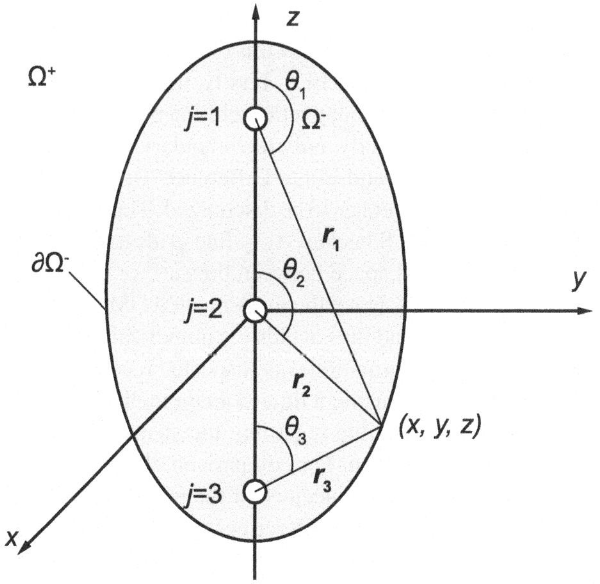

Fig. 7.3 Concept of distributed multipoles to compute the scattered and interior electromagnetic fields as used in the GMT

$$\hat{\nu} \times (\mathbf{E}^{\mathrm{sca}} - \mathbf{E}^{\mathrm{int}}) = \hat{\nu} \times \mathbf{E}^{\mathrm{inc}},$$
$$\hat{\nu} \times (\mathbf{H}^{\mathrm{sca}} - \mathbf{H}^{\mathrm{int}}) = \hat{\nu} \times \mathbf{H}^{\mathrm{inc}}, \tag{7.22}$$

at so-called matching points. In the current model the matching points are distributed equidistantly along the polar and azimuthal angle of the boundary $\partial\Omega^-$ of the spheroid. In (7.22), $\hat{\nu}$ is the outward unit vector field, and $\mathbf{E}^{\mathrm{inc}}$ and $\mathbf{H}^{\mathrm{inc}}$ are the incident electric and magnetic field intensities given by (7.16).

In principle the number of matching points has at least to be equal to the number of unknown coefficients in the multipole expansions. This is referred to as *Point matching (PM)* method. Ludwig [24], however, recommends to used two to three times as many matching points as unknown coefficients to increase the stability of the method. This leads to an overdetermined system of equations that is solved in a *least-squares* sense so that

$$\min\left\{\|\mathbf{Ax} - \mathbf{b}\|_2^2,\ \mathbf{x} \in \mathbb{C}^{N_c}\right\}. \tag{7.23}$$

The latter method is commonly named *Generalized point matching (GPM)* method.

The matrix $\mathbf{A}$ is the system matrix containing the SVWF, the vector $\mathbf{x}$ contains the unknown expansion coefficients, and $\mathbf{b}$ represents the incident electromagnetic fields. The quantity N_c is the number of unknown coefficients.

Finally, after solving (7.23) to determine the unknown expansion coefficients, (7.19) and (7.20) are used to compute the scattered electric field intensity along the path of the electron. Next, the field is integrated numerically to compute the electron energy loss probability $P(\omega)$ according to (7.11). The process is then repeated for the designated range of incident wavelengths to compute the whole EEL spectrum.

As already mentioned at the beginning of this chapter, the GMT has some important advantages for the simulation of EELS. Firstly, the use of analytic expressions for the scattered and interior electromagnetic field intensities ensures accurate and smooth numerical solutions. Secondly, only the boundary of the nanoparticle has to be discretized. In Finite Element and Finite Difference Time Domain Methods the volume in- and outside the scatterer has to be discretized. This is especially disadvantageous in the simulation of EELS because very fine grids have to be used to ensure an accurate solution in the small space between the surface of the scatterer and the trajectory of the electron, cf. [25]. Next, the implemented GMT method works exclusively in the frequency domain and thus avoids the numerical Fourier transformation of the results. The Fourier transformation was suspected recently to cause numerical errors in the simulation of EELS using a time-domain method, cf. [25]. Finally, the implemented GMT method is flexible regarding the shape of the nanoparticle and the trajectory of the incident electron. For comparison, the Null-Field Method with Discrete Sources is restricted in the trajectory of the incident electron by the smallest circumscribing sphere around the scatterer because the field expressions are not valid in the specific space between the surface of the nanoparticle and the smallest circumscribing sphere, see [26]. Hence, the scattered electric field intensity can not be computed along any trajectory that intersects the smallest circumscribing sphere. Only recently Doicu and Wriedt [1], and Forestiere et al. [7], presented a method in the frame of the T-matrix method to compute the near field in the space between the scatterer and the smallest circumscribing sphere. Contrary, no geometric restrictions arise within the GMT approach.

7.4 Validation and Numerical Results

7.4.1 Electromagnetic Scattering of Spheroidal Dielectric Nanoparticles

The implemented GMT is validated and tested by simulating classical electromagnetic scattering by spherical and spheroidal nanoparticles and comparisons to analytical and numerical solutions.

The particles, placed in plain free space, have semidiameters a and c, and relative permittivity ε_r. A uniform plane wave propagating in the positive z-direction with wavelength λ_0 and amplitude $E = 1\,\mathrm{Vm}^{-1}$ is used as illumination. The plane wave is polarized in the xz-plane.

For perfectly spherical particles only one exterior and one interior multipole is placed in the center of the nanoparticles. The multipole order was chosen between $M = 4$ and $M = 10$ ($M = M_{\mathrm{ext}} = M_{\mathrm{int}}$) depending on the optical size a/λ_0 of the scatterer. For spheroidal particles up to three exterior and interior multipole origins are distributed equidistantly along the z-axis. The multipole orders are chosen between $M = 4$ and $M = 5$.

The obtained differential scattering cross sections (DSCS) are compared to Prahl's Mie Scattering Calculator [33] for spherical particles, and to the Superellipsoid Scattering Tool (SScatT) [38] for spheroidal particles. Figure 7.4 shows the normalized DSCS for two representative optical particle sizes and relative permittivities. It is seen that the GMT gives accurate results, with maximal relative errors below 1%, for classical electromagnetic scattering by dielectric particles.

Figure 7.5 shows the DSCS for two spheroidal nanoparticles with aspect ratios $c/a = 0.5$ and $c/a = 1.5$, and relative permittivity $\varepsilon = 1.5$. Again, the results are in excellent agreement with the numerical reference method. The average relative errors are below 5%. This demonstrates the ability of the implemented method to provide accurate results for representative physical parameters. Comparisons for intermediate optical sizes, relative permittivities, and aspect ratios are presented in [19].

7.4.2 Low-Loss EEL Spectra of Spherical Dielectric Nanoparticles

The ability of the implemented method to compute EEL spectra is validated by computing EEL spectra of spherical nanoparticles for different materials and comparisons of the results against available analytic solutions by Garc̀ia de Abajo [8, 9, 11]. The spherical particles are again placed in plain free space, and the influence of any supporting structure is neglected. The radius of the particles is $a = 10\,\text{nm}$. The kinetic energy of the incident electrons is $E_0 = 50\,\text{keV}$, and the impact parameter is $\mathbf{b} = (10.5\,\text{nm}, 0\,\text{nm})$. The relative permittivities of gold (Au) [17], silver (Ag) [17], aluminum (Al) [34], silicon (Si) [32], silica (SiO_2) [32], and alumina (Al_2O_3) [32] are obtained by interpolating experimental data given in the references stated above. As in the case of electromagnetic scattering of perfect spheres, one exterior and one interior multipole origin is placed in the center of the spherical particles. The multipole order was chosen between $M = 3$ and $M = 5$, depending on the material.

It is seen from Fig. 7.6 that it is sufficient to integrate the scattered electric field intensity from $-5a$ to $5a$ along the trajectory of the electron. The integration is

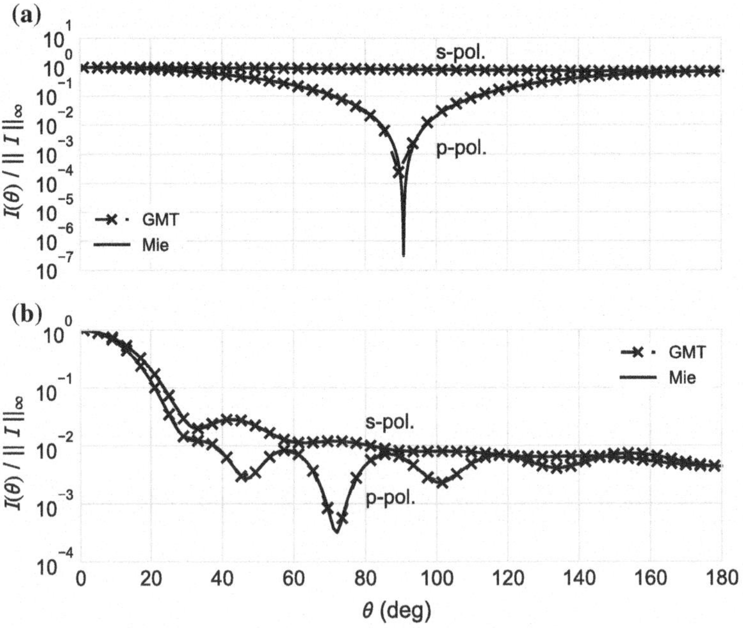

Fig. 7.4 Normalized DSCS of perfectly spherical nanoparticles of radius $a = 10\,\text{nm}$. The relative permittivities are **a** $\varepsilon = 1.5$, and **b** $\varepsilon = 1.5 + 10$ i. The incident wave has wavelengths **a** $\lambda_0 = 100\,\text{nm}$, and **b** $\lambda_0 = 10\,\text{nm}$, corresponding to size parameters **a** $k_0 a = 0.63$, and **b** $k_0 a = 6.28$, and is polarized in the xz-plane

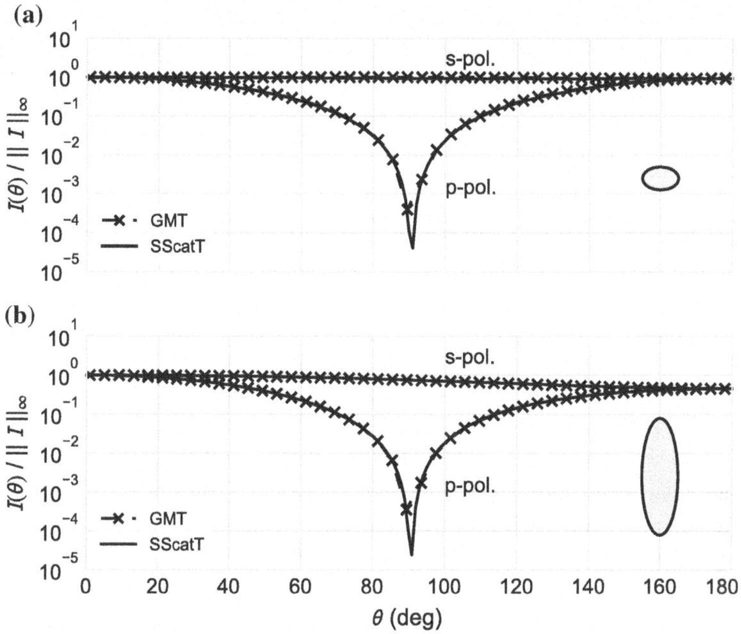

Fig. 7.5 Normalized DSCS of spheroidal nanoparticles of semidiameters **a** $a = 10\,\text{nm}$ and $c = 5\,\text{nm}$, and **b** $a = 10\,\text{nm}$ and $c = 15\,\text{nm}$. The relative permittivity is $\varepsilon = 1.5$. The incident wave has wavelength $\lambda_0 = 100\,\text{nm}$, corresponding to size parameters of an equivolume sphere **a** $k_0 r = 0.50$, and **b** $k_0 r = 0.72$, and is polarized in the xz-plane

performed using Shampine's [36] adaptive Gauss–Kronrod quadrature procedure with relatively tight tolerances of 10^{-15}. Further, it is beneficial to use a denser distribution of sampling points along the path of the electron close to the scatterer because the scattered field oscillates more significantly near the scatterer than far away from it. Also, we observed better results if the matching points along the surface of the scatterer are distributed equidistantly contrary to a uniform angular distribution of the matching points.

Figures 7.7, 7.8 and 7.9 show the normalized EEL spectra of the spherical nanoparticles for the six different materials. All EEL spectra are normalized with respect to their maximum value for comparison. In general, the implemented method provides excellent qualitative results. Especially for Au (Fig. 7.7a), Ag (Fig. 7.7b), Si (Fig. 7.8b), and Al_2O_3 (Fig. 7.9a), the GMT results match excellently with the analytical solution in the whole range under consideration. In the case of SiO_2 the peak around $\lambda_0 = 120\,\text{nm}$ agrees well with the analytical solution, however, the relative amplitudes of the peaks in the range from 70 to 100 nm do not match with the reference. Nevertheless, the position of the peaks compares fairly well, and an appropriate change of the numerical parameters will most likely improve the computed solution in the specific range. Thus, it is concluded that the GMT in principle is able to compute very accurate EEL spectra of spherical nanoparticles for different materials.

7.4.3 Low-Loss EEL Spectra of Spheroidal Nanoparticles

After validation of the implemented method for spherical nanoparticles, the program is used to predict the EEL spectra of prolate spheroidal nanoparticles. Figure 7.10 shows the EEL spectrum of a slightly prolate spheroidal Al particle with semidiameters $a = 10\,\text{nm}$ and $c = 12\,\text{nm}$ excited by a 50 keV electron. The frequency-dependence of the dielectric function is modeled by the Drude model (7.13) with $\hbar\omega_p = 15.8\,\text{eV}$ and $\hbar\eta = 0.6\,\text{eV}$ [10]. In total, three exterior and three interior multipoles are distributed equidistantly along the z-axis.

The results computed with the implemented method correspond to the DGTD [25] results for $\lambda_0 \leq 100\,\text{nm}$ and $\lambda_0 \geq 130\,\text{nm}$. In between, the results show a significant discrepancy. This is also observed for different aspect ratios c/a, up to 1.5, and different materials. An appropriate change of the numerical parameters such as number and order of the multipoles, and number and distribution of the matching points along the boundary of the spheroid does not improve the solution, but might cause an instability of the solution. Also, using the rules of Moreno et al. [28] to dis-

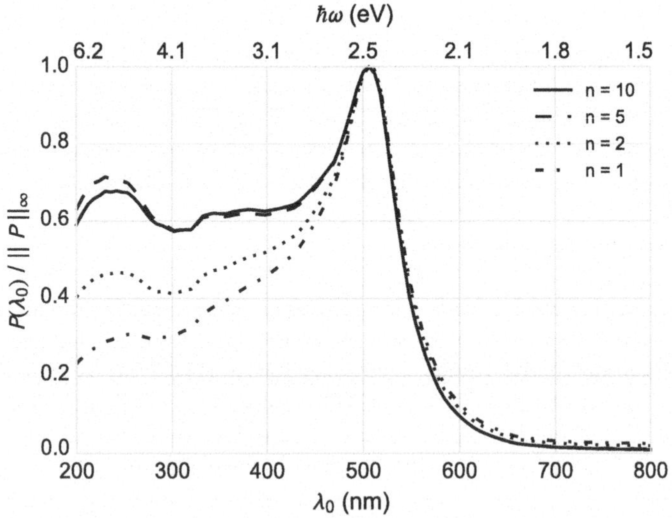

Fig. 7.6 Normalized EEL spectra of 50 keV electrons passing a spherical gold (Au) nanoparticle of radius $a = 10\,\text{nm}$ for varying lengths of the integration path. The integration was performed from $-na$ to na. Impact parameter $\mathbf{b} = (10.5\,\text{nm}, 0\,\text{nm})$

tribute the multipole origins did not remove the instability. Hence, further extensive numerical experiments have to be conducted to identify the source of instability, and to provide empirical rules of thumb to set the numerical parameters according to the configuration.

Nevertheless, it is seen from Figs. 7.8a and 7.10 that the peaks in the EEL spectrum depend on the geometry of the nanoparticle. This would enable the inversion of EEL spectra to characterize individual nanoparticles.

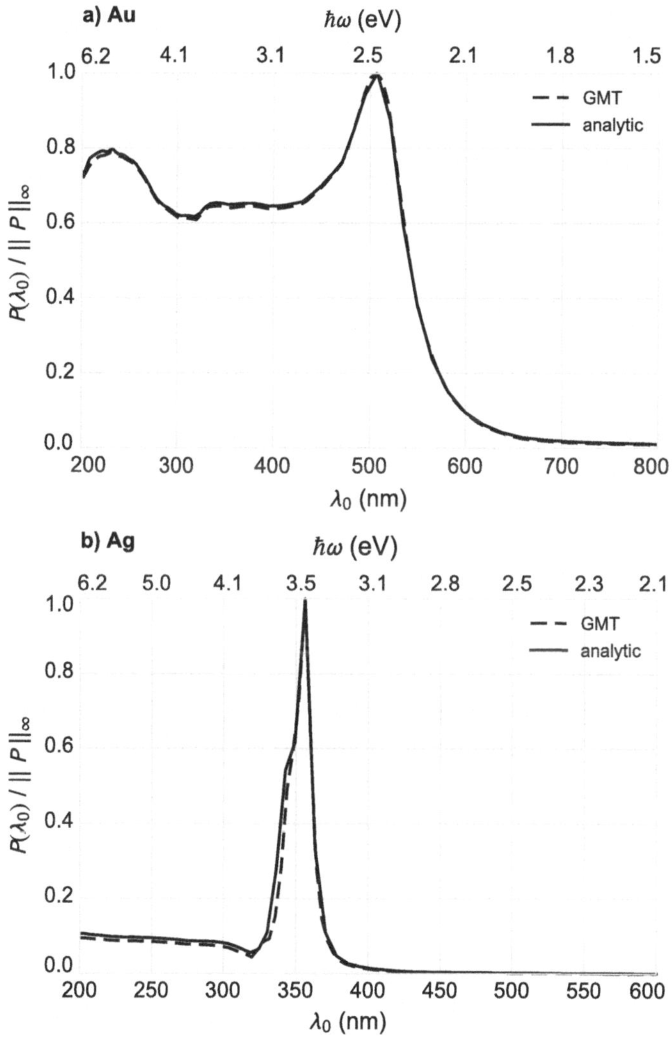

Fig. 7.7 Normalized EEL spectra of 50 keV electrons passing spherical nanoparticles of radius $a = 10\,\text{nm}$. Impact parameter $\mathbf{b} = (10.5\,\text{nm}, 0\,\text{nm})$. Analytical solution from [8, 9, 11]

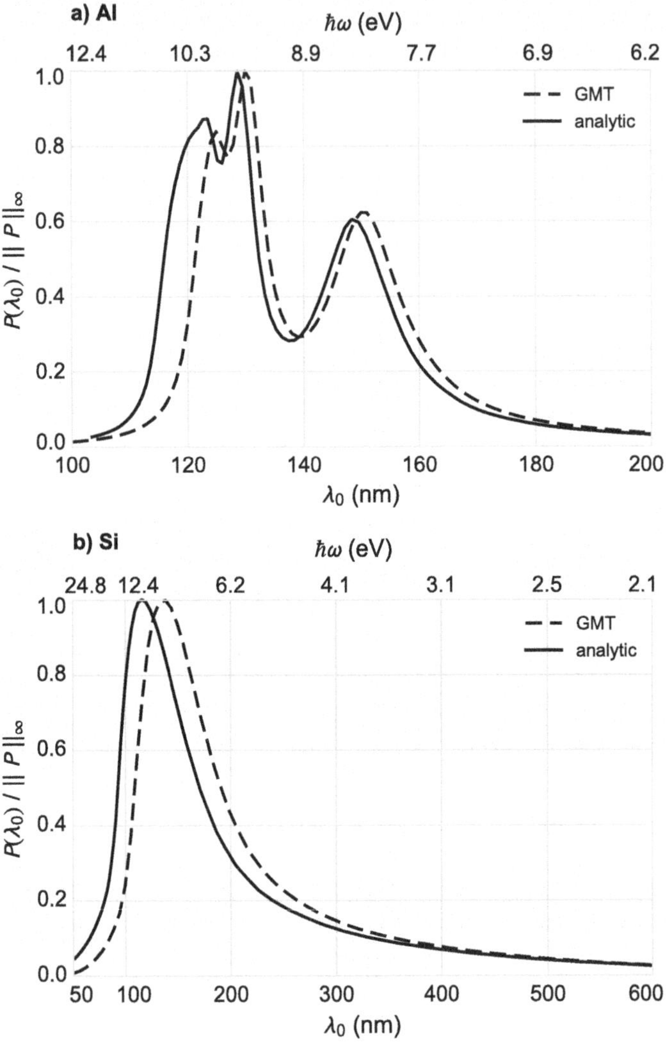

Fig. 7.8 Normalized EEL spectra of 50 keV electrons passing spherical nonaparticles of radius $a = 10\,\text{nm}$. Impact parameter $\mathbf{b} = (10.5\,\text{nm}, 0\,\text{nm})$. Analytical solution from [8, 9, 11]

Finally, Fig. 7.11 shows the absolute scattered and interior electric field intensities around the slightly prolate spheroidal Al nanoparticle in the xy-plane ($\varphi = 0°$). The dotted line indicates the particle, and the dashed line illustrates the trajectory of the electron. A strong near-field enhancement is seen around the metallic particle. This effect is used in Surface Enhanced Raman Spectroscopy [12] to measure the loading of metallic surfaces. In addition, it is seen that the implemented method

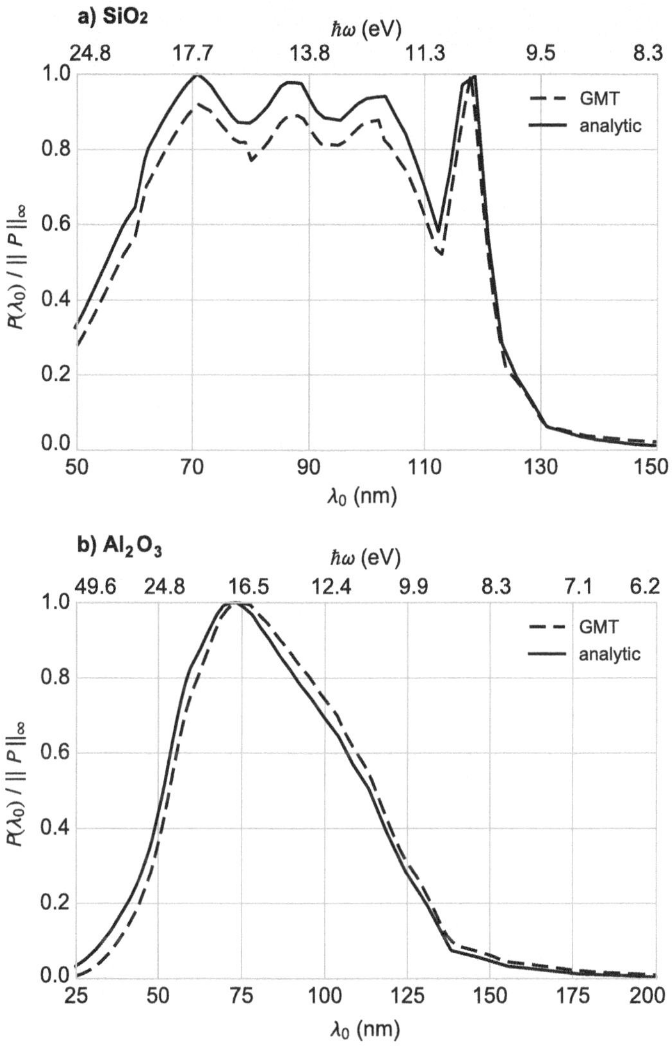

Fig. 7.9 Normalized EEL spectra of 50 keV electrons passing spherical nonaparticles of radius $a = 10\,\text{nm}$. Impact parameter $\mathbf{b} = (10.5\,\text{nm}, 0\,\text{nm})$. Analytical solution from [8, 9, 11]

provides smooth near fields that are easily numerically integrated along the path of the electron.

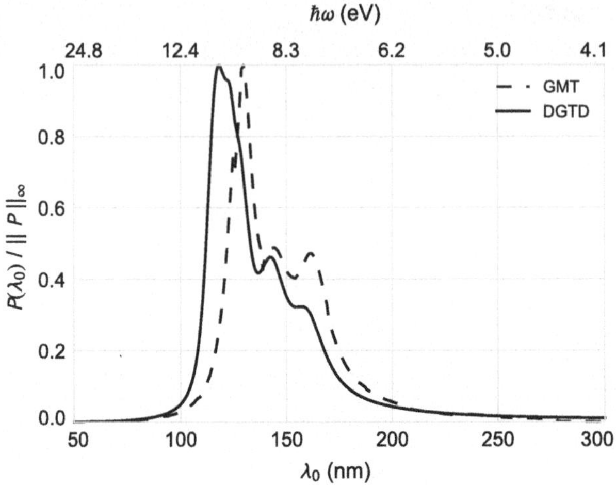

Fig. 7.10 EEL spectrum of 50 keV electrons passing a prolate spheroidal Al nanoparticle of semidiamters $a = 10$ nm and $c = 12$ nm computed with the GMT and the DGTD method. Impact parameter $\mathbf{b} = (10.5\,\text{nm}, 0\,\text{nm})$

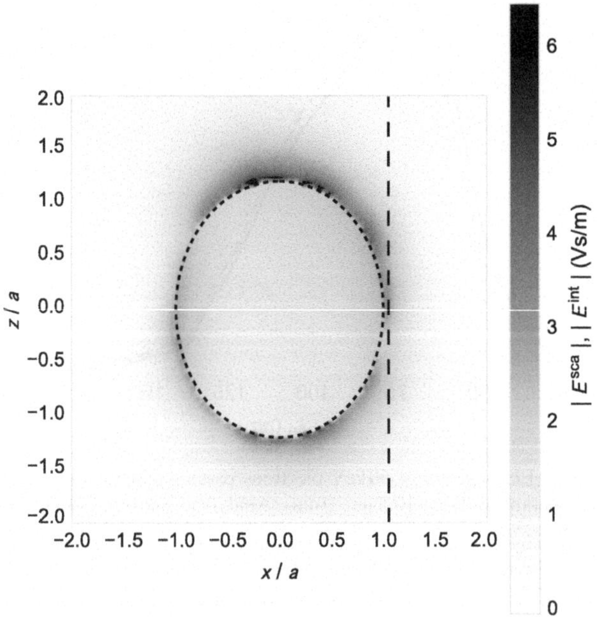

Fig. 7.11 Electric near-field intensity around a prolate spheroidal Al nanoparticle of semidiameters $a = 10$ nm and $c = 12$ nm (dashed line). The arrow indicates the trajectory of the 50 keV electron with impact parameter $\mathbf{b} = (10.5\,\text{nm}, 0\,\text{nm})$. The wavelength associated with the shown energy loss is $\lambda_0 = 162$ nm ($\hbar\omega = 7.62$ eV)

7.5 Summary and Conclusions

In this chapter we presented a semianalytical method based on the GMT to compute classical electromagnetic scattering and EEL spectra of spheroidal dielectric nanoparticles, exclusively in the frequency domain. Hence, no Fourier transformation of the results is needed. Given the appropriate numerical parameters, such as multipole order and distribution, and number and distribution of the matching points, the implemented method is able to provide very accurate results. Furthermore, the method accepts either experimental data or dielectric functions modeled by Drude–Lorentz type equations.

The implemented method is validated extensively by considering classical electromagnetic scattering and EELS by dielectric spheroidal nanoparticles. The computed results agree well with available analytical and numerical solutions. Consequently, the presented method is able to provide accurate EEL spectra that help to interpret experimental data, and to design innovative plasmonic devices. Furthermore, the model forms a fundamental basis for the inversion of experimental EEL spectra to characterize individual nanoparticles.

Future research includes addressing the stability of the implemented method and taking into account the interaction between several nanoparticles and supporting structures.

References

1. A. Doicu, T. Wriedt, Near-field computation using the null-field method. J. Quant. Spectrosc. Radiat. Transf. **111**, 466–473 (2010)
2. A. Doicu, Y.A. Eremin, T. Wriedt, *Acoustic and Electromagnetic Scattering Analysis Using Discrete Sources* (Academic Press, London, 2000)
3. M. Dressel, G. Grüner, *Electrodynamics of Solids: Optical Properties of Electrons in Matter* (Cambridge University Press, Cambridge, 2002)
4. P. Drude, Zur Elektronentheorie der Metalle. Annalen der Physik **306**, 566–613 (1900)
5. P. Drude, Zur Elektronentheorie der Metalle; II. Teil. Galvanomagnetische und thermomagnetische Effecte. Annalen der Physik **308**, 369–402 (1900)
6. L.B. Felsen, N. Marcuvitz, *Radiation and Scattering of Waves* (Prentice-Hall, Englewood Cliffs, 1973)
7. C. Forestiere, G. Iadarola, L.D. Negro, G. Miano, Near field calculation based on the T-matrix method with discrete sources. J. Quant. Spectrosc. Radiat. Transf. **112**, 2384–2394 (2011)
8. F.J. García de Abajo, Electron energy loss and cathodoluminescence in spheres, http://nanophotonics.csic.es/static/widgets/sphere/index.html. Accessed 13 Sept 2012
9. F.J. García de Abajo, Relativistic energy loss and induced photon emission in the interaction of a dielectric sphere with an external electron beam. Phys. Rev. B: Condens. Matter. Mater. Phys. **59**, 3095–3107 (1999)
10. F.J. García de Abajo, Optical excitations in electron microscopy. Rev. Mod. Phys. **82**, 209–275 (2010)
11. F.J. García de Abajo, A. Howie, Relativistic electron energy loss and electron-induced photon emission in inhomogeneous dielectrics. Phys. Rev. Lett. **80**, 5180–5183 (1998)

12. A. Gopinath, S.V. Boriskina, B.M. Reinhard, L.D. Negro, Deterministic aperiodic arrays of metal nanoparticles for surface-enhanced Raman scattering (SERS). Opt. Express **17**, 3741–3753 (2009)
13. C. Hafner, Beiträge zur Berechnung der Ausbreitung elektromagnetischer Wellen in zylindrischen Strukturen mit Hilfe des 'Point-Matching-Verfahrens'. Ph.D. dissertation, Department of Information Technology and Electrical Engineering, ETH Zürich, Zürich, CHE, 1980
14. C. Hafner, *The Generalized Multipole Technique for Computational Electromagnetics* (Artech House, Boston, 1990)
15. C. Hafner, L. Bomholt, *The 3D Electromagnetic Wave Simulator* (Wiley, Chichester, 1993)
16. M.F. Iskander, A. Lakhtakia, C.H. Durney, A new iterative procedure to solve for scattering and absorption by dielectric objects. Proc. IEEE **70**, 1361–1362 (1982)
17. P.B. Johnson, R.W. Christy, Optical constants of noble metals. Phys. Rev. B: Condens. Matter. Mater. Phys. **6**, 4370–4379 (1972)
18. F. Kahnert, Numerical methods in electromagnetic scattering theory. J. Quant. Spectrosc. Radiat. Transf. **79–80**, 775–824 (2003)
19. L. Kiewidt, Numerical inversion of EELS data for the estimation of quantum spectra of nanoparticles, Thesis, Department of Production Engineering, University of Bremen, Bremen, GER, M.Sc, 2012
20. L. Kiewidt, M. Karamehmedović, C. Matyssek, W. Hergert, L. Mdler, T. Wriedt, Numerical simulation of electron energy loss spectroscopy using a generalized multipole technique. Ultramicroscopy **133**, 101–108 (2013)
21. P. Leidenberger, P. Schnle, C. Hafner, Simulations of a dielectric slot tip for scanning near-field microwave microscope. J. Comput. Theor. Nanosci. **8**, 1556–1563 (2011)
22. Y. Leviatan, P.G. Li, A.T. Adams, J. Perini, Single-post inductive obstacle in rectangular waveguide. IEEE Trans. Microw. Theory Tech. **31**, 806–811 (1983)
23. A.C. Ludwig, A comparison of spherical wave boundary value matching versus integral equation scattering solulions for a perfectly conducting body. IEEE Trans. Antennas Propag. **34**, 857 (1986)
24. A. Ludwig, The generalized multipole technique. Comput. Phys. Commun. **68**, 306–314 (1991)
25. C. Matyssek, J. Niegemann, W. Hergert, K. Busch, Computing electron energy loss spectra with the Discontiuous Galerkin Time-Domain method. Photonics Nanostructures: Fundam. Appl. **9**, 367–373 (2011)
26. C. Matyssek, V. Schmidt, W. Hergert, T. Wriedt, The T-Matrix method in electron energy loss and cathodoluminescence spectroscopy calculations for metallic nano-particles. Ultramicroscopy **117**, 46–52 (2012)
27. S. Mohsen Raeis Zadeh Bajestani, M. Shahabadi, N. Talebi, Analysis of plasmon propagation along a chain of metal nanospheres using the generalized multipole technique. J. Opt. Soc. Am. B **28**, 937–943 (2011)
28. E. Moreno, D. Erni, C.H.H. Vahldieck, Multiple multipole method with automatic multipole setting applied to the simulation of surface plasmons in metallic nanostructures. J. Opt. Soc. Am. A **19**, 101–111 (2002)
29. J.C. Nédélec, *Acoustic and Electromagnetic Equations: Integral Representations for Harmonic Problems* (Springer, New York, 2001)
30. M. Nishimura, S. Takamatsu, H. Shigesawa, A numerical analysis of electromagnetic scattering of perfect conducting cylinders by means of discrete singularity method improved by optimization process. Electron. Commun. Jpn. I **67**, 75–81 (1984)
31. N. Piller, O. Martin, Extension of the generalized multipole technique to three-dimensional anisotropic scatterers. Opt. Lett. **23**, 579–581 (1998)
32. E.D. Palik (ed.), *Handbook of Optical Constants of Solids* (Academic Press, San Diego, 1998)
33. S. Prahl, Mie Scattering Calculator, http://omlc.ogi.edu/calc/mie_calc.html. Accessed 9 Aug 2012
34. A.D. Rakić, Algorithm for the determination of intrinsic optical constants of metal films: application to aluminum. Appl. Opt. **34**, 4755–4767 (1995)

35. A.D. Rakić, A.B. Djurii, J.M. Elazar, M.L. Majewski, Optical properties of metallic films in vertical-cavity optoelectric devices. Appl. Opt. **37**, 5271–5283 (1998)
36. L.F. Shampine, Vectorized adaptive quadrature in MATLAB. J. Comput. Appl. Math. **211**, 131–140 (2008)
37. T. Wriedt (ed.), *Generalized Multipole Techniques for Electromagnetic and Light Scattering* (Elsevier, Amsterdam, 1999)
38. T. Wriedt, Using the T-matrix method for light scattering computations by non-axisymmetric particles: superellipsoids and realistically shaped particles. Part. Part. Syst. Charact. **19**, 256–268 (2002)

Chapter 8
Introduction to Yasuura's Method of Modal Expansion with Application to Grating Problems

Akira Matsushima, Toyonori Matsuda and Yoichi Okuno

Abstract In this chapter we introduce the theory of the Yasuura's method based on modal expansion and explain the methods of numerical computation in detail for several grating problems. After a sample problem we discuss the methods for solving two types of problems that require additional knowledge and steps, that is, scattering by a dielectric cylinder and diffraction by a grating. Some numerical results are shown to give an evidence of an experimental rule for the number of linear equations in formulating the least-squares problem that determines the modal coefficients. After confirming the rule we show a couple of examples of practical interest, i.e., scattering by a relatively deep metal grating, plasmon surface waves on a metal grating placed in conical mounting, scattering by a metal surface modulated in two directions, and scattering by periodically located dielectric spheres. To provide supplementary explanations of particular problems, four appendices are given; H-wave scattering from a cylinder, the normal equation and related topics, conical diffraction by a dielectric grating, and comparison of modal functions and the algorithm of the smoothing procedures.

A. Matsushima (✉)
Kumamoto University, Kumamoto 860-8555, Japan
e-mail: matsua@cs.kumamoto-u.ac.jp

T. Matsuda
National Institute of Technology, Kumamoto College,
Kumamoto 861-1102, Japan
e-mail: tmatsu@kumamoto-nct.ac.jp

Y. Okuno
South China Normal University, Guangzhou 510006, China
e-mail: okuno@gpo.kumamoto-u.ac.jp

T. Wriedt and Y. Eremin (eds.), *The Generalized Multipole Technique for Light Scattering*, Springer Series on Atomic, Optical, and Plasma Physics 99,
https://doi.org/10.1007/978-3-319-74890-0_8

8.1 Introduction

In Chap. 6 of the last edition [27] we have introduced Yasuura's method of modal expansion from two main points of view: one was the relation with the fictitious or equivalent source methods; and another was the employment of smoothing procedures (SP's) [10, 24, 25, 31, 41, 42] to obtain rapidly converging solutions. We needed the first point to have the method recognized as one of the modal expansion methods having firm theoretical foundations and a wide range of application. While in the second point we tried to explain our tool to cope with the problem of slow convergence. Because we had been working with the separated solutions as modal functions, we were often troubled by their poor approximation power. Accordingly, Yasuura et al. hit upon an idea of the SP, which works to accelerate the convergence of solutions by reducing the higher-order oscillations on the boundary. The SP, hence, is an important step in solving a 2-D problem[1] where the cross section of the obstacle is strongly deformed from the coordinate curves of a separable system of coordinates.

In the last edition we have included: (**1**) the theory and the method of numerical execution of the original form of Yasuua's method, which we call the conventional Yasuura's method (CYM) today; (**2**) Yauura's method with a smoothing or a singular-smoothing procedure (YMSP or YMSSP); and (**3**) numerical examples obtained mainly by the YMSP and YMSSP. In the present chapter, however, we decided to omit a greater part of SP-related topics in view of the recent trend in computational electromagnetics. That is, the methods for 3-D as well as 2-D analysis of structures made of a dielectric are required in various areas. Instead of removing the SP, we include a detailed explanation on the solution process by the CYM. We hope this helps those who are interested in solving their problems by using Yasuura's methods, CYM, YMSP, and YMSSP. Because the process with the SP's are almost in common with that of the CYM, the detailed introduction of the CYM execution process would be useful not only for the CYM users but also for those who intend to employ the YMSP or YMSSP.

The contents of this chapter are as follows: In Sect. 8.2 we first introduce the theory of the CYM briefly and explain the method of numerical computation in detail taking a sample problem. Then, we move on to the methods for solving two types of problems that require additional knowledge and steps: (**1**) scattering by a dielectric cylinder; and (**2**) diffraction by a grating. In Sect. 8.3 we show some numerical results. The aim of Sect. 8.3.1 is to give an evidence of an experimental rule for the number of sampling points or, in general, the number of linear equations in formulating the least-squares problem that determines the modal coefficients. Computational results show the number should be twice as many as the number of unknown modal coefficients. After confirming the rule we show a couple of examples of practical interest in Sects. 8.3.2–8.3.5: Scattering by a relatively deep metal grating, Plasmon surface waves on a metal grating placed in conical mounting, Scattering by a metal surface modulated in two directions, and Scattering by periodically located dielectric

[1]The reason why we set a limit "2-D" is that the SP, in the present form, is available only in 2-D problems. This is because we employ an indefinite integral to realize a low-pass spatial filter.

spheres. Section 8.4 is a conclusion where we state some additional remarks. Finally, four appendices follow mainly providing supplementary explanations of particular problems: **1** H-wave scattering from a cylinder; **2** The normal equation and related topics; **3** Conical diffraction by a dielectric grating; and **4** Comparison between two types of modal functions and a brief introduction to the algorithm with the SP.

8.2 Yasuura's Method of Modal Expansion

In this section we introduce the foundations of Yasuura's method of modal expansion. We start by formulating a sample problem: plane wave scattering by a perfectly-conducting (PC) cylinder, the problem from which we can learn the essential part of the method together with important concepts and ideas in Yasuura's method.

8.2.1 Scattering by a Perfectly-Conducting Cylinder

The geometry of the sample problem is shown in Fig. 8.1. The closed curve C is the cross section and S_e is the exterior infinite region of C. We denote a point in S_e by $\mathbf{r}\,(r, \theta)$; and one on C by an arc-length s along C measured counterclockwise from a fixed point s_0. S_{e0} is an arbitrary closed region that is entirely inside S_e. Let the incident plane wave be polarized in z and

$$\mathbf{E}^{\mathrm{i}}(\mathbf{r}) = \mathbf{u}_z F(\mathbf{r}) = \mathbf{u}_z \exp[-ikr\cos(\theta - \iota)], \tag{8.1}$$

where $\mathbf{u}_z$ is a unit vector in z-direction, ι is the angle of incidence shown in Fig. 8.1, and $k = 2\pi/\lambda = \omega/c$ is the wavenumber of the incident field. The $e^{i\omega t}$ time dependence is assumed. This case of polarization is called E-wave,[2] which is one of the two basic polarizations. We will deal with an E-wave problem in this section and summarize important results of an H-wave case, which is another basic polarization, in Appendix 1.

In the present problem a surface current flows in the z-direction exciting a scattered wave polarized in z again:

$$\mathbf{E}^{\mathrm{s}}(\mathbf{r}) = \mathbf{u}_z \Psi(\mathbf{r}). \tag{8.2}$$

Other non-zero components of the scattered wave, $\mathbf{H}^{\mathrm{s}}(\mathbf{r}) = \mathbf{u}_x H_x^{\mathrm{s}}(\mathbf{r}) + \mathbf{u}_y H_y^{\mathrm{s}}(\mathbf{r})$, can be obtained by[3]

[2]It is also termed Transverse-Electric (TE) wave, which means the electric field is orthogonal to the xy-plane. While in the H-wave (or TM-wave) the magnetic field has the z-component alone.

[3]The component is called a leading field if it gives other nonzero components as in (8.3). Note that the derivation of $\mathbf{H}^{\mathrm{s}}$ by (8.3) is a proper procedure because the sequence of our approximate

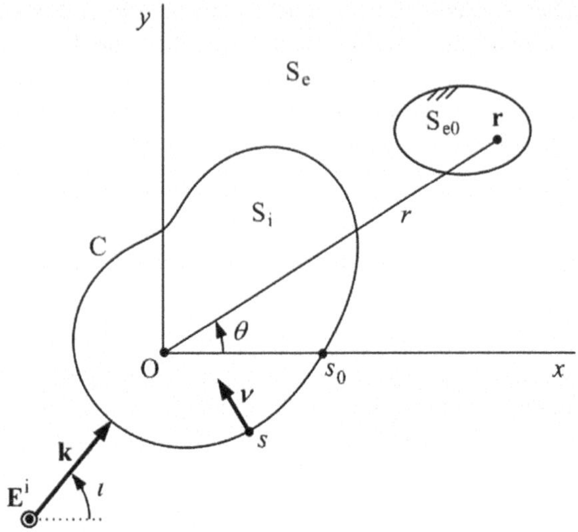

Fig. 8.1 Geometry of the sample problem. C is a closed curve of length C representing the cross section of a cylindrical obstacle. S_e is the exterior infinite region of C; and S_i is the interior region, which we need in Sects. 8.2.2.1 and 8.2.4

$$\mathbf{H}^s(\mathbf{r}) = \frac{i}{\omega\mu_0}\nabla\Psi(\mathbf{r}) \times \mathbf{u}_z, \tag{8.3}$$

where $\nabla = (\partial/\partial x, \partial/\partial y, 0)$ is the 2-D nabla operator. Hence, our target is $\Psi(\mathbf{r})$ and we can state our sample problem as:

Problem 1 E-wave, PC. Find the scattered electric field $\Psi(\mathbf{r})$ that satisfies:

(D1) The 2-D Helmholtz equation in S_e

$$\nabla^2\Psi(\mathbf{r}) + k^2\Psi(\mathbf{r}) = 0 \quad (\mathbf{r} \in S_e), \tag{8.4}$$

(D2) The 2-D radiation condition at infinity

$$\sqrt{r}\left(\frac{\partial\Psi(\mathbf{r})}{\partial r} + ik\Psi(\mathbf{r})\right) \to 0 \quad (r \to \infty), \tag{8.5}$$

(D3) The boundary condition

$$\Psi(s) = f(s) \equiv -F(s) \quad (s \in C, \text{ i.e., } 0 \le s \le C). \tag{8.6}$$

Here, $\nabla^2 = \partial^2/\partial x^2 + \partial^2/\partial y^2$ denotes the 2-D Laplacian. The condition given by (8.6) is called Dirichlet's or the first-kind boundary condition.

---

solutions converges to the true solution uniformly in wider sense in the exterior region S_e as we will see later.

8.2.2 Modal Functions, Approximate Solution, and Least-Squares Boundary Matching

Here we introduce the analytical part of Yasuura's method [38–40]. Because it is one of the modal expansion methods, we need: **(i)** definition of a set of modal functions; **(ii)** a method to construct an approximate solution; and **(iii)** the sense in which the solution approximates the boundary condition. Let us see these points below.

8.2.2.1 Definition of the Set of Modal Functions

Modal functions for the sample problem are solutions of Helmholtz's equation (8.4) satisfying some additional requirements. Here, we define a set of modal functions $\{\varphi_m(\mathbf{r}) : m = 1, 2, \ldots\}$ as a countable set that satisfy the following three requirements:

(M1) Each $\varphi_m(\mathbf{r})$ satisfies the Helmholtz equation in S_e;
(M2) Each $\varphi_m(\mathbf{r})$ meets the 2-D radiation condition;
(M3) Both the set of boundary values $\{\varphi_m(s) : m = 1, 2, \ldots\}$ and the set of normal derivatives $\{\partial\varphi_m(s)/\partial\nu : m = 1, 2, \ldots\}$ are complete (or total)[4] in the function space $\mathbf{H} = L^2(\mathrm{C})$ consisting of all the square-integrable functions defined on the boundary C.

The first two requirements are natural and easy to understand; but the third is rather complicated and needs explanation. Here, we would like to call readers' attention to the fact that (**M3**) is a little different from the original requirement given in [39], which seems to be lacking in concreteness than the statement above. We have modified the original statement to require completeness of the boundary values.

Now, let us see a couple of examples first to facilitate the understanding. Then, we will give additional explanations for this issue throughout this section.

Example 1 The set of radiative separated solutions

$$\varphi_m(\mathbf{r}) = H_m^{(2)}(kr)\exp(im\theta) \quad (m = 0, \pm 1, \pm 2, \ldots), \tag{8.7}$$

where $H_m^{(2)}(kr)$ is the second kind Hankel function of order m and the coordinate origin should be inside S_i, the complimentary region of S_e.[5]

[4]We hope the readers consult a treatise on *Functional Analysis*, e.g. [14], in case of need.

[5]If C is a circle centered at the origin, it is apparent that the sets of boundary values and normal derivatives are both complete because the members of each set are nothing other than the Fourier bases. Even in case if C is not a circle, the sets are still complete because of Example 2: let L be a circle centered at the origin and take the Fourier bases for $f_m(t)$ in (8.8), then we get a set of separated solutions.

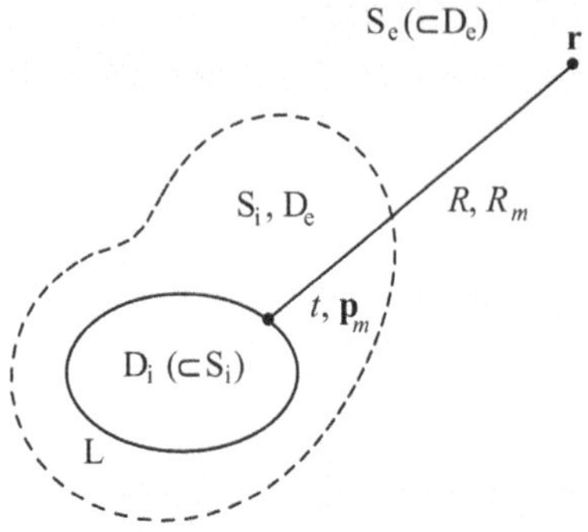

Fig. 8.2 Relevant to definition of the set of modal functions. The cross section C is shown by a dashed curve; L is another closed curve inside C

Example 2 Let L be a smooth closed curve that is entirely inside $\mathrm{S_i}$ and $\mathrm{D_e}$ be an exterior infinite region of L. As shown in Fig. 8.2, $\mathrm{S_e}$ is a subregion of $\mathrm{D_e}$; and $\mathrm{D_i}$, the complementary region of $\mathrm{D_e}$, is a subregion of $\mathrm{S_i}$. Now, let an enumerable set of functions $\{f_m(s) : m = 1, 2, \ldots\}$ be complete in the function space $L^2(\mathrm{L})$. Then, the set of potential functions defined in $\mathrm{D_e}$ with $f_m(t)$'s as double-layer density functions on L

$$\varphi_m(\mathbf{r}) = -\int_{\mathrm{L}} f_m(t)\,\frac{\partial\psi(kR)}{\partial\nu_t}\,dt \quad (\mathbf{r} \in \mathrm{D_e};\ R = \overline{t\mathbf{r}};\ m = 1, 2, \ldots) \tag{8.8}$$

is a set of modal functions in $\mathrm{D_e}$ provided that k does not coincide with a member of $\{k_{\mathrm{H}}(\mathrm{D_i})\}$, the set of eigenvalues of the homogeneous H-wave (Neumann) problem in $\mathrm{D_i}$.[6] Here, R is the distance between t and $\mathbf{r}$, $\psi(kR) = H_0^{(2)}(kR)/4\mathrm{i}$ is the free-space Green's function, and $\partial/\partial\nu_t$ denotes normal derivative at t. Note that the ensemble of single-layer potentials can also be the set of modal functions provided $k \notin \{k_{\mathrm{E}}(\mathrm{D_i})\}$, the set of eigenvalues of homogeneous E-wave (Dirichlet) problem in $\mathrm{D_i}$.

Example 3 Monopole fields whose poles $\mathbf{p}_m$ are located on L

$$\varphi_m(\mathbf{r}) = H_0^{(2)}(kR_m) \quad (\mathbf{r} \in \mathrm{D_e};\ R_m = \overline{\mathbf{p}_m\mathbf{r}};\ m = 1, 2, \ldots, M) \tag{8.9}$$

form a set of modal functions in $\mathrm{D_e}$ when we let $M \to \infty$ while letting $\overline{\mathbf{p}_m\mathbf{p}_{m+1}} \to 0$ provided there is no internal resonance in $\mathrm{D_i}$ [26, 30].

Example 4 The set of multiple-multipole fields whose poles $\mathbf{p}_m$ are on L

$$\begin{gathered}\varphi_{mn}(\mathbf{r}) = H_n^{(2)}(kR_m)\exp(\mathrm{i}n\theta_m) \\ (\mathbf{r} \in \mathrm{D_e};\ R_m = \overline{\mathbf{p}_m\mathbf{r}};\ m = 1, 2, \ldots, M;\ n = 0, \pm 1, \pm 2, \ldots)\end{gathered} \tag{8.10}$$

is also an example of modal functions.

[6] This is not a strong exception because we can modify the contour L (and hence $\mathrm{D_i}$) slightly to avoid the coincidence. Example 2 is a key theorem of generation of complete sets, which has been proven by Yasuura and Itakura [39] as an analogy of Runge's (or Runge-Walsh's) theorem known in *Theory of Complex Functions*.

8.2.2.2 Construction of an Approximate Solution

To define an approximate solution, we first choose a set of modal functions from among possible candidates. Let us take the set of separated solutions in the following analysis. This is because the set of separated solutions is one of the most familiar functions and each member has physical meaning.[7] Thus we can define an approximate solution as a *finite summation* of the outgoing separated solutions (8.7) with unknown coefficients:

$$\Psi_N(\mathbf{r}) = \sum_{m=-N}^{N} A_m(M)\,\varphi_m(\mathbf{r}). \tag{8.11}$$

Here, $A_m(M)$ means that the A_m coefficient depends on $M = 2N + 1$, the number of modal functions employed.[8] Because of the definition of modal functions, the approximate solution already satisfies the requirements (**D1**) and (**D2**). The A_m coefficients, hence, should be determined so that the solution meets the boundary condition in a sense of approximation. Let us call this procedure *boundary matching* and keep in mind that the sense of approximation in boundary matching determines a method of solution.

We employ the least-squares approximation in Yasuura's method, i.e., minimization of mean-squares boundary residual. We will see in Sect. 8.2.2.3 that this is a promising way in boundary matching provided the completeness of the set of boundary values (**M3**) is guaranteed.

8.2.2.3 Least-Squares Boundary Matching

We employ integral representations of the solutions to explain the method of solution including convergence of the approximate solutions. For this purpose let us define the Green's function of our problem, $G(\mathbf{r}, \mathbf{r}')$, satisfying Helmholtz's equation with a unit source at $\mathbf{r}$, radiation condition with respect to $\mathbf{r}'$, and a homogeneous boundary condition[9]

[7] Unfortunately, separated solutions are not very efficient in a problem where C is strongly modulated from a circle (or, in general, a coordinate surface of the system of coordinates employed). As an example, we show a comparison between types of modal functions: the separated solutions (8.7) and monopole fields (8.9) in Appendix 4.

[8] This dependence is natural because the boundary values of modal functions, in general, do not form an orthogonal set in **H**. This type of summation is usually called a flexible summation. Note that the approximate solution is defined in a finite summation of modal functions. By considering a sequence of finite-sum solutions, we can avoid the constraint of the convergence area of an infinite series solution. Yasuura's original papers [38–40] has been written from this point of view. Reference [9] includes an interpretation of the difference between series and sequence solutions.

[9] $G(\mathbf{r}, \mathbf{r}')$ is a total electric field observed at $\mathbf{r}'$ ($\neq \mathbf{r}$) when a unit line source is placed at $\mathbf{r}$ in Fig. 8.1. Note that employment of the Green function satisfying (8.12) is for convenience and is not essential: The whole theory has been established in [38–40], where the free-space Green function alone was used.

$$G(\mathbf{r}, s) = 0 \quad (\mathbf{r} \in \mathrm{S_e};\ s \in \mathrm{C}). \tag{8.12}$$

Using the Green's formula to $\Psi(\mathbf{r}')$ and $G(\mathbf{r}, \mathbf{r}')$, we have

$$\Psi(\mathbf{r}) = -\int_{s=0}^{C} \partial_\nu G(\mathbf{r}, s)\Psi(s)\,ds = -\int_{s=0}^{C} \partial_\nu G(\mathbf{r}, s) f(s)\,ds \quad (\mathbf{r} \in \mathrm{S_e}). \tag{8.13}$$

Here, ∂_ν denotes the normal derivative at s and the second equality comes from (**D3**). Besides, we get a similar representation for the approximate solution $\Psi_N(\mathbf{r})$. Subtracting (8.13) from the representation of $\Psi_N(\mathbf{r})$ side by side, we have

$$\Psi_N(\mathbf{r}) - \Psi(\mathbf{r}) = -\int_{s=0}^{C} \partial_\nu G(\mathbf{r}, s)[\Psi_N(s) - f(s)]\,ds \quad (\mathbf{r} \in \mathrm{S_e}). \tag{8.14}$$

Although (8.14) is a formal representation, we can deduce useful results starting from it.

Let the observation point $\mathbf{r}$ be inside the closed region $\mathrm{S_{e0}}$ in Fig. 8.1. Then, $\partial_\nu G(\mathbf{r}, s)$ is a continuous function of s because there is a non-zero distance between s and $\mathbf{r}$. Taking the absolute value of both sides of (8.14) and applying Cauchy–Schwarz's inequality to the right-hand side, we obtain

$$\left|\Psi_N(\mathbf{r}) - \Psi(\mathbf{r})\right| \le \sqrt{\int_{s=0}^{C} |\partial_\nu G(\mathbf{r}, s)|^2\,ds}\ \left\|\Psi_N - f\right\| \quad (\mathbf{r} \in \mathrm{S_e}). \tag{8.15}$$

Here, $\|f\|$ stands for the Euclidean norm of a function $f(s)$ defined by

$$\|f\| = \left[\int_{s=0}^{C} |f(s)|^2\,ds\right]^{1/2}. \tag{8.16}$$

Because the integrand on the right of (8.15) is a continuous function of $\mathbf{r}$, the integral, as a function of $\mathbf{r}$, has a maximum inside the closed region $\mathrm{S_{e0}}$:

$$G(\mathrm{S_{e0}}) = \max_{\mathbf{r} \in \mathrm{S_{e0}}} \sqrt{\int_{s=0}^{C} |\partial_\nu G(\mathbf{r}, s)|^2\,ds} \quad (\mathrm{S_{e0}} \subset \mathrm{S}). \tag{8.17}$$

Thus we have an estimation

$$|\Psi_N(\mathbf{r}) - \Psi(\mathbf{r})| \le G(\mathrm{S_{e0}})\left\|\Psi_N - f\right\| \quad (\mathbf{r} \in \mathrm{S_{e0}} \subset \mathrm{S_e}), \tag{8.18}$$

which means that the maximal absolute error in $\mathrm{S_{e0}}$ cannot exceed the product of the mean-squares boundary residual and a factor of proportionality $G(\mathrm{S_{e0}})$. Note that the latter depends on the region $\mathrm{S_{e0}}$ but does not depend on $\mathbf{r}$.

Now, let us remember the completeness (**M3**) of the set of boundary values of modal functions. Because the given boundary value $f(s)$ is a member of $\mathbf{H} = L^2(\mathrm{C})$, for given any positive number ε, there is a positive integer N_0 such that

$$\left\| \Psi_N - f \right\| < \varepsilon \quad (N > N_0). \tag{8.19}$$

That is, there exists a sequence of boundary values of the approximate solutions $\{\Psi_0(s), \Psi_1(s), \Psi_2(s), \ldots\}$ that converges to the true boundary value $f(s)$ in the mean-squares sense:

$$\left\| \Psi_N - f \right\| \to 0 \quad (N \to \infty). \tag{8.20}$$

Referring to (8.18), we can conclude that the corresponding sequence of approximate solutions $\{\Psi_0(\mathbf{r}), \Psi_1(\mathbf{r}), \Psi_2(\mathbf{r}), \ldots\}$ converges to $\Psi(\mathbf{r})$ uniformly in the closed region $\mathrm{S_{e0}}$[10]: for given any positive number ε, there is a positive integer $N_0(\mathrm{S_{e0}}, \varepsilon)$ such that

$$|\Psi_N(\mathbf{r}) - \Psi(\mathbf{r})| < \varepsilon \quad (\mathbf{r} \in \mathrm{S_{e0}} \subset \mathrm{S_e};\ N > N_0(\mathrm{S_{e0}}, \varepsilon)). \tag{8.21}$$

We can get such a sequence by solving repeatedly the following least-squares problem (LSP) stated in the function space **H**.

LSP 1: E-wave, PC. Find the coefficients $A_m(M)$ $(m = 0, \pm 1, \ldots, \pm N;\ M = 2N + 1)$ that minimize the normalized mean-squares boundary residual

$$E_N = \frac{\left\| \Psi_N - f \right\|^2}{\| f \|^2} = \frac{1}{\| f \|^2} \left\| \sum_{m=-N}^{N} A_m(M) \varphi_m - f \right\|^2. \tag{8.22}$$

Note that the least-squares boundary matching means a relaxation of the boundary condition because (8.6) implies $\|\Psi - f\| = 0$; but the converse is not always true. The smoothing procedure (SP), which we mentioned in Introduction, is an extension of the relaxation idea: we minimize $\| \int (\Psi - f)\, ds \|$ instead of $\|\Psi - f\|$; and extinction of the latter is stronger than vanishing of the former [10, 24, 31, 41, 42]. Although the Yasuura's method with the SP is a strong tool for 2-D problems, we shall not get deeply in this subject.

8.2.3 *Method of Numerical Solution*

Because computers cannot handle continuous functions, we need (**i**) method of discretization of **LSP 1** and (**ii**) method of solution to the discretized problem.[11]

[10]This kind of convergence is called uniform convergence in wider sense in $\mathrm{S_e}$.

[11]Until the middle of 80s we employed normal equations (NE) in solving **LSP 1**. Now we solve the problem using the method in Sect. 8.2.3.2. We state the reason why we stopped using the NE and attach some comments in Appendix 2.

8.2.3.1 Method of Discretization

To discretize the problem we first locate J ($\geq M$) sampling points on C, assign a set of integers from 0 through J to them, and get a numbered set of sampling points $\{s_0, s_1, \ldots, s_J\}$. Because C is a closed curve, we give two numbers, 0 and J, to the point $s = 0$. Two methods are usually used in locating the points:

1. An equal division of the boundary C: In most applications we can recommend to use the points given by

$$s_j = \frac{jC}{J} \quad (j = 0, 1, 2, \ldots, J) \tag{8.23}$$

   without any reservation in theory. If we take this method, we may have to solve an (transcendental) (8.23) in locating the points.
2. An equal division with respect to a coordinate variable: For example, if C is represented as $r = r(\theta)$, it must be convenient to use the discretization

$$\theta_j = \frac{j2\pi}{J} \quad (j = 0, 1, 2, \ldots, J). \tag{8.24}$$

   This choice, however, means a variable transformation in (8.13) and in other integrals on C and will lead us solving a weighted least-squares problem unexpectedly. Users should notice this and be careful in applying this method of location in a problem where the boundary C is strongly deformed from a circle.[12]

Having located the sampling points on C, we can define discretized forms of the functions $f(s)$, $\varphi_m(s)$, and so on:

$$\mathbf{f} = [f(s_1)\ f(s_2)\ \cdots\ f(s_J)]^{\mathrm{T}} \tag{8.25}$$

and

$$\boldsymbol{\varphi}_m = [\varphi_m(s_1)\ \varphi_m(s_2)\ \cdots\ \varphi_m(s_J)]^{\mathrm{T}}. \tag{8.26}$$

Here, the superscript T denotes a transposed vector or matrix and the discretized forms are J-dimensional complex-valued column vectors. Next, we define a $J \times M$ matrix by

$$\Phi = \left[\boldsymbol{\varphi}_{-N}\ \boldsymbol{\varphi}_{-N+1}\ \cdots\ \boldsymbol{\varphi}_N\right] = \begin{bmatrix} \varphi_{-N}(s_1) & \varphi_{-N+1}(s_1) & \cdots & \varphi_N(s_1) \\ \varphi_{-N}(s_2) & \varphi_{-N+1}(s_2) & \cdots & \varphi_N(s_2) \\ \vdots & \vdots & \ddots & \vdots \\ \varphi_{-N}(s_J) & \varphi_{-N+1}(s_J) & \cdots & \varphi_N(s_J) \end{bmatrix} \tag{8.27}$$

[12]On the other hand, there is a possibility to make possible use of the weighting function accompanying the variable transformation. For example, a Schwarz–Christoffel-type transformation works to remove the singularity of Green's function in a problem of an edged cross section [23].

which is usually termed a Jacobian matrix. Finally, defining an M-dimensional solution vector

$$\mathbf{A} = \left[A_{-N}(M)\ A_{-N+1}(M)\ \cdots\ A_N(M)\right]^{\mathrm{T}}, \tag{8.28}$$

we can represent a discretized form of an approximate solution on C in vector-matrix notation

$$\boldsymbol{\Psi}_N = \sum_{m=-N}^{N} A_m(M)\boldsymbol{\varphi}_m = \boldsymbol{\Phi}\mathbf{A}. \tag{8.29}$$

Thus, we have an approximation to the mean-squares boundary residual in (8.22):

$$E_{NJ} = \frac{\|\boldsymbol{\Phi}\mathbf{A} - \mathbf{f}\|^2}{\|\mathbf{f}\|^2}. \tag{8.30}$$

Here, $\|\mathbf{f}\|$ denotes a Euclidean norm of a J-dimensional complex-valued vector $\mathbf{f}$. Because C is a closed curve and $f(s_0) = f(s_J)$, etc., (8.30) can be understood as a trapezoidal-rule approximation of (8.22). Now, we can state a discretized form of **LSP 1** as follows:

DLSP 1: E-wave, PC. Find the solution vector $\mathbf{A}$ that minimizes the numerator of (8.30).

Here arises an important issue of the number of sampling points[13]: How many J do we need? If we answer to this question in generality, we should say: It depends. However, employing the results of examination in Sect. 8.3.1, we can state an experimental rule:

$$J \doteq 2(2N + 1) = 2M. \tag{8.31}$$

Here, the symbol $\doteq$ means that the number on the right-hand side is usually *sufficient in finding the scattered field*. This might be considerably smaller than what uninitiates expect because **DLSP 1** with the number J of (8.31) does not seem to be a good approximation of **LSP 1**. This is because an inner product $(C/J)\mathbf{f}^{\dagger}\mathbf{g}$ implicitly included in the norm on the right of (8.30) cannot be a precise approximation of (f, g) in (8.22) if either $f(\mathbf{r})$ or $g(\mathbf{r})$ is a higher-order space harmonic. Nevertheless, **DLSP 1** with (8.31) gives an approximate solution having converged with respect to J. We know this welcome nature of **DLSP 1** since we started solving the scattering problem on a computer in early 70s. At that time the method in Sect. 8.2.3.2 was not known widely and we solved the problem using a normal equation (NE; see Appendix 2). We found (8.31) was effective even in using the NE where the inner products $(C/J)\mathbf{f}^{\dagger}\mathbf{g}$ appeared explicitly as the matrix elements. In the method of solution that we introduce next, we do not have to calculate these inner products. That is one of the advantages of the method.

---

[13]It is more reasonable to ask "How many linear equations do we need?" This is because (i) we get two equations at one sampling point in a 2-media problem (see Sect. 8.2.4); and (ii) we should understand (8.31) as a relation between the numbers of equations J and unknowns M.

8.2.3.2 Solution Method to the Discretized Problem

To solve the least-squares problem in the J-dimensional vector space, we employ orthogonal decomposition of the Jacobian matrix: the singular-value decomposition (SVD) and the QR decomposition (QRD) [15]. They have the following features:

- The SVD informs us of the character of the Jacobian matrix through singular values. This is helpful in designing and testing a process of numerical solution, in particular, choice of modal functions, number and location of sampling points, etc. Instead, the computational complexity, in both memory and time, is bigger than that of the QRD.
- The QRD needs less computation than the SVD and solves the problem provided no rank deficiency occurs.[14]

Hence, we recommend the use of the SVD for designing and testing the discretized least-squares problem. After the problem is established, application of the QRD is appropriate. Let us see how to use these decompositions in examining and solving **DLSP 1**.

Utilization of the SVD

Applying the SVD, we get a decomposition of the Jacobian matrix in the form

$$\Phi = \mathrm{U}\Sigma\mathrm{V}^{\dagger}, \tag{8.32}$$

where U ($J \times J$) and V ($M \times M$) are unitary matrices, and $\dagger$ denotes Hermitian conjugation: $\mathrm{V}^{\dagger} = \bar{\mathrm{V}}^{\mathrm{T}}$. Σ is a stack of an $M \times M$ diagonal matrix and a $(J - M) \times M$ zero matrix. The diagonal elements of Σ, σ_m, are non-negative and are called the singular values of Φ. Arranging the M singular values in the order of decreasing magnitude, we have $\sigma_1 \geq \sigma_2 \geq \cdots \geq \sigma_M$ $(M = 2N + 1)$. Let us call σ_1 and σ_M by $\sigma_{\max}$ and $\sigma_{\min}$ because this order of σ_m does not necessarily agree with the order of modal functions. The following items are widely known and accepted:

- The singular values are non-negative square roots of the eigenvalues of a positive semidefinite Hermitian matrix $\Phi^{\dagger}\Phi$: $\sigma_m(\Phi) = \sqrt{\lambda_m(\Phi^{\dagger}\Phi)}$. And, vanish of the smallest singular value, $\sigma_{\min} = 0$, means $\det \Phi^{\dagger}\Phi = 0$. Because $\Phi^{\dagger}\Phi$ is the coefficient matrix of the NE (8.106) in Appendix 2, this is a serious problem: the least-squares problem does not have a unique solution. Although $\sigma_{\min} = 0$ in strict sense seldom occurs in practice, very tiny $\sigma_{\min}$ is not rare and causes substantial rank deficiency.
- The ratio of the maximum singular value to the minimum

$$\mathrm{cond}(\Phi) = \frac{\sigma_{\max}}{\sigma_{\min}} \tag{8.33}$$

[14] In addition, the solution by a QRD program, usually, is not inferior in accuracy to one by an SVD program. This may be because of the greater computational complexity of the SVD.

defines the condition number of Φ, which shows the degree of numerical difficulty in solving the least-squares problem with the Jacobian matrix Φ. In general, a problem with a small cond(Φ) is easy to solve and is termed well-conditioned; while one with a huge cond(Φ) is difficult and called ill-conditioned. In this connection an empirical rule is known: if the reciprocal of cond(Φ) is of the same order as or smaller than the machine epsilon[15] of the system of floating-point numbers, effective rank of Φ might be less than M and **DLSP 1** may not be solved properly.

Although our main purpose to employ the SVD is to check the nature of Φ, we can solve **DLSP 1** in the following way:

(a) Modifying $\|\Phi \mathbf{A} - \mathbf{f}\|^2$ by insertion of (8.32), we have

$$\|\Phi \mathbf{A} - \mathbf{f}\|^2 = \|\mathrm{U}^\dagger(\Phi \mathbf{A} - \mathbf{f})\|^2 = \|\Sigma \mathrm{V}^\dagger \mathbf{A} - \mathrm{U}^\dagger \mathbf{f}\|^2 = \|\Sigma \mathbf{B} - \mathbf{d}\|^2. \quad (8.34)$$

Here, we have used that the matrices U and V are unitary and that a unitary transformation does not change the norm of a vector. Also, note that the last equal sign defines the vectors **B** and **d**.

(b) We get the solution to **DLSP 1** from

$$B_m = \frac{d_m}{\sigma_m} \quad (m = 1, 2, \ldots, M \ (= 2N + 1)) \quad (8.35)$$

and the squared norm by

$$\|\Phi \mathbf{A} - \mathbf{f}\|^2 = \sum_{j=M+1}^{J} |d_j|^2. \quad (8.36)$$

Utilization of the QRD

Employment of the QRD leads us to a decomposition of the form

$$\Phi = \mathrm{Q}\tilde{\mathrm{R}} = \mathrm{Q}\begin{bmatrix} \mathrm{R} \\ 0 \end{bmatrix}, \quad (8.37)$$

where Q is a $J \times J$ unitary matrix and $\tilde{\mathrm{R}}$ is a stack of $M \times M$ upper triangular matrix R and a $(J - M) \times M$ zero matrix. Having the decomposition (8.37), we can solve **DLSP 1** by the following procedure:

(a) Inserting (8.37) into $\|\Phi \mathbf{A} - \mathbf{f}\|^2$, we have

[15] The machine epsilon, EPS, is the minimum positive number that satisfies $1 + \mathrm{EPS} > 1$ in the floating-point system employed.

$$\|\Phi\mathbf{A} - \mathbf{f}\|^2 = \|Q^\dagger(\Phi\mathbf{A} - \mathbf{f})\|^2 = \|\tilde{R}\mathbf{A} - Q^\dagger\mathbf{f}\|^2 \equiv \left\| \begin{bmatrix} \mathrm{R}\mathbf{A} \\ 0 \end{bmatrix} - \begin{bmatrix} \mathbf{d} \\ \mathbf{z} \end{bmatrix} \right\|^2, \tag{8.38}$$

where the last equality defines the vectors **d** and **z**.

(b) We obtain the solution by solving

$$\mathrm{R}\mathbf{A} = \mathbf{d}. \tag{8.39}$$

Because R is triangular, we need only back substitution to solve (8.39). The residual norm is given by

$$\|\Phi\mathbf{A} - \mathbf{f}\|^2 = \|\mathbf{z}\|^2. \tag{8.40}$$

8.2.4 Application to Dielectric or Metal Obstacles

This section introduces the Yasuura's method applied to problems with dielectric or metal obstacles [35, 43–45]. Although metals have unique nature, we here regard a metal as a dielectric with a complex permittivity depending on the frequency. Therefore we consider a material whose permittivity and refractive index are given by complex numbers ε and $n = \sqrt{\varepsilon/\varepsilon_0}$. Usually the material is penetrable and there is a non-zero transmitted field in $\mathrm{S_i}$, the complementary region of $\mathrm{S_e}$. Thus we have two unknown functions $\Psi_\mathrm{i}(\mathbf{r})$ ($\mathbf{r} \in \mathrm{S_i}$) and $\Psi_\mathrm{e}(\mathbf{r})$ ($\mathbf{r} \in \mathrm{S_e}$); and we need two boundary conditions to determine the two unknown functions. The continuity of tangential components of the electric and magnetic field satisfies the necessity.

8.2.4.1 E-Wave Scattering by a Cylindrical Obstacle Made of a Dielectric

Let us assume that the obstacle in Fig. 8.1 is made of a dielectric and that an E-wave is incident. The electric field in $\mathrm{S_e}$ is a sum of the incident and the scattered wave: $\mathbf{u}_z(F + \Psi_\mathrm{e})(\mathbf{r})$; while the field in the interior region $\mathrm{S_i}$ is the transmitted field $\mathbf{u}_z\Psi_\mathrm{i}(\mathbf{r})$. They are the solutions of Helmholtz's equation in each region:

$$\begin{cases} \left(\nabla^2 + k^2\right)\Psi_\mathrm{e}(\mathbf{r}) = 0 & (\mathbf{r} \in \mathrm{S_e}), \\ \left(\nabla^2 + (nk)^2\right)\Psi_\mathrm{i}(\mathbf{r}) = 0 & (\mathbf{r} \in \mathrm{S_i}). \end{cases} \tag{8.41}$$

The exterior solution, in addition, should meet the radiation condition (8.5). The continuity of tangential components of electric and magnetic fields requires the boundary conditions[16]

[16] To get the latter we set $\mathbf{u}_\nu \times (\mathbf{H}_\mathrm{e} - \mathbf{H}_\mathrm{i}) = 0$. Insertion of $\mathbf{H}_\mathrm{e} = (\mathrm{i}/\omega\mu)\nabla E_\mathrm{e} \times \mathbf{u}_z$ etc. finds the desired relation. Here, $E_\mathrm{e} = F + \Psi_\mathrm{e}$ stands for the total electric field in $\mathrm{S_e}$.

$$\begin{cases} \Psi_\mathrm{e}(s) - \Psi_\mathrm{i}(s) = f(s) \equiv -F(s), \\ \dfrac{\partial \Psi_\mathrm{e}(s)}{\partial \nu} - \dfrac{\partial \Psi_\mathrm{i}(s)}{\partial \nu} = g(s) \equiv -\dfrac{\partial F(s)}{\partial \nu}. \end{cases} \tag{8.42}$$

Thus we get a boundary-value problem for $\Psi_\mathrm{e}(\mathbf{r})$ and $\Psi_\mathrm{i}(\mathbf{r})$:

Problem 2 E-wave, dielectric. Find the electric fields $\Psi_\mathrm{e}(\mathbf{r})$ and $\Psi_\mathrm{i}(\mathbf{r})$ that satisfy (8.41), (8.5), and (8.42).

Note that in dealing with an H-wave problem, the second line of (8.42) should include the refractive index n or permittivity ε (see (8.100) in Appendix 1).

8.2.4.2 Modal Functions and Approximate Solutions

We need two sets of modal functions to solve the problem, one is for $\Psi_\mathrm{e}(\mathbf{r})$ and another is for $\Psi_\mathrm{i}(\mathbf{r})$. Let us call them the exterior and interior modal functions and represent them as $\{\varphi_{\mathrm{e}m}(\mathbf{r})\}$ and $\{\varphi_{\mathrm{i}m}(\mathbf{r})\}$. They should satisfy the requirements below, which are almost in common with the conditions from (**M1**) through (**M3**) given in Sect. 8.2.2.1

(**MD1**) Each member of the set of exterior modal functions satisfies the Helmholtz equation in $\mathrm{S_e}$ and meets the radiation condition at infinity.

(**MD2**) Each member of the set of interior modal functions satisfies the Helmholtz equation in $\mathrm{S_i}$.

(**MD3**) The sets of boundary values $\{\varphi_{\mathrm{e}m}(s)\}$ and $\{\varphi_{\mathrm{i}m}(s)\}$, and the sets of normal derivatives $\{\partial\varphi_{\mathrm{e}m}(s)/\partial\nu\}$ and $\{\partial\varphi_{\mathrm{i}m}(s)/\partial\nu\}$ are all complete in the function space **H**.

Here, we take the sets of separated solutions again because they are familiar to many people working with boundary-value problems. Then, the exterior and interior modal functions are:

$$\begin{cases} \varphi_{\mathrm{e}m}(\mathbf{r}) = H_m^{(2)}(kr)\exp(im\theta), \quad \varphi_{\mathrm{i}m}(\mathbf{r}) = J_m(nkr)\exp(im\theta) \\ \qquad (m = 0, \pm 1, \pm 2, \ldots). \end{cases} \tag{8.43}$$

Here, $J_m(nkr)$ stands for the Bessel function of order m. Then, we can define approximate solutions in $\mathrm{S_e}$ and $\mathrm{S_i}$ as

$$\begin{cases} \Psi_{\mathrm{e}N}(\mathbf{r}) = \displaystyle\sum_{m=-N}^{N} A_{\mathrm{e}m}(M)\varphi_{\mathrm{e}m}(s) \quad (\mathbf{r} \in \mathrm{S_e}), \\ \Psi_{\mathrm{i}N}(\mathbf{r}) = \displaystyle\sum_{m=-N}^{N} A_{\mathrm{i}m}(M)\varphi_{\mathrm{i}m}(s) \quad (\mathbf{r} \in \mathrm{S_i}). \end{cases} \tag{8.44}$$

They satisfy the Helmholtz equation in each region and $\Psi_{\mathrm{e}N}$ meets the radiation condition.

8.2.4.3 Error Estimation and Least-Squares Boundary Matching

After some analytical work we get error estimations similar to (8.18)[17]:

$$|\Psi_{eN}(\mathbf{r}) - \Psi_{e}(\mathbf{r})| \le G_{e1}(S_{e0}) \left\| \frac{\partial \Psi_{eN}}{\partial \nu} - \frac{\partial \Psi_{iN}}{\partial \nu} - g \right\| + G_{e2}(S_{e0}) \left\| \Psi_{eN} - \Psi_{iN} - f \right\| \quad (\mathbf{r} \in S_{e0} \subset S_{e}) \tag{8.45}$$

and

$$|\Psi_{iN}(\mathbf{r}) - \Psi_{i}(\mathbf{r})| \le G_{i1}(S_{i0}) \left\| \frac{\partial \Psi_{eN}}{\partial \nu} - \frac{\partial \Psi_{iN}}{\partial \nu} - g \right\| + G_{i2}(S_{i0}) \left\| \Psi_{eN} - \Psi_{iN} - f \right\| \quad (\mathbf{r} \in S_{i0} \subset S_{i}). \tag{8.46}$$

Here, S_{e0} and S_{i0} are arbitrary closed regions in S_e and S_i, and G_{pq} (p = e, i; q = 1, 2) are positive constants depending on S_{e0} and S_{i0}.

We can prove that: provided the sets of modal functions satisfy the requirement (**MD3**), there exists a sequence of pairs of approximate solutions

$$\begin{bmatrix} \Psi_{e0}(\mathbf{r}) \\ \Psi_{i0}(\mathbf{r}) \end{bmatrix}, \begin{bmatrix} \Psi_{e1}(\mathbf{r}) \\ \Psi_{i1}(\mathbf{r}) \end{bmatrix}, \ldots, \begin{bmatrix} \Psi_{eN}(\mathbf{r}) \\ \Psi_{iN}(\mathbf{r}) \end{bmatrix}, \ldots \tag{8.47}$$

whose boundary values and normal derivatives satisfy

$$E_N \equiv \frac{\left\| \Psi_{eN} - \Psi_{iN} - f \right\|^2}{\|f\|^2} + \frac{\left\| \partial \Psi_{eN}/\partial \nu - \partial \Psi_{iN}/\partial \nu - g \right\|}{\|g\|^2} \to 0 \quad (N \to \infty). \tag{8.48}$$

The sequence (8.47), hence, converges to the true solutions of the problem uniformly in wider sense in S_e and S_i:

$$\Psi_{pN}(\mathbf{r}) \to \Psi_{p}(\mathbf{r}) \quad (N \to \infty;\ \mathrm{p} = 1, 2;\ \text{uniformly in } S_{p0}). \tag{8.49}$$

Members of such a sequence can be found by solving the least-squares problem:

LSP 2: E-wave, dielectric. Find the modal coefficients $\{A_{pm}(M) : m = 0, \pm 1, \ldots, \pm N\}$ (p = e, i) that minimize the normalized mean-square error E_N defined in (8.48).

It is worth to note the following matters: Because the convergence in (8.48) is a consequence of completeness of the four sets of boundary functions in a product space $\mathbf{H} \times \mathbf{H}$, the choice of denominators in (8.48), $\|f\|^2$ and $\|g\|^2$, is no more than a

[17] We cannot include the derivation of equations from (8.45) through (8.49) because it takes much space. Interested readers can find the details in [35, 43–45]. The paper by Petit and Cadilhac [33] is also helpful.

convention to get non-dimensional quantities or to unify the units.[18] Speaking from a computational point of view, however, the ratio $\|f\|^2/\|g\|^2$ may have an effect on the condition number of **LSP 2** and, sometimes it is effective to introduce a parameter γ $(0 < \gamma < 1)$ to modify the definition of E_N as

$$E_N \equiv \gamma \frac{\left\|\Psi_{eN} - \Psi_{iN} - f\right\|^2}{\|f\|^2} + (1-\gamma)\frac{\left\|\partial\Psi_{eN}/\partial\nu - \partial\Psi_{iN}/\partial\nu - g\right\|^2}{\|g\|^2}. \quad (8.50)$$

The parameter should be determined by optimization to get a permissible condition number.

8.2.4.4 Notes on the Method of Numerical Computation

In above formulation we have $2M = 2(2N+1)$ unknowns. If we apply the rule in Sect. 8.3.1 (and also in Sect. 8.2.3.1), we need $2 \times 2M = 4M$ linear equations. The number of sampling points required for the $4M$ equations, however, is $2M$ again. This is because we have two equations at each sampling point: the first and the second equation of (8.42).

Let us follow the method of discretization in Sect. 8.2.3.1. Locating $J\ (= M = 2(2N+1))$ sampling points on C, we define J-dimensional vectors

$$\begin{cases} \mathbf{f} = [f(s_1)\ f(s_2)\ \cdots\ f(s_J)]^{\mathrm{T}}, \\ \mathbf{g} = [g(s_1)\ g(s_2)\ \cdots\ g(s_J)]^{\mathrm{T}}, \end{cases} \quad (8.51)$$

$$\begin{cases} \boldsymbol{\varphi}_{em} = [\varphi_{em}(s_1)\ \varphi_{em}(s_2)\ \cdots\ \varphi_{em}(s_J)]^{\mathrm{T}}, \\ \boldsymbol{\varphi}_{im} = [\varphi_{im}(s_1)\ \varphi_{im}(s_2)\ \cdots\ \varphi_{im}(s_J)]^{\mathrm{T}} \end{cases} \quad (8.52)$$

and

$$\begin{cases} \partial_\nu\boldsymbol{\varphi}_{em} = [\partial_\nu\varphi_{em}(s_1)\ \partial_\nu\varphi_{em}(s_2)\ \cdots\ \partial_\nu\varphi_{em}(s_J)]^{\mathrm{T}}, \\ \partial_\nu\boldsymbol{\varphi}_{im} = [\partial_\nu\varphi_{im}(s_1)\ \partial_\nu\varphi_{im}(s_2)\ \cdots\ \partial_\nu\varphi_{im}(s_J)]^{\mathrm{T}}, \end{cases} \quad (8.53)$$

where the mode-number m runs from $-N$ to N.

Next, we construct four $J \times M$ matrices

$$\begin{cases} \Phi_{11} = \left[\boldsymbol{\varphi}_{e,-N}\ \boldsymbol{\varphi}_{e,-N+1}\ \cdots\ \boldsymbol{\varphi}_{e,N}\right], \\ \Phi_{12} = \left[\boldsymbol{\varphi}_{i,-N}\ \boldsymbol{\varphi}_{i,-N+1}\ \cdots\ \boldsymbol{\varphi}_{i,N}\right] \end{cases} \quad (8.54)$$

and

$$\begin{cases} \Phi_{21} = \left[\partial_\nu\boldsymbol{\varphi}_{e,-N}\ \partial_\nu\boldsymbol{\varphi}_{e,-N+1}\ \cdots\ \partial_\nu\boldsymbol{\varphi}_{e,N}\right], \\ \Phi_{22} = \left[\partial_\nu\boldsymbol{\varphi}_{i,-N}\ \partial_\nu\boldsymbol{\varphi}_{i,-N+1}\ \cdots\ \partial_\nu\boldsymbol{\varphi}_{i,N}\right]. \end{cases} \quad (8.55)$$

[18] The use of intrinsic impedance is also possible and is widely employed. That is: find the coefficients by minimization of $|\text{error in } \mathbf{E}|^2 + Z_0^2|\text{error in } \mathbf{H}|^2$. Here, Z_0 is the intrinsic impedance of vacuum or surrounding material. We use this formulation in Appendix 3.

Arranging the four matrices, we get a $2J \times 2M$ Jacobian matrix

$$\Phi = \begin{bmatrix} p\Phi_{11} & p\Phi_{12} \\ q\Phi_{21} & q\Phi_{22} \end{bmatrix}. \tag{8.56}$$

Here,

$$p = \frac{\gamma}{\mathbf{f}^{\dagger}\mathbf{f}}, \quad q = \frac{1-\gamma}{\mathbf{g}^{\dagger}\mathbf{g}} \tag{8.57}$$

are normalizing constants with the parameter γ appeared in (8.50).[19] Finally, defining a $2M$-dimensional solution vector

$$\mathbf{A} = \begin{bmatrix} \mathbf{A}_{\mathrm{e}} \\ \mathbf{A}_{\mathrm{i}} \end{bmatrix}, \tag{8.58}$$

where

$$\begin{cases} \mathbf{A}_{\mathrm{e}} = \left[A_{\mathrm{e},-N}(M)\ A_{\mathrm{e},-N+1}(M)\ \cdots\ A_{\mathrm{e},N}(M)\right]^{\mathrm{T}}, \\ \mathbf{A}_{\mathrm{i}} = \left[A_{\mathrm{i},-N}(M)\ A_{\mathrm{i},-N+1}(M)\ \cdots\ A_{\mathrm{i},N}(M)\right]^{\mathrm{T}} \end{cases} \tag{8.59}$$

are M $(= 2N + 1)$ dimensional column vectors. Thus, we can state a discretized problem as:

DLSP 2: E-wave, dielectric. Find the solution vector $\mathbf{A}$ that minimizes the discretized form of normalized boundary residual

$$E_{NJ} = \left\| \Phi\mathbf{A} - \begin{bmatrix} p\mathbf{f} \\ q\,\mathbf{g} \end{bmatrix} \right\|^2 = \left\| \begin{matrix} p\Phi_{11}\mathbf{A}_{\mathrm{e}} + p\Phi_{12}\mathbf{A}_{\mathrm{i}} - p\mathbf{f} \\ q\Phi_{21}\mathbf{A}_{\mathrm{e}} + q\Phi_{22}\mathbf{A}_{\mathrm{i}} - q\mathbf{g} \end{matrix} \right\|^2. \tag{8.60}$$

8.2.5 Application to Gratings

Here we consider the problem of plane-wave diffraction by a grating and state the points of difference from scattering by a cylindrical obstacle. The book edited by Petit [32] includes a nice introduction to Yasuura's method applied to grating problems as of late 70s.

8.2.5.1 Diffraction by a PC Grating

Figure 8.3 shows the cross section of a grating, an incident wave, and the system of coordinates. The cross section C is periodic in X with a period d and the surface is

[19] If the compensation by γ is not necessary, we can set $p = 1/\mathbf{f}^{\dagger}\mathbf{f}$ and $q = 1/\mathbf{g}^{\dagger}\mathbf{g}$ or use the intrinsic impedance.

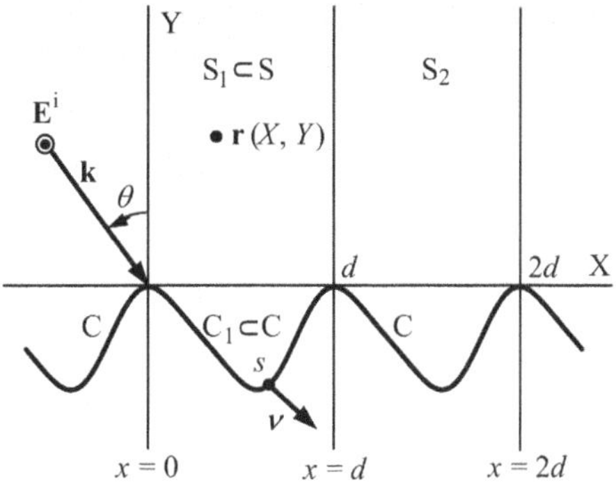

Fig. 8.3 Diffraction by a PC grating

uniform in Z. The semi-infinite region S over C is a vacuum and the region below C is occupied by a PC. We assume C is represented by a single-valued smooth function

$$\mathrm{C}:\ y = \eta(x), \tag{8.61}$$

where $\eta(x)$ is periodic in x, $\eta(x + d) = \eta(x)$, and (x, y) denotes a point on C.

Let an electromagnetic wave having an electric field

$$\mathbf{u}_Z F(\mathbf{r}) = \mathbf{u}_Z \exp(-ikX \sin\theta + ikY \cos\theta) \tag{8.62}$$

is incident on the grating. This case of polarization is termed E-wave, TE wave, or s-polarization.[20] Here, $\mathbf{r} = (X, Y)$ is a point in S, $\mathbf{u}_Z$ is a unit vector in Z, and θ is the angle of incidence shown in Fig. 8.3. The diffracted electric field has only a Z-component, which we describe by $\Psi(\mathbf{r})$. $\Psi(\mathbf{r})$ is the solution of the following problem.

Problem 3 E-wave, PC grating. Find $\Psi(\mathbf{r})$ that satisfies the conditions below:

(GD1) The 2-D Helmholtz equation in S;
(GD2) A radiation condition in Y that $\Psi(\mathbf{r})$ propagates or attenuates in positive Y;
(GD3) A periodicity condition

$$\Psi(X + d, Y) = \exp(-ikd \sin\theta)\, \Psi(X, Y); \tag{8.63}$$

(GD4) The boundary condition

$$\Psi(x, \eta(x)) = f(x) \equiv -F(x, \eta(x)). \tag{8.64}$$

[20] s stands for senkrecht (German), which means the electric field is perpendicular to the plane of incidence, the plane spanned by $\mathbf{u}_Y$ (grating normal) and the incident wavevector.

Conditions (**GD1**) and (**GD4**) are common to the case of cylindrical obstacle in Sect. 8.2.1, while (**GD2**) is quite different from (**D2**) and (**GD3**) is a new requirement. These differences come from the pseudo-periodic nature of the problem. Because a grating has a periodic structure, and because we have assumed a plane-wave incidence, the phenomena at (X, Y) and $(X + d, Y)$ are almost the same; the only discrepancy can be seen in the phase difference (8.63). Hence, if we divide S by vertical lines $X = 0, \pm d, \pm 2d, \ldots$ as shown in Fig. 8.3, the diffracted fields in neighboring strip regions are the same except for the phase shift. This is a characteristic feature of a grating problem called quasi- or pseudo-periodicity and explains why the 1-D radiation condition appears in (**GD2**). In solving a grating problem, hence, we can assume that the observation point $\mathbf{r} = (X, Y)$ is inside the first strip region S_1 $(0 < X \leq d;\ Y \geq \eta(X))$ shown in Fig. 8.3.

8.2.5.2 Modal Functions, Approximate Solution, and Key Points in the Solution Method

Here again we choose separated solutions as modal functions. The separated solutions satisfying the periodicity are known as Floquet modes. We take the Floquet modes satisfying the radiation condition (**GD2**)

$$\varphi_m(\mathbf{r}) = \exp(-\mathrm{i}\alpha_m X - \mathrm{i}\beta_m Y) \quad (m = 0, \pm 1, \pm 2, \ldots) \tag{8.65}$$

as the set of modal functions, where

$$\alpha_m = k \sin\theta + \frac{2m\pi}{d}, \quad \beta_m = \sqrt{k^2 - \alpha_m^2} \quad (\mathrm{Re}\,\beta_m \geq 0,\ \mathrm{Im}\,\beta_m \leq 0). \tag{8.66}$$

The term $k \sin\theta$ in α_m is for the periodicity, the definition of β_m implies the Helmholtz equation, and the sign of β_m (positive or negative imaginary) is for the radiation condition.

We construct an approximate solution following the way we took in Sect. 8.2.2.2:

$$\Psi_N(\mathbf{r}) = \sum_{m=-N}^{N} A_m^{\mathrm{E}}(M)\,\varphi_m(\mathbf{r}). \tag{8.67}$$

This solution satisfies conditions (**GD1**), (**GD2**), and (**GD3**). Hence, the A_m^{E} coefficients[21] should be determined so that the solution satisfies the boundary condition (**GD4**) approximately. Let us see briefly the least-squares boundary matching works to yield a sequence of solutions converging to the true solution.

Some analysis starting from an assumption that $\mathbf{r}$ is inside a closed region S_{10} $(\subset S_1)$ leads us to an estimation

[21] The superscript E denotes that the coefficients concern the E-wave. Later we will also use the superscripts H, TE, and TM in accordance with polarizations.

$$|\Psi_N(\mathbf{r}) - \Psi(\mathbf{r})| \leq G(\mathrm{S}_{10})\|\tilde{\Psi}_N - \tilde{f}\| \quad (\mathbf{r} \in \mathrm{S}_{10} \subset \mathrm{S}_1). \tag{8.68}$$

Here, G is a positive constant depending on the closed region S_{10} and the quantities with tildes, e.g. $\tilde{f}$, mean periodic functions derived from the pseudo-periodic functions:

$$\tilde{f}(s) = \exp(\mathrm{i}\alpha_0 x) f(x, y) = -\exp(-\mathrm{i}\beta_0 y), \tag{8.69}$$

$$\tilde{\Psi}_N(s) = \exp(\mathrm{i}\alpha_0 x)\Psi_N(x, y) = \sum_{m=-N}^{N} A_m^{\mathrm{E}}(M)\,\tilde{\varphi}_m(x, y), \tag{8.70}$$

and

$$\tilde{\varphi}_m(x, y) = \exp(\mathrm{i}\alpha_0 x)\varphi_m(x, y) = \exp\left(-\frac{2m\pi \mathrm{i}x}{d} - \mathrm{i}\beta_m y\right). \tag{8.71}$$

The norm of a function $g(s)$ defined on C_1, the first period of C, is defined by

$$\|g\| = \left[\int_{s=0}^{C} |g(s)|^2 ds\right]^{1/2}, \tag{8.72}$$

where C denotes the length of C_1. Thus we have a least-squares problem:

LSP 3: E-wave, PC grating. Find the A_m^{E} coefficients that minimize the numerator of the normalized mean-square error

$$E_N = \frac{\left\|\tilde{\Psi}_N - \tilde{f}\right\|^2}{\left\|\tilde{f}\right\|^2}. \tag{8.73}$$

The modification of the boundary values to define the periodic functions is the key point in the solution of grating problems. Introducing the modification, we can establish a correspondence between one period of the grating surface C_1 and the cross section of a cylindrical obstacle C in Sect. 8.2.1.[22] The method of numerical solution for **LSP 3** is similar to that in Sect. 8.2.3. To solve the problem of diffraction by a grating made of dielectric or metal we can combine the method in this section with that in Sect. 8.2.4. Guidance to the problem of conical diffraction can be found in Appendix 3.

[22] If we employ the SP, this correspondence is essentially important because we need periodicity of the functions defined on the boundary. In using Yasuura's method without the SP, we can say the following points: (**1**) If we get the solution through the NE, this modification is not necessary because it is done automatically in calculating the inner products; (**2**) While if we employ the QRD or SVD: (**2.i**) The modification may accelerate the convergence of the solutions because the target function and the modal functions are periodically continuous after modification; (**2.ii**) And, a quadrature by parts (or rectangular-rule) approximation is equivalent to a trapezoidal-rule in numerical integrations.

8.3 Numerical Examples

In this section we show some results of numerical computations obtained by the methods in the last section. First, we examine the nature of the Jacobian matrices taking grating problems as examples to show the validity of the experimental rule (8.31). Meanwhile we add some comments that are useful in applying the method. Then we give the results of four problems of practical interest.

8.3.1 Rule on the Number of Sampling Points

We have solved the problem of diffraction by a grating made of PC and by one made of BK7 optical glass varying the number of sampling points or of linear equations. The results support our experimental rule. In addition, we have made a comparison between the two methods of locating the sampling points, (8.23) and (8.24), introduced in Sect. 8.2.3.1 and found little difference in the rage $J \geq 2M$ for the problem parameters employed in numerical analysis.

8.3.1.1 A PC Grating

We consider the grating shown in Fig. 8.3 and assume that the cross section C is given by[23]

$$\mathrm{C} : y = H\left(\cos\frac{2\pi x}{d} - 1\right). \tag{8.74}$$

We assume also that an E- or H-polarized plane wave is incident at $\theta = 0$ (normal incidence). Other physical parameters are: $d = 556\,\mathrm{nm}$, $H/d = 0.15$, and $\lambda = 500\,\mathrm{nm}$.[24] The computational parameters are: the number of truncation $N = 20$; the total number of modal functions $M = 41$; and the number of sampling points J is in the range $M \leq J \leq 4M$. This means that the number of unknown coefficients is M and the number of linear equations is between M and $4M$.

The first example, Fig. 8.4, shows the convergence of the solution and related parameters in the E-wave. The curves in Fig. 8.4a includes the maximum and minimum singular value, the condition number cond(Φ), $E_{20\,J}$ of (8.30), and an error on the power balance

$$e_{NJ} = 1 - \sum_{\text{prop}} \rho_m = 1 - \sum_{\beta_m > 0} \frac{\beta_m}{\beta_0} |A_m^{\mathrm{E}}(M, J)|^2, \tag{8.75}$$

---

[23] Note that the bias setting (we used -1 here) has an effect on the accuracy of numerical computation when the grating is deep.

[24] Although the use of normalization by wavelength (i.e., $kd = 2\pi d/\lambda$ etc.) is convenient in handling a problem with a PC obstacle, we employ real length here.

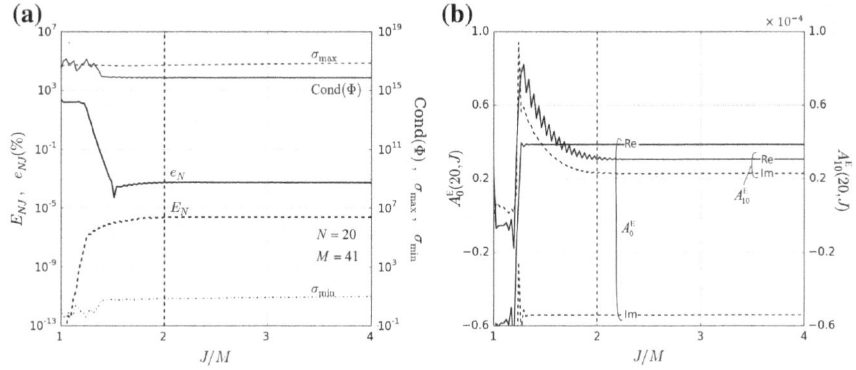

Fig. 8.4 Convergence of the solutions with respect to J for a PC grating in the E-wave. $\theta = 0$, $d = 556\,\text{nm}$, $H/d = 0.15$, and $\lambda = 500\,\text{nm}$: **a** Cond(Φ) and errors; **b** $A_0^{\text{E}}(20, J)$ and $A_{10}^{\text{E}}(20, J)$

where $\sum_{\text{prop}}$ and $\sum_{\beta_m>0}$ mean the summation in respect to the propagating orders.[25] We observe these quantities are approaching final values with increasing J; and have converged for $J \geq 2M$.[26] Figure 8.4b illustrates the convergence of $A_0^{\text{E}}(20, J)$ and $A_{10}^{\text{E}}(20, J)$ coefficients. The former has converged before reaching $J = 2M$; while the latter is with small ripples until $J = 2.2M$. We, however, can neglect this oscillation in finding the diffracted wave because the mode with $m = 10$ is evanescent and cannot be observed at a point apart from the grating surface.

The second set of figures, Fig. 8.5, displays the same thing for the H-wave. The curves in Fig. 8.5a show the max and min singular value, cond(Φ), $E_{20\,J}$, and $e_{20\,J}$. While in Fig. 8.5b we show the convergence of $A_0^{\text{H}}(20, J)$ and $A_{10}^{\text{H}}(20, J)$. We observe all the quantities have converged substantially in the range $J \geq 2M$.

The third example, Fig. 8.6, shows the convergence of solutions: N dependence of the normalized mean-square error E_N and energy error e_N of E- and H-wave solutions. The rule $J = 2M$ is applied. Because the surface modulation is moderate in the problem, we get precise solutions with 10^{-6} or 10^{-4} percent energy error easily for both E- and H-wave problem.[27] It is worth to mention that a modal coefficient—e.g. $A_m^{\text{E}}(M)$, as a function of M, converges to a final value: $A_m^{\text{E}}(M) \to A_m^{\text{E}}$ $(M \to \infty)$. This convergence, however, is not uniform with respect to m.

---

[25] ρ_m is referred to as the (reflection) efficiency of the mth order.

[26] When the number of truncation is small (e.g., $N \leq 10$), we sometimes observe a phenomenon that the condition number continues to decrease slightly beyond $J = 2M$ due to tiny increment of $\sigma_{\min}$.

[27] We should notice, however, that the accuracy of an H-wave solution is lower than that of an E-wave solution by one or two digits. This is observed generally; and was Yasuura's motivation of introducing the SP. His idea came from the fact that a Neumann problem for an electrostatic potential is equivalent to a Dirichlet problem for a stream function. The prototype of the SP, hence, was called *an algorithm using the stream function in a wave field*.

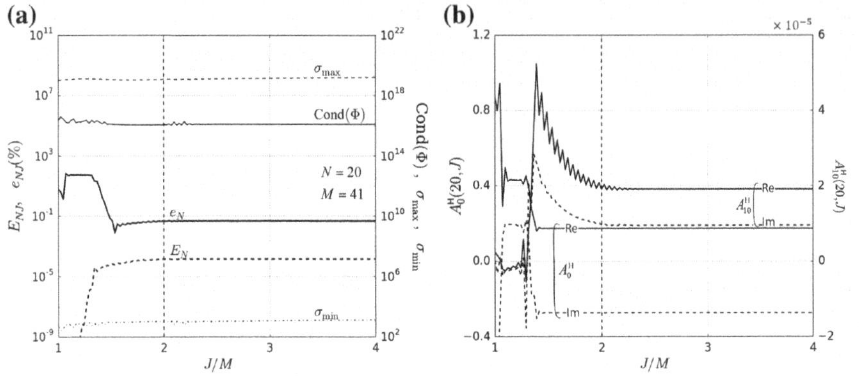

Fig. 8.5 Convergence of the solutions with respect to J for a PC grating in the H-wave. Optical parameters are the same as those in Fig. 8.4: **a** Cond(Φ) and errors; **b** $A_0^{\mathrm{H}}(20, J)$ and $A_{10}^{\mathrm{H}}(20, J)$

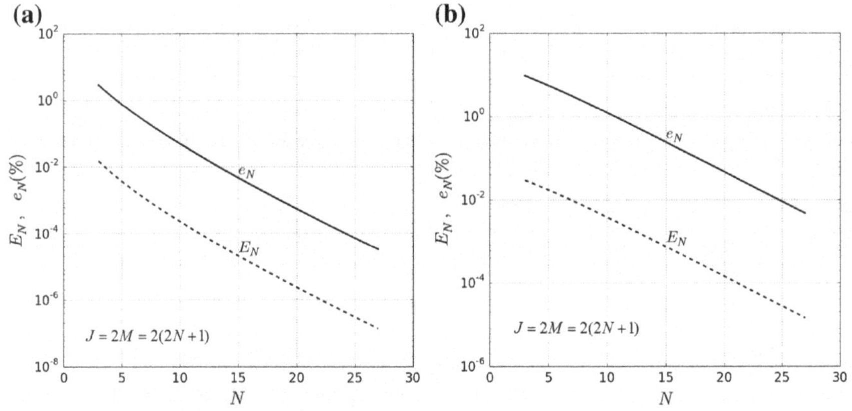

Fig. 8.6 Convergence of the solutions with respect to N for a PC grating. Optical parameters are the same as those in Fig. 8.4: **a** E-wave; **b** H-wave

8.3.1.2 A BK7 Optical Glass Grating

Here we examine the case of a dielectric grating made of an optical glass BK7 whose refractive index is 1.5139 [1]. Other parameters are the same as in Sect. 8.3.1.1. In the present problem we have transmitted fields ($\mathbf{E}^{\mathrm{t}}$ and $\mathbf{H}^{\mathrm{t}}$) in the region V_2 below the grating surface in addition to the reflected fields ($\mathbf{E}^{\mathrm{r}}$ and $\mathbf{H}^{\mathrm{r}}$) over the grating V_1. We, hence, define approximate solutions following (8.44) in Sect. 8.2.4. That is, we employ Floquet modes in V_1 and V_2 and construct approximations of leading fields in each region in the form of finite linear combinations of the Floquet modes. Let the number of truncation be N. Then, we have $2(2N + 1) = 2M$ unknown coefficients in total.

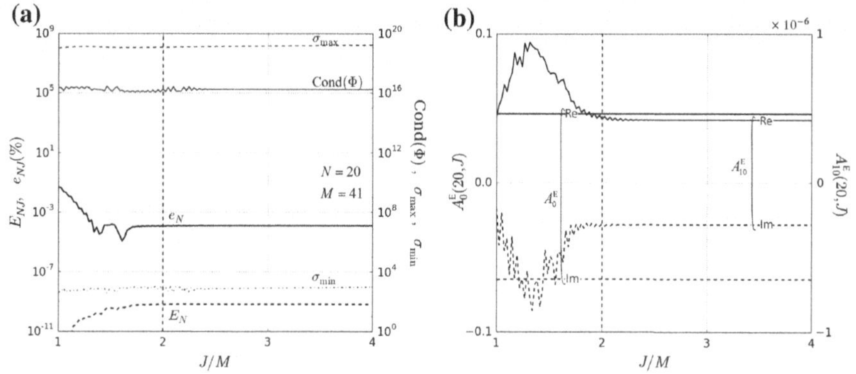

Fig. 8.7 Convergence of the solutions with respect to J for a BK7 grating in the E-wave. $\theta = 0$, $d = 556\,\text{nm}$, $H/d = 0.15$, and $\lambda = 500\,\text{nm}$: **a** Cond(Φ) and errors; **b** $A_0^{\text{E}}(20, J)$ and $A_{10}^{\text{E}}(20, J)$

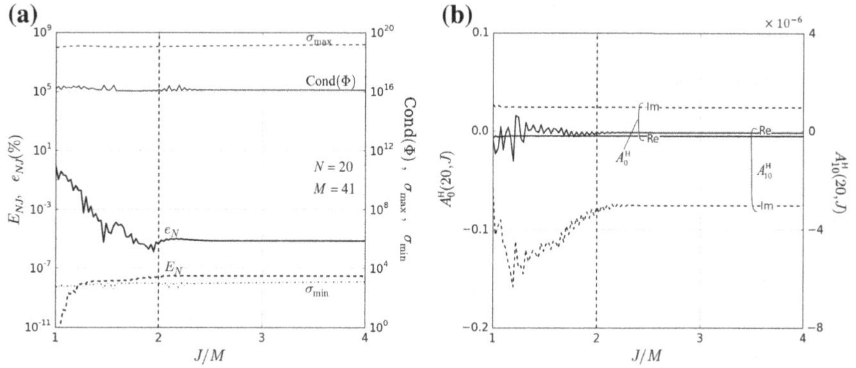

Fig. 8.8 Convergence of the solutions with respect to J for a BK7 grating in the H-wave. Optical parameters are the same as those in Fig. 8.7: **a** Cond(Φ) and errors; **b** $A_0^{\text{H}}(20, J)$ and $A_{10}^{\text{H}}(20, J)$

Figure 8.7 shows the convergence of the solution and related parameters as functions of the number of sampling points J in the E-wave and $N = 20$. We observe that all the errors and the parameters have converged in the range $J \geq 2M = 2(2N + 1)$ except for small ripples. Figure 8.8 shows the same thing in the H-wave. This means that the number of linear equations in the least-squares problem can be twice as many as the number of unknowns ($2 \times 2M = 4(2N + 1)$; see Sect. 8.2.4.4).

Figure 8.9 illustrates the N dependence of the errors of the solutions. We get precise solutions with 10^{-5} percent energy error easily on a personal computer.

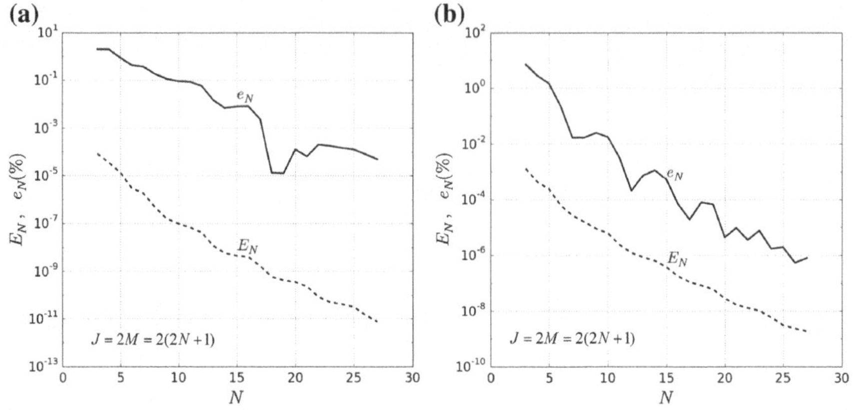

Fig. 8.9 Convergence of the solutions with respect to N for a BK7 grating. Optical parameters are the same as those in Fig. 8.7: **a** E-wave; **b** H-wave

8.3.2 Scattering by Relatively Deep Gratings

Yasuura's method, when combined with the partition of the groove region, can solve the problem of diffraction from a deep grating with a depth-to-period ratio beyond unity. In the conventional Yasuura's method without partition, this ratio is said to be about 0.5 and a little less than 0.4 in the E- and H-wave cases, respectively. In the present subsection, some numerical results are given for the scattering by relatively deep gratings using a combination of up-and down-going Floquet modal functions [22].

The period and height of the sinusoidal profile are d and $2H$, respectively, as shown in Fig. 8.10. At first we deal with a perfectly conducting grating as a fundamental problem where the electromagnetic fields exist only in the vacuum region. The semi-infinite region over the grating surface is divided into an upper half plane U_0 and a groove region a fictitious boundary (a horizontal line). The latter is further divided into shallow horizontal layers $\mathrm{U}_1, \mathrm{U}_2, \ldots, \mathrm{U}_Q$ again by fictitious boundaries.

An approximate solution in U_0, that is $\Psi_{0N}(\mathbf{r})$, is defined in a usual manner as (8.67), while the solutions in U_q $(q = 1, 2, \ldots, Q)$ include not only the up-going but also the down-going modal functions as

$$\Psi_{qN}(\mathbf{r}) = \sum_{m=-N}^{N} \left[A_{qm}^{+}(N)\,\varphi_m^{+}(\mathbf{r} - \mathbf{u}_Y y_q) + A_{qm}^{-}(N)\,\varphi_m^{-}(\mathbf{r} - \mathbf{u}_Y y_{q-1})\right], \quad (8.76)$$

where $\varphi_m^{\pm}(\mathbf{r}) = \exp(\mathrm{i}\alpha_m X \pm \mathrm{i}\beta_m Y)$, and the plane $Y = y_q$ is the boundary between U_q and U_{q+1}. Thus the total number of unknown coefficients is $(2N+1)(2Q+1)$. These coefficients should be determined in order that the solutions meet the boundary condition (**GD4**) and an additional set of boundary conditions on the Q fictitious boundaries:

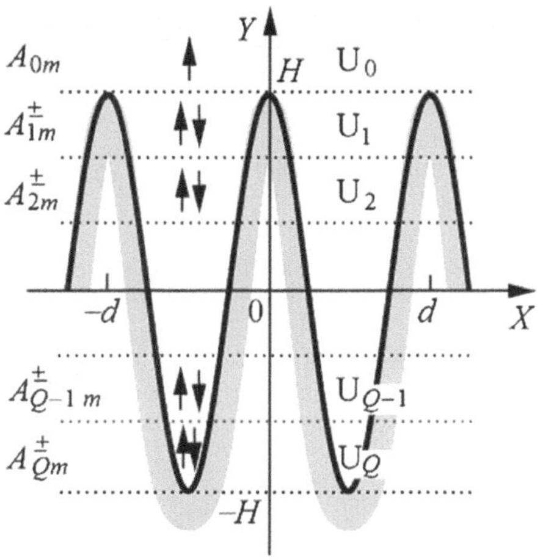

Fig. 8.10 Cross section of a perfectly conducting (PC) or dielectric sinusoidal grating. In the case of PC grating, partition into $Q+1$ layers is done only in the vacuum region

$$\begin{cases} \left(F\delta_{q0} + \Psi_q\right)\big|_{Y=y_q+0} = \Psi_{q+1}\big|_{Y=y_q-0}, \\ \left.\dfrac{\partial\left(F\delta_{q0} + \Psi_q\right)}{\partial Y}\right|_{Y=y_q-0} = \left.\dfrac{\partial \Psi_{q+1}}{\partial Y}\right|_{Y=y_q-0}, \end{cases} \tag{8.77}$$

where δ_{q0} is Kronecker's delta. The mean-square error is defined in the same form as (8.73), but the integration range in the norm (8.72) must include not only the grating surface but also the fictitious boundaries.

Let us check the convergence of the results obtained by the present method. Figure 8.11 shows the variation of the normalized mean-square error and the energy error as functions of the number of truncation N for both E- and H-wave incidence. As is observed in these figures, the mean-square error decreases as N increases. An approximate solution with 0.1 percent energy error is accomplished at $N = 14$ for an E-wave. In the H-wave case convergence of solutions is not so fast as in the E-wave case. We attain to one percent energy error at $N = 23$ in that case of polarization.

Figure 8.12 shows comparison of reflection efficiency for a perfectly conducting grating as functions of the incident angle at E-wave incidence. The numbers (N, Q) are (15, 4), (15, 5), and (30, 20) as $H/d = 0.31$, 0.4, and 1.066, respectively. The curves and symbols represent the present results and the results by the integral equation method [46]. We find good agreement between the results. For dielectric gratings, partition must be made not only in the vacuum region but also in the dielectric one. As a result, numbers of unknown modal coefficients and boundary conditions become doubled compared with the previous case.

Figure 8.13 shows comparison of transmission efficiency for a dielectric grating as functions of the incident angle at H-wave incidence. The numbers (N, Q) are (11,4). The curves and symbols represent the present results and the results by the finite element method [20]. We find that the results agree with each other except for the grazing limit.

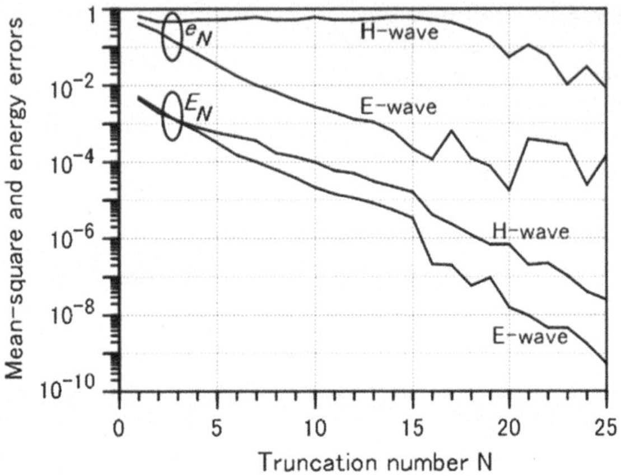

Fig. 8.11 Normalized mean-square error E_N and energy error e_N as functions of the truncation number N for a PC grating with $2H/d = 1$, $d/\lambda = 0.5$, and $\theta = 30°$

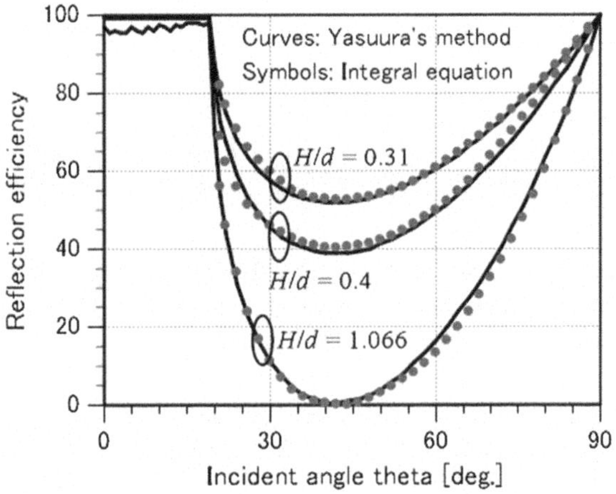

Fig. 8.12 Reflection efficiency in percent for a PC grating with $d/\lambda = 0.75$ at E-wave incidence. Comparison with the integral equation method [46]

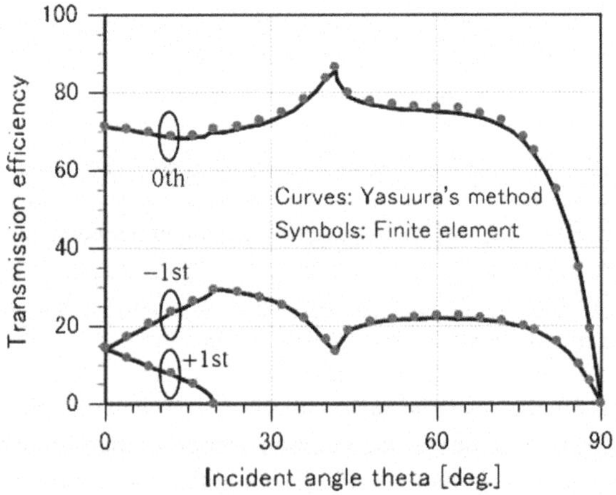

Fig. 8.13 Transmission efficiency in percent for a dielectric grating with $\varepsilon_r = 4$, $2H/d = 1$, and $d/\lambda = 0.6$ at H-wave incidence. Comparison with the finite element method [20]

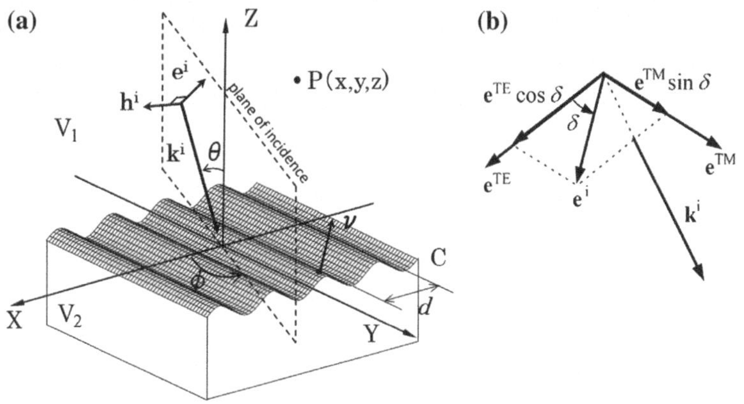

Fig. 8.14 Diffraction by a dielectric grating: **a** Conical mounting; **b** Definition of δ

Although there are a couple of methods that are capable of solving the problems of extremely deep gratings, the present results make sense because they show a limit of a conventional modal-expansion approach when using the Floquet modes as basis functions.

8.3.3 Plasmon Surface Waves Excited on a Metal Grating Placed in Conical Mounting

We show some numerical results in regard to plasmon surface waves excitation on a metal grating placed in conical mounting [29]. Conical mounting is an optical arrangement in which the plane of incidence is not perpendicular to grooves of a grating as shown in Fig. 8.14. Readers can find detailed description of problem in Appendix 3. We here illustrate the results obtained by the method explained there.

We deal with a sinusoidal silver grating whose surface profile is given by $z = H\sin(2\pi x/d)$. The upper region V_1 over the grating surface is assumed to be vacuum with a refractive index $n_1 = 1$ and the grating is made of silver with a complex refractive index n_2. As an incident light we consider an electromagnetic plane wave, which is specified by the wavenumber in vacuum ($k = 2\pi/\lambda$), the polar angle (θ) between the wavevector and the grating normal, and the azimuthal angle (ϕ) between the X axis and the plane of incidence.

The diffracted fields in the conical mounting are decomposed into a TE and a TM component which mean that the relevant electric and magnetic field are perpendicular to the plane of incidence. The efficiency of the mth-order diffracted mode in V_1, hence, is represented as $\rho_m = \rho_m^{\mathrm{TE}} + \rho_m^{\mathrm{TM}}$. Here, ρ_m^{TE} or ρ_m^{TM} is the efficiency of the

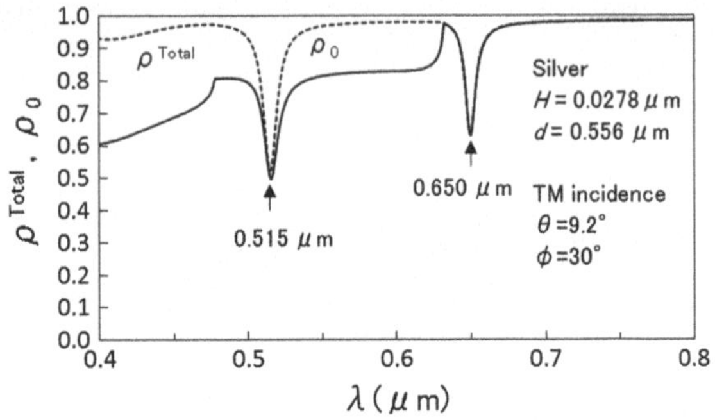

Fig. 8.15 ρ_0 and ρ^{Total} as functions of wavelength λ

TE- or TM-component of the mth-order diffracted mode.[28] In the numerical examples below we deal with a shallow grating made of silver with a period $d = 0.556\,\mu\text{m}$ and an amplitude $H = 0.0278\,\mu\text{m}$. Yasuura's method provides sufficiently reliable results for the problem of such a grating at the truncation number of the approximate solutions $N = 10$.

Figure 8.15 shows the efficiency of the 0th-order diffracted mode ρ_0 and the total diffraction efficiency ρ^{Total} as functions of wavelength λ.[29] The incident light is in the TM incidence—a polarization angle $\delta = \pi/2$ in (8.114) of Appendix 3—where the magnetic field is perpendicular to the plane of incidence. The polar angle and the azimuthal angle are chosen as $\theta = 9.2°$ and $\phi = 30°$. As a complex refractive index of silver n we take the interpolated values for the experimental data in the literature [8]. In the figure we observe partial absorption of incident light at $\lambda = 0.515\,\mu\text{m}$ and $\lambda = 0.650\,\mu\text{m}$ as dips in the total efficiency curve.[30] As we will see later, the dips are associated with plasmon resonance absorption, which is caused by coupling of surface plasmons with an evanescent mode diffracted by the grating [21, 34].

Figure 8.16 shows the 0th-order efficiency ρ_0 and the TE and TM component ρ_0^{TM} and ρ_0^{TE} as functions of θ with a fixed azimuthal angle $\phi = 30°$. The wavelength is chosen as $\lambda = 0.650\,\mu\text{m}$ and a refractive index is $n_2 = 0.07 - \text{i}4.2$. Remaining parameters are the same as those of Fig. 8.15. We observe in Fig. 8.16 partial absorption of incident light at $\theta = 9.2°$, we call it a resonance angle, as a dip in the ρ_0 curve. In addition we notice that ρ_0^{TM} takes a minimal value at the resonance angle, but ρ_0^{TE} increases there to the contrary. This illustrates the enhancement of TM-TE mode conversion [5] that a TM component of the incident light is strongly converted into a TE component of the 0th-order diffracted light when plasmon resonance absorption occurs in a metal grating in conical mounting.

[28]The efficiencies are given by $\rho_m^{\text{TE}} = (\gamma_{1m}/\gamma_{10})\,|A_{1m}^{\text{TE}}|^2$ and $\rho_m^{\text{TM}} = (\gamma_{1m}/\gamma_{10})\,|A_{1m}^{\text{TM}}|^2$ where γ_{1m} is the propagation constant in the Z-direction of the mth-order propagating mode ($\text{Re}\,(\gamma_{1m}) \geq 0$) concerning the upper region V_1, and A_{1m}^{TE} and A_{1m}^{TM} are the expansion coefficients of the approximate solutions defined in (8.120) of Appendix 3.

[29]ρ^{Total} is a summation of ρ_m over the propagating orders.

[30]$1 - \rho^{\text{Total}}$ represents the ratio of the absorbed light power by a metal grating to the incident light power.

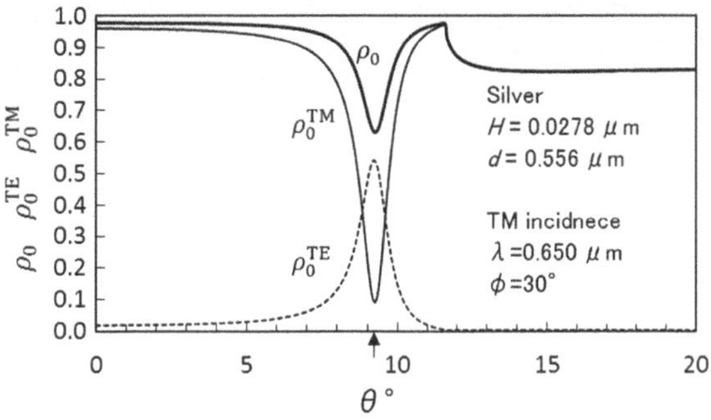

Fig. 8.16 ρ_0, ρ_0^{TE} and ρ_0^{TM} as functions of θ

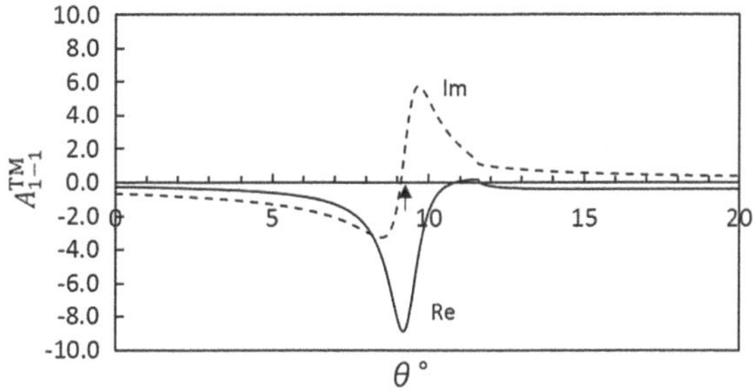

Fig. 8.17 A_{1-1}^{TM} as functions of θ

In Fig. 8.17 we show the expansion coefficient of the -1st-order TM vector modal function A_{1-1}^{TM} defined in (8.120) of Appendix 3 as a function of θ. The parameters in the figure are the same as those of Fig. 8.16 where the -1st-order mode is evanescent. The solid curve in Fig. 8.17 represents the real part of the expansion coefficient and the dashed curve is the imaginary part. From this result we observe the resonance property of the expansion coefficient A_{1-1}^{TM} at the angle of incidence $\theta = 9.2°$ and confirm that the TM component of the -1st-order evanescent mode couples with surface plasmons at the resonance angle. We thus demonstrate that plasmon resonance absorption is associated with coupling of surface plasmons with an evanescent mode diffracted by a metal grating.

We note that the excitation of surface plasmons is largely affected by the azimuthal angle ϕ. Figure 8.18 shows the plasmon resonance absorption for several ϕ's under the same parameters as those of Fig. 8.16. We observe that the resonance angle varies with ϕ as shown in Fig. 8.18. This means direction of propagation depends on ϕ, the direction in which the plasmon surface wave propagates. The azimuthal angle ϕ has also large influence on the enhancement of TM-TE mode conversion through plasmon resonance absorption. For example, a TM component of the 0th-order diffracted mode almost vanishes at the resonance absorption at $\phi = 45°$, but a TE component becomes to be 0.7 there.

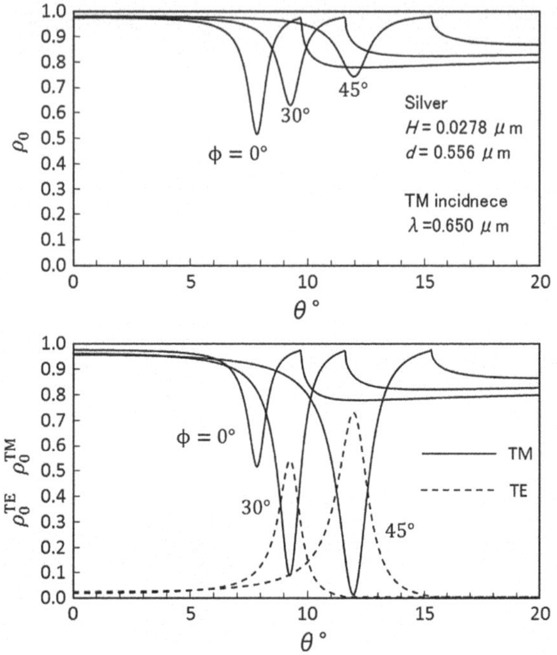

Fig. 8.18 Dependence of TE-TM mode conversion on ϕ

8.3.4 Scattering by a Metal Bigrating

In this subsection we deal with a 3-D problem: diffraction by a metal bigrating whose surface profile is periodically corrugated in two directions. We briefly describe the formulation of Yasuura's method for solving the problem by a metal bigrating and then show numerical results of plasmon resonance absorption in the grating [16].

We consider a bisinusoidal metal grating shown in Fig. 8.19. The surface profile of the grating is given by

$$\eta(x, y) = H\left[\sin\left(\frac{2\pi x}{d}\right) + \sin\left(\frac{2\pi y}{d}\right)\right]. \tag{8.78}$$

The upper region V_1 over the grating surface S_0 is vacuum with a refractive index $n_1 = 1$ and the region V_2 below the grating surface consists of a lossy metal with a complex refractive index n_2. The permeability of the metal is assumed to be μ_0.

The incident light is an electromagnetic plane wave

$$\begin{bmatrix} \mathbf{E}^{\mathrm{i}} \\ \mathbf{H}^{\mathrm{i}} \end{bmatrix}(\mathbf{r}) = \begin{bmatrix} \mathbf{e}^{\mathrm{i}} \\ \mathbf{h}^{\mathrm{i}} \end{bmatrix}\exp(-\mathrm{i}\mathbf{k}^{\mathrm{i}} \cdot \mathbf{r}). \tag{8.79}$$

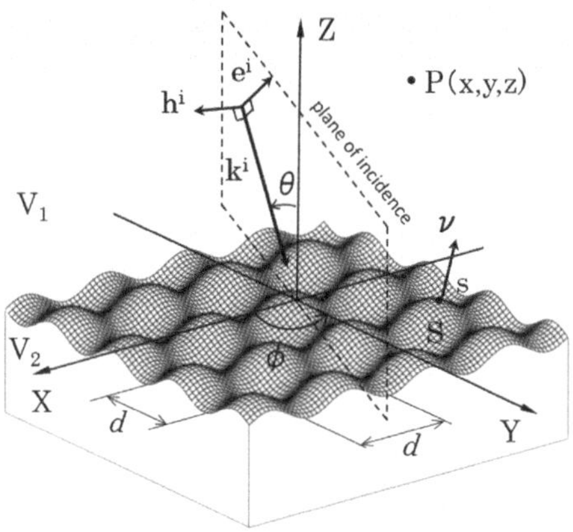

Fig. 8.19 A bisinusoidal grating and coordinate system

Here, $\mathbf{r}$ is the position vector for an observation point, $\mathbf{k}^{\mathrm{i}}$ is the wavevector of the incident wave, and $\mathbf{h}^{\mathrm{i}} = (1/\omega\mu_0)\,\mathbf{k}^{\mathrm{i}} \times \mathbf{e}^{\mathrm{i}}$. The wavevector is given by

$$\mathbf{k}^{\mathrm{i}} = (\alpha, \beta, -\gamma) \tag{8.80}$$

with $\alpha = n_1 k \sin\theta\cos\phi$, $\beta = n_1 k \sin\theta \sin\phi$, and $\gamma = n_1 k \cos\theta$. Here, $k\ (= 2\pi/\lambda)$ is the wavenumber in vacuum, and θ is the polar angle between the Z axis and the incident wavevector, and ϕ is the azimuthal angle between the X axis and the plane of incidence.

We denote the diffracted electric and magnetic fields by $\mathbf{E}^{\mathrm{d}}_{\ell}(\mathrm{P})$, $\mathbf{H}^{\mathrm{d}}_{\ell}(\mathrm{P})$ in the regions V_ℓ $(\ell = 1, 2)$. Here we explain briefly Yasuura's method for finding the diffracted fields. We first introduce TE and TM vector modal functions defined in the region V_ℓ $(\ell = 1, 2)$:

$$\begin{cases} \boldsymbol{\varphi}^{\mathrm{TE}}_{\ell mn}(\mathbf{r}) = \mathbf{e}^{\mathrm{TE}}_{\ell mn} \exp(-\mathrm{i}\mathbf{k}_{\ell mn}\cdot\mathbf{r}), \quad \mathbf{e}^{\mathrm{TE}}_{\ell mn} = \dfrac{\mathbf{k}_{\ell mn}\times\mathbf{u}_Z}{|\mathbf{k}_{\ell mn}\times\mathbf{u}_Z|}, \\ \boldsymbol{\varphi}^{\mathrm{TM}}_{\ell mn}(\mathbf{r}) = \mathbf{e}^{\mathrm{TM}}_{\ell mn} \exp(-\mathrm{i}\mathbf{k}_{\ell mn}\cdot\mathbf{r}), \quad \mathbf{e}^{\mathrm{TM}}_{\ell mn} = \dfrac{\mathbf{e}^{\mathrm{TE}}_{\ell mn}\times\mathbf{k}_{\ell mn}}{|\mathbf{e}^{\mathrm{TE}}_{\ell mn}\times\mathbf{k}_{\ell mn}|} \\ \quad (m,\ n = 0, \pm1, \pm2, \ldots). \end{cases} \tag{8.81}$$

Here, $\mathbf{u}_Z$ is a unit vector in the Z-direction and $\mathbf{k}_{\ell mn}$ $(\ell = 1, 2)$ is the wavevector of the (m, n)th-order diffracted wave:

$$\mathbf{k}_{1mn} = (\alpha_m, \beta_n, \gamma_{1mn}), \quad \mathbf{k}_{2mn} = (\alpha_m, \beta_n, -\gamma_{2mn}) \tag{8.82}$$

with

$$\begin{cases} \alpha_m = \alpha + \dfrac{2m\pi}{d}, \quad \beta_n = \beta + \dfrac{2n\pi}{d}, \quad \gamma_{\ell m} = \sqrt{(n_\ell k)^2 - (\alpha_m^2 + \beta_n^2)} \\ (\operatorname{Re}\gamma_{\ell mn} \ge 0,\ \operatorname{Im}\gamma_{\ell mn} \le 0). \end{cases} \tag{8.83}$$

We form approximate solutions for the diffracted electric and magnetic fields:

$$\begin{bmatrix} \mathbf{E}^{\mathrm{d}}_{\ell N} \\ \mathbf{H}^{\mathrm{d}}_{\ell N} \end{bmatrix}(\mathbf{r}) = \sum_{m,n=-N}^{N} A^{\mathrm{TE}}_{\ell mn} \begin{bmatrix} \boldsymbol{\varphi}^{\mathrm{TE}}_{\ell mn} \\ \boldsymbol{\psi}^{\mathrm{TE}}_{\ell mn} \end{bmatrix}(\mathbf{r}) + \sum_{m,n=-N}^{N} A^{\mathrm{TM}}_{\ell mn} \begin{bmatrix} \boldsymbol{\varphi}^{\mathrm{TM}}_{\ell mn} \\ \boldsymbol{\psi}^{\mathrm{TM}}_{\ell mn} \end{bmatrix}(\mathbf{r}) \quad (\ell = 1, 2) \tag{8.84}$$

with

$$\boldsymbol{\psi}^{\mathrm{q}}_{\ell mn}(\mathbf{r}) = \frac{1}{\omega\mu_0} \mathbf{k}_{\ell \mathrm{mn}} \times \boldsymbol{\varphi}^{\mathrm{q}}_{\ell mn}(\mathbf{r}) \quad (\mathrm{q} = \mathrm{TE}, \mathrm{TM}). \tag{8.85}$$

The expansion coefficients $A^{\mathrm{TE}}_{\ell mn}$, $A^{\mathrm{TM}}_{\ell mn}$ are determined so that the approximate solutions $\mathbf{E}^{\mathrm{d}}_{\ell N}(\mathrm{P})$, $\mathbf{H}^{\mathrm{d}}_{\ell N}(\mathrm{P})$ satisfy the boundary conditions in a weighted least-squares sense. To do this, we minimize the mean-square error

$$\begin{aligned} E_N = &\int_{\mathrm{S}} \left|\boldsymbol{\nu} \times \left(\mathbf{E}^{\mathrm{d}}_{1N} + \mathbf{E}^{\mathrm{i}} - \mathbf{E}^{\mathrm{d}}_{2N}\right)(s)\right|^2 dS \\ &+ Z_0^2 \int_{\mathrm{S}} \left|\boldsymbol{\nu} \times \left(\mathbf{H}^{\mathrm{d}}_{1N} + \mathbf{H}^{\mathrm{i}} - \mathbf{H}^{\mathrm{d}}_{2N}\right)(s)\right|^2 dS, \end{aligned} \tag{8.86}$$

where S is one period cell of the grating surface S_0, $\boldsymbol{\nu}$ is a unit normal vector to the grating surface, and Z_0 is an intrinsic impedance of the medium of V_1.

The mean-square error E_N is discretized by applying a two-dimensional trapezoidal rule where the number of divisions in the X- and Y-directions is chosen to be $J = 2(2N+1)$. The discretized LSP with $24(2N+1)^2 \times 4(2N+1)^2$ Jacobian is solved by QRD.

The diffraction efficiency ρ_{mn} of the (m,n)th-order mode $(\gamma_{1m} \geq 0)$ in V_1 is given by

$$\rho_{mn} = \rho^{\mathrm{TE}}_{mn} + \rho^{\mathrm{TM}}_{mn}, \tag{8.87}$$

where the efficiency of the (m,n)th-order TE or TM mode is given by

$$\rho^{\mathrm{TE}}_{mn} = \frac{\gamma_{1m}}{\gamma} |A^{\mathrm{TE}}_{1mn}|^2, \quad \rho^{\mathrm{TM}}_{mn} = \frac{\gamma_{1m}}{\gamma} |A^{\mathrm{TM}}_{1mn}|^2. \tag{8.88}$$

We show the plasmon resonance absorption in a bisinusoidal grating made of silver [12]. We consider a shallow bisinusoidal grating with a corrugation depth $H = 0.0075\,\mu\mathrm{m}$ and a period $d = 0.556\,\mu\mathrm{m}$. The wavelength of the incident light is chosen as $\lambda = 0.650\,\mu\mathrm{m}$ where only the (0, 0)th-order diffracted mode propagates. We take $n_2 = 0.07 - \mathrm{i}4.2$ as the refractive index of silver at this wavelength.

Figure 8.20 shows the diffraction efficiency of the (0, 0)th-order diffracted mode ρ_{00} as functions of the polar angle θ when the azimuthal angle $\phi = 30°$ is fixed. In the efficiency curve we observe four dips A, B, C, and D at which incident light power is strongly absorbed by the grating. The dips are associated with absorption that is caused by the coupling of surface plasmons with an evanescent mode diffracted by

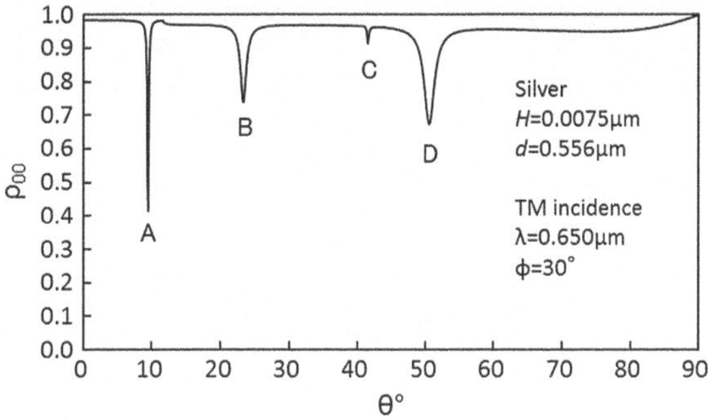

Fig. 8.20 ρ_{00} as functions of θ

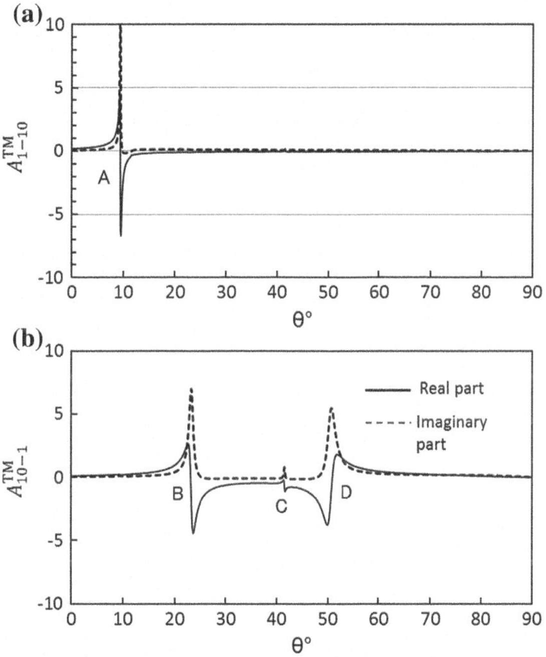

Fig. 8.21 Expansion coefficients as functions of θ: **a** $A^{\mathrm{TM}+}_{1-10}$; **b** $A^{\mathrm{TM}+}_{10-1}$

a bisinusoidal silver grating. This is confirmed from Fig. 8.21 where the expansion coefficients (a) A^{TM}_{1-10} and (b) A^{TM}_{10-1} are plotted as functions of θ under the same parameters as in Fig. 8.20. The solid curves in Fig. 8.21 represent the real part of the expansion coefficient and the dashed curves are the imaginary part. In Fig. 8.21a, a resonance property of the A^{TM}_{1-10} curve at $\theta = 9.5°$, i.e., a dip A, illustrates that the TM component of the $(-1, 0)$ evanescent mode couples with surface plasmons at a dip A. From the resonant property of the A^{TM}_{10-1} curve in Fig. 8.21b we confirm that dips B and D are associated with the coupling of the $(0, -1)$ evanescent mode with surface plasmons. Similarly, we can show a dip C is caused by coupling of the $(-1, -1)$ evanescent mode.

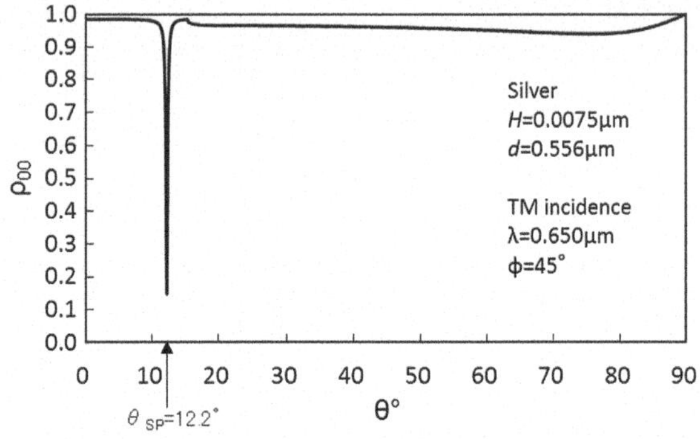

Fig. 8.22 Simultaneous resonance absorption in a bisinusoidal silver grating

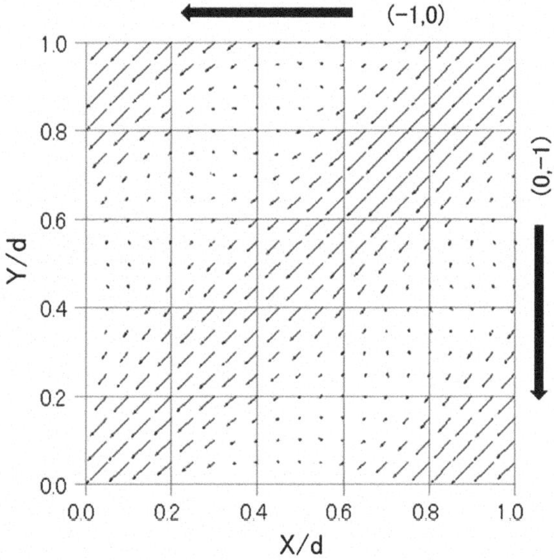

Fig. 8.23 Power flows of the total field when simultaneous resonance absorption occurs

When an incident light with $\phi = 45°$ illuminates a bisinusoidal grating at the specific angle of θ, i.e., the resonance angle, two surface plasmon waves are excited and propagate in directions symmetric with respect to the plane of incidence. The absorption associated with the two surface plasmon waves is called simultaneous resonance absorption [12]. Figure 8.22 shows an example of the simultaneous resonance absorption where the $(-1, 0)$th- and $(0, -1)$st-order evanescent modes couple simultaneously with two surface plasmon waves at the same polar angle $\theta = 12.2°$. The two surface plasmon waves excited simultaneously on the grating surface interact with each other and the interference of the surface plasmon waves causes the standing wave in the vicinity of the grating surface. This is confirmed in Fig. 8.23, where the X and Y components of Poynting's vector $\mathbf{S}$ on the surface $0.01d$ above the one-unit cell of the grating surface are plotted as the vector (S_X, S_Y).

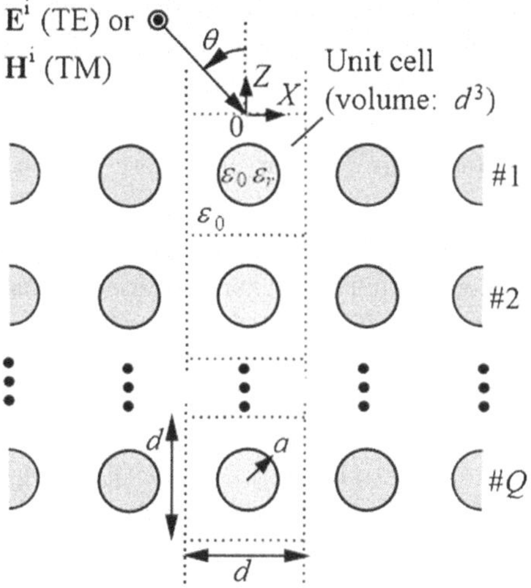

Fig. 8.24 Scattering by dielectric spheres. The structure is periodic both parallel to and perpendicularly to the page

8.3.5 *Scattering by Periodically Located Spheres*

Some numerical results are given for the scattering by dielectric spheres located periodically in three directions [17]. This kind of structure is a fundamental model of photonic crystals having properties of electromagnetic or optical band gaps.

As shown in Fig. 8.24, the structure is composed by stacking cubic unit cell regions with a volume d^3, each of which includes a sphere with radius a and relative permittivity ε_r. The number of spheres is infinity along the both X and Y axes, and the two-dimensionally infinite periodic structures are stacked to compose finite Q layers in the Z direction. At present we limit ourselves to the case where either electric or magnetic field of the incident plane wave is perpendicular to the page, allowing us to use only one incident angle θ.

In the upper and lower semi-infinite spaces, the approximate wave functions $(\mathbf{E}_{0N}(\mathbf{r}), \mathbf{H}_{0N}(\mathbf{r}))$ and $(\mathbf{E}_{Q+1\,N}(\mathbf{r}), \mathbf{H}_{Q+1\,N}(\mathbf{r}))$ are expressed in terms of modal coefficients $A_{0mn}^{\mathrm{TE,TM}}(N)$ and $A_{Q+1\,mn}^{\mathrm{TE,TM}}(N)$, respectively. The set of modal functions here is the same as that employed in Sect. 8.3.4 for the two-dimensional periodic structures. On the other hand, for the fields in the areas of periodically distributed spheres, a set of vector spherical wave functions $\{\mathbf{m}_{mn}^{\mathrm{e,h}}(\mathbf{r}), \mathbf{n}_{mn}^{\mathrm{e,h}}(\mathbf{r})\}$ is used to write the approximate wave functions. In the cube region of the layer $\sharp q$, they are expressed by

$$\left\{ \begin{array}{l} \begin{bmatrix} \mathbf{E}_{qN}(\mathbf{r}) \\ Z_0\mathbf{H}_{qN}(\mathbf{r}) \end{bmatrix} = \displaystyle\sum_{n=1}^{3N} \sum_{m=-n}^{n} \begin{bmatrix} \mathbf{m}_{mn}^{\mathrm{e}}(\mathbf{r}_q) & \mathbf{n}_{mn}^{\mathrm{e}}(\mathbf{r}_q) \\ -i\mathbf{n}_{mn}^{\mathrm{h}}(\mathbf{r}_q) & -i\mathbf{m}_{mn}^{\mathrm{h}}(\mathbf{r}_q) \end{bmatrix} \begin{bmatrix} A_{qmn}^{\mathrm{TE}}(N) \\ A_{qmn}^{\mathrm{TM}}(N) \end{bmatrix} \\ (q = 1, 2, \ldots, Q), \end{array} \right. \tag{8.89}$$

where Z_0 is the intrinsic impedance of vacuum and $\mathbf{r}_q = (r_q, \theta_q, \phi_q)$ is a position vector with its origin placed at the center of the qth sphere on the Z axis. Note that the truncation number is selected as $3N$ in order to maintain the balance with the half spaces from the viewpoint of the degree of approximation. The spherical wave functions $\{\mathbf{m}_{mn}^{\mathrm{e,h}}(\mathbf{r}_q), \mathbf{n}_{mn}^{\mathrm{e,h}}(\mathbf{r}_q)\}$ are written by combination of the spherical Bessel functions of the nth order, the associated Legendre functions $P_n^{|m|}(\cos\theta_q)$, and the exponential (trigonometric) functions $\exp(im\phi_q)$.[31] The functions with respect to $\mathbf{r}_q$ are constructed beforehand so that they automatically satisfy the continuity conditions for E_θ, E_ϕ, H_θ, and H_ϕ over the spherical surfaces $r_q = a$ [17]. As a result, the present problem is reduced to the determination of the modal coefficients such that the remaining boundary conditions on the horizontal planes

$$\begin{cases} \mathbf{u}_Z \times (\mathbf{E}_q, \mathbf{H}_q) = \mathbf{u}_Z \times (\mathbf{E}_{q+1}, \mathbf{H}_{q+1}) \\ \qquad \text{(between the layers } \sharp q \text{ and } \sharp q+1;\ q = 0, 1, 2, \ldots, Q) \end{cases} \tag{8.90}$$

and the periodicity conditions on the vertical planes

$$\begin{cases} \mathbf{u}_X \times (\mathbf{E}_q, \mathbf{H}_q) \exp(ikd \sin\theta)\big|_{X=-d/2+0} = \mathbf{u}_X \times (\mathbf{E}_q, \mathbf{H}_q)\big|_{X=d/2-0}, \\ \mathbf{u}_Y \times (\mathbf{E}_q, \mathbf{H}_q)\big|_{Y=-d/2+0} = \mathbf{u}_Y \times (\mathbf{E}_q, \mathbf{H}_q)\big|_{Y=d/2-0} \\ \qquad (q = 1, 2, \ldots, Q) \end{cases} \tag{8.91}$$

should be satisfied on the faces of the unit cells in the sense of least-squares. In the boundary conditions (8.91), we count the upper and lower half spaces by the numbers $\sharp 0$ and $\sharp Q+1$, respectively.

Figure 8.25 shows the normalized mean-square error and energy error as functions of the truncation number N. We find that both errors decrease monotonically when N increases. The period d is 0.8 times as the wavelength of the incident wave λ ($=2\pi/k$). Since the wavelength in dielectric material is shorter than that in the air, we need large N for big spheres. However, even at $a/d = 0.3$, these errors become less than 1% if $N \geq 4$.

Figure 8.26 is drawn to observe the effect of increasing the layer number on the band of total transmission and total reflection. For the single layer at $Q = 1$, we find two reflection points at $d/\lambda \approx 0.77$ and 0.91. When the layer is increased, these points are changed to reflection bands.

Figure 8.27 presents the reflected power for each mode as a function of incident angle for a 4-layered structure. We observe the power is totally reflected when θ is less than about 40°. This property disappears for larger θ due to the emergence of the $(-1, 0)$th higher order modes having a cutoff angle $\theta = 46°$.

---

[31] The vector $\mathbf{m}_{mn}^{\mathrm{e,h}}(\mathbf{r}_q)$ is perpendicular to the r_q axis, whereas $\mathbf{n}_{mn}^{\mathrm{e,h}}(\mathbf{r}_q)$ has an r_q component. That is, the superscript TE (TM) in (8.89) means transverse electric (transverse magnetic) with respect to r_q.

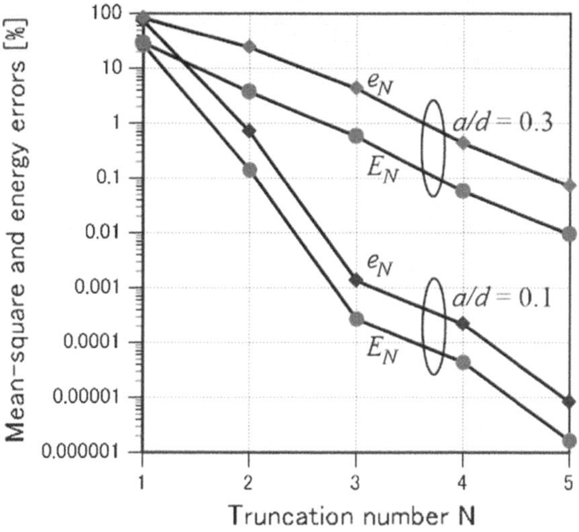

Fig. 8.25 Normalized mean-square error E_N and energy error e_N as functions of truncation number N; $Q = 2$, $\varepsilon_r = 10$, $d/\lambda = 0.8$, and $\theta = 0$

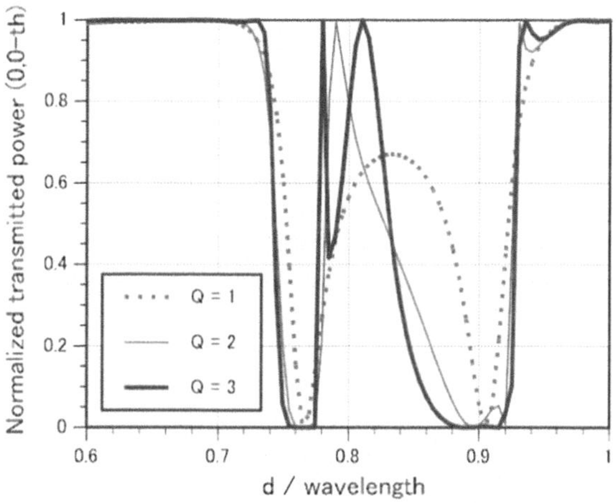

Fig. 8.26 Normalized transmitted powers as functions of wavelength; $a/d = 0.25$, $\varepsilon_r = 12$, and $\theta = 0$

We should note that introduction of sequential accumulation in the process of QR decomposition reduces the computation time from $O(Q^3)$ to $O(Q^1)$ and the memory requirement from $O(Q^2)$ to $O(Q^1)$, with Q being a number of sphere layers. See [17] for the detailed data.

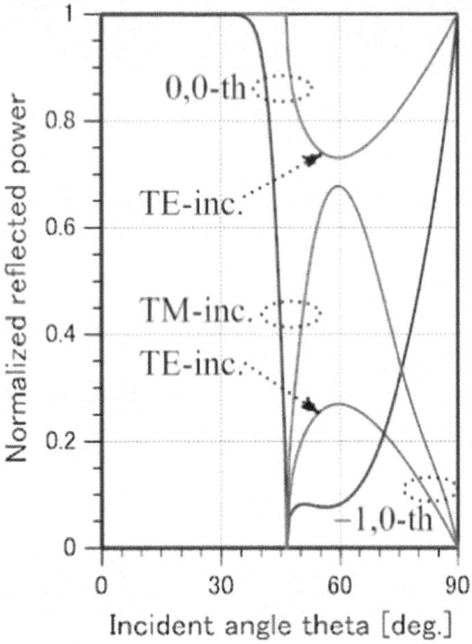

Fig. 8.27 Normalized reflected powers as functions of θ; $Q = 4$, $a/d = 0.25$, $\varepsilon_r = 12$, and $d/\lambda = 0.58$

8.4 Conclusions

Because of the reasons we have stated in Sect. 8.1, we reviewed Yasuura's method of modal expansion attaching importance to the process of solution by the CYM: choice of modal functions; a finite-sum approximate solution; least-squares boundary matching; location and number of sampling points; and solution method for the LSP. In addition, we included guidances for handling dielectric obstacles and gratings placed in planer or conical mounting. Still more, we gave a comparison between separated solutions and monopole fields in approximation power.

As for applications to 3D, we have only two grating problems in Sects. 8.3.4 and 8.3.5. Because we have been working in diffraction gratings, we do not have appropriate examples that show the effectiveness of the CYM in 3D scattering problems. However, our former colleagues have solved the problems using the CYM and published their results [11, 13]. Speaking from a theoretical point of view, they have employed the set of multipole functions as the modal functions whose completeness has been proven by Calderón [6].

We hope that the contents of this chapter would be useful for researchers and engineers who need reliable methods for solving electromagnetic boundary-value problems.

Acknowledgements The authors thank Mr. BenWen Chen and Mr. Rui Gong, Centre for Optical and Electromagnetic Research, South China Academy of Advanced Optoelectronics, South China Normal University for preparing the figures in Sect. 8.3.1 including numerical computations.

One of the authors (A.M.) wish to express his thanks to Japan Society for Promotion of Science (JSPS) for partial support to the work in Sects. 8.3.2 and 8.3.5 under Grant Number JP15K06023 (KAKENHI).

Another one of the authors (Y.O.) is grateful to Prof. S. He, COER-SCNU, COER-ZJU, and JORSEP-KTH for his continuous help and encouragement.

Appendix 1: H-Wave Scattering by a PC Cylinder

Let us consider a problem where an H-wave (TM-wave) is incident to the obstacle shown in Fig. 8.1. That is, the incident wave is polarized in the xy-plane so that the incident magnetic field has only a z-component

$$\mathbf{H}^{\mathrm{i}}(\mathbf{r}) = \mathbf{u}_z F(\mathbf{r}) = \mathbf{u}_z \exp[-ikr\cos(\theta - \iota)]. \tag{8.92}$$

The scattered magnetic field has only a z-component

$$\mathbf{H}^{\mathrm{s}}(\mathbf{r}) = \mathbf{u}_z \Psi(\mathbf{r}) \tag{8.93}$$

which is a leading field of the problem. Thus, we have

Problem 1': H-wave, PC. Find $\Psi(\mathbf{r})$ that satisfies:

(N1) The 2-D Helmholtz equation in $\mathrm{S_e}$;
(N2) The 2-D radiation condition at infinity;
(N3) The boundary condition

$$\partial_\nu \Psi(s) = g(s) \equiv -\partial_\nu F(s) \quad (s \in \mathrm{C}). \tag{8.94}$$

Here, ∂_ν denotes a normal derivative at s. Equation (8.94) is called Neumann's or the second-kind boundary condition.

Employing the Green's (or Neumann's) function of this boundary-value problem satisfying a homogeneous boundary condition

$$\partial_\nu N(\mathbf{r}, s) = 0 \quad (\mathbf{r} \in \mathrm{S};\, s \in \mathrm{C}), \tag{8.95}$$

we get a formal representation similar to (8.13)

$$\Psi_N(\mathbf{r}) - \Psi(\mathbf{r}) = -\int_{s=0}^{C} N(\mathbf{r}, s)\,[\partial_\nu \Psi_N(s) - g(s)]\, ds \quad (\mathbf{r} \in \mathrm{S}). \tag{8.96}$$

Here, Ψ_N denotes an approximate solution defined by

$$\Psi_N(\mathbf{r}) = \sum_{m=-N}^{N} B_m(M)\varphi_m(\mathbf{r}). \tag{8.97}$$

After a discussion similar to that in Sect. 8.2.2.3, we have a least-squares problem for the H-wave problem:

LSM 1': H-wave, PC. Find the coefficients $B_m(M)$ $(m = 0, \pm 1, \ldots, \pm N)$ that minimize the mean-squares boundary residual

$$E_N = \frac{\|\partial_\nu \Psi_N - g\|^2}{\|g\|^2} = \frac{1}{\|g\|^2} \left\| \sum_{m=-N}^{N} B_m(M) \partial_\nu \varphi_m - g \right\|^2 . \tag{8.98}$$

We can solve this problem on a computer following the procedure in Sect. 8.2.3. Approximations to other nonzero components can be found by

$$\mathbf{E}_N^{\mathrm{s}}(\mathbf{r}) = \frac{1}{\mathrm{i}\omega\varepsilon_0} \nabla \Psi_N(\mathbf{r}) \times \mathbf{u}_z . \tag{8.99}$$

It is worth noting that in an H-wave scattering from a dielectric obstacle, the boundary condition (8.42) should be altered slightly. Let $\mathbf{H}^{\mathrm{s}}(\mathbf{r}) = \mathbf{u}_z \Psi_{\mathrm{e}}(\mathbf{r})$, and $\mathbf{H}^{\mathrm{t}}(\mathbf{r}) = \mathbf{u}_z \Psi_{\mathrm{i}}(\mathbf{r})$, then we have

$$\begin{cases} \Psi_{\mathrm{e}}(s) - \Psi_{\mathrm{i}}(s) = f(s) \equiv -F(s) \\ \partial_\nu \Psi_{\mathrm{e}} - n^{-2} \partial_\nu \Psi_{\mathrm{i}}(s) = g(s) \equiv -\partial_\nu F(s), \end{cases} \tag{8.100}$$

where the second line means the electric-field continuity and $n^2 = \varepsilon/\varepsilon_0$.

Appendix 2: Solution of LSP 1 by a Normal Equation and Related Topics

Although we do not use a normal equation in numerical analysis, we look over the solution method by the equation because it is an important theoretical tool in working with a least-squares problem. Let us define an inner product between two functions in $\mathbf{H} = L^2(0, C)$ by

$$(f, g) = \int_{s=0}^{C} \overline{f(s)} g(s)\, ds, \tag{8.101}$$

then we find that $\|f\| = \sqrt{(f, f)}$. Employing these relations, we modify (8.22) to obtain

$$E_N = \sum_{m=-N}^{N} \sum_{n=-N}^{N} \overline{A_m} (\varphi_m, \varphi_n) A_n - \sum_{m=-N}^{N} \overline{A_m} (\varphi_m, f) - \sum_{n=-N}^{N} (f, \varphi_n) A_n + \|f\|^2 . \tag{8.102}$$

The predictable M is not shown.

Now we define a subspace of **H**, Φ_N, spanned by the boundary values of a finite number of modal functions $\{\varphi_0(s), \varphi_{\pm 1}, \ldots, \varphi_{\pm N}\}$. An element of Φ_N can be represented as

$$\Psi_N(s) = \sum_{n=-N}^{N} A_n \, \varphi_n(s). \tag{8.103}$$

Apparently, there is a minimum value of E_N, which is a squared distance between $f(s)$ and a point in Φ_N.[32] The minimum is achieved when (8.103) agrees with the foot of a perpendicular line from $f(s)$ to the surface of Φ_N. The necessary and sufficient condition for this is that: The A_m coefficients are the solutions of the set of linear equations

$$\sum_{n=-N}^{N} (\varphi_m, \varphi_n) \, A_n = (\varphi_m, f) \quad (m = 0, \pm 1, \ldots, \pm N). \tag{8.104}$$

This is referred to as the normal equation (NE) of **LSP 1** and is a formal solution to the problem.[33]

Next, let us consider the minimization from a computational point of view. That is, we try to find the A_m coefficients using the sampled values of boundary functions; and the functions are represented by J-dimensional complex-valued vectors **f**, $\boldsymbol{\varphi}_m$, and $\boldsymbol{\Psi}_N$ as in Sect. 8.2.3. This leads us to **DLSP 1**. We know the orthogonal decompositions are useful tools for solving the problem. However, setting them aside, we here consider a NE based on **DLSP 1**. Because the Jacobian matrix Φ is $J \times M$ $(J > M)$, the set of linear equations

$$\Phi \mathbf{A} = \mathbf{f} \tag{8.105}$$

is over-determined and does not have a usual solution. However, if we multiply (8.105) by $\Phi^\dagger$ from the left, we have

$$\mathrm{H}\mathbf{A} = \mathbf{b}, \tag{8.106}$$

where

$$\mathrm{H} = \Phi^\dagger \Phi \tag{8.107}$$

is an $M \times M$ positive-definite Hermitian matrix provided Φ is full rank. And,

$$\mathbf{b} = \Phi^\dagger \mathbf{f} \tag{8.108}$$

---

[32]If $f \in \Phi_N$, then $E_N = 0$. This, however, cannot occur in practice: For example, even in the case of scattering from a circular cylinder made of a PC, we need an infinite series to represent a rigorous solution because the boundary value has a form $\exp[-ika\cos(\theta_s - \iota)]$. In addition, note that Φ_N is closed.

[33]We get (8.104) by setting $(\varphi_m, \Psi_N - f) = 0$ $(m = 0, \pm 1, \ldots, \pm N)$; or from $\partial E_N / \partial \overline{A}_m = 0$.

is an M-dimensional right-hand side. Usually, (8.106) is referred to as the NE of **DLSP 1** and has been employed as a standard method of solution for a long time.

Obviously, (8.106) is an approximation of (8.104). For example, an (m, n)th element of the coefficient matrix of (8.104) can be represented as

$$(\varphi_m, \varphi_n) = \int_{s=0}^{C} \overline{\varphi_m(s)}\varphi_n(s)\,ds \simeq \frac{C}{J}\sum_{j=1}^{J} \overline{\varphi_m(s_j)}\varphi_n(s_j) = \frac{C}{J}\,\boldsymbol{\varphi}_m^{\dagger}\boldsymbol{\varphi}_n. \tag{8.109}$$

The right-hand side of (8.109) is the (m, n)th entry of H multiplied by the line element C/J. Hence, (8.104) and (8.106) are essentially the same thing, and they have common weak points in numerical computations. Widely-accepted key observations are:

- The NE is rigorous, in principle, and can be employed in theoretical considerations;
- The NE combined with Gaussian elimination (diagonal pivoting assumed) is equivalent to the (modified) Schmidt QRD except for the next two items;
- The NE may lose information in constructing $\mathrm{H} = \Phi^{\dagger}\Phi$, and this process is time consuming usually;
- The NE is dominated by the condition number of H that is square of the original condition number: $\mathrm{cond}(\mathrm{H}) = [\mathrm{cond}(\Phi)]^2$.

The last item means (8.104) and (8.106) are more sensitive to computational errors than **LSP 1** and **DLSP 1**. Therefore, the NE's are more difficult to solve on a computer than the original least-squares problems. We, hence, do not recommend the use of (8.104) or (8.106). Even if we are working in the case where the inner products in (8.104) can be calculated analytically, we should not employ (8.104) because of the last item.

Before closing this Appendix, we would like to state a couple of comments on (8.105). Apparently, J cannot be less than M because (8.105) is indeterminate for $J < M$. If we set $J = M$, we have a point-matching method (PMM) or a collocation method. The method is known to be effective if the contour C coincides with a part of a coordinate curve of a system of coordinates in which Helmholtz's equation is separable; and that the modal functions are the separated solutions in that system. Convergence of the PMM solution is related to the validity of the Rayleigh hypothesis [2, 3, 18].

In Yasuura's method we usually set $J = 2M$ as we see in Sect. 8.3.1. That is, we employ $2M$ linear equations to determine M unknown coefficients. This may be understood as a small device or improvement of the PMM. However, this produces good results such as proof of convergence, wide range of application, and so on with little increase of computational complexity as a reasonable cost.

Appendix 3: Conical Diffraction by a Grating

In Sect. 8.2.5 we dealt with diffraction by a grating, where all the field components were functions of two variables (X and Y) and two independent cases of polarization [E-wave (TE, s) and H-wave (TM, p)] existed. In addition, the directions of propagating diffraction-orders were parallel to the plane of incidence. These were possible because: (**1**) the grating surface was uniform in Z; and (**2**) the plane of incidence was in parallel to the direction of periodicity $\mathbf{u}_X$. Here, we concisely examine the problem of a lossless dielectric grating in which the second condition is not satisfied, i.e., the plane of incidence makes a nonzero angle ϕ with the positive X-direction as shown in Fig. 8.14a. We will see that

- The field components are functions of X, Y, and Z, but the dependence on Y—the direction of uniformity—is limited;
- The two cases of polarization are not independent, i.e., both TE and TM diffracted waves exist for TE (or TM) incidence[34];
- The direction of propagating orders lie on the surface of a cone whose vertex agrees with the coordinate origin O; the direction of the zeroth mode is on the plane of incidence at the same time.

Because of the third characteristic, this arrangement ($\phi \neq 0$) is called conical mounting and the term *conical diffraction* is used. In this connection, the arrangement in Sect. 8.2.5 is termed planar mounting.

Let the incident wave be

$$\begin{bmatrix} \mathbf{E}^{\mathrm{i}} \\ \mathbf{H}^{\mathrm{i}} \end{bmatrix}(\mathbf{r}) = \begin{bmatrix} \mathbf{e}^{\mathrm{i}} \\ \mathbf{h}^{\mathrm{i}} \end{bmatrix} \exp(-\mathrm{i}\mathbf{k}^{\mathrm{i}} \cdot \mathbf{r}). \tag{8.110}$$

Here, $\mathbf{e}^{\mathrm{i}}$ and $\mathbf{h}^{\mathrm{i}}$ are electric- and magnetic-field amplitude, which are related by

$$\mathbf{h}^{\mathrm{i}} = \frac{1}{\omega\mu_0}\mathbf{k}^{\mathrm{i}} \times \mathbf{e}^{\mathrm{i}} \tag{8.111}$$

and $\mathbf{k}^{\mathrm{i}}$ is the incident wavevector defined by

$$\mathbf{k}^{\mathrm{i}} = (n_1 k \sin\theta \cos\phi, n_1 k \sin\theta \sin\phi, -n_1 k \cos\theta) \equiv (\alpha, \beta, -\gamma) \tag{8.112}$$

with θ being the polar angle between the wavevector $\mathbf{k}^{\mathrm{i}}$ and the grating normal $\mathbf{u}_Z$. We decompose the incident wave into a TE(s)- and a TM(p)-component, where TE (or TM) means that the electric (or magnetic) field of the relevant incident wave is perpendicular to the plane of incidence. To do this, we define two unit vectors that span a plane orthogonal to the incident wavevector

---

[34] Assume a PC surface-relief grating with a TE-wave incidence, for simplicity, and imagine the surface current induced. It apparently has a Z-oriented ingredient, which excites a TM-wave component.

$$\mathbf{e}^{\mathrm{TE}} = (\sin\phi, -\cos\phi, 0), \quad \mathbf{e}^{\mathrm{TM}} = (\cos\theta\cos\phi, \cos\theta\sin\phi, \sin\theta). \tag{8.113}$$

They give the directions of the incident electric fields that are in the TE- and TM-polarization.[35] Thus the decomposition is

$$\mathbf{e}^{\mathrm{i}} = \mathbf{e}^{\mathrm{TE}}\cos\delta + \mathbf{e}^{\mathrm{TM}}\sin\delta, \tag{8.114}$$

where δ is a polarization angle shown in Fig. 8.14b. $\delta = 0$ and $\pi/2$ mean TE- and TM-incidence. Hence, an incident wave has three angular parameters: ϕ, θ, and δ.

We consider the problem to seek the diffracted electric and magnetic field in the semi-infinite regions V_1 and V_2 over and below the grating surface $\mathrm{S_G}$.

Problem 4 conical, dielectric grating. Find the solutions that satisfy the following requirements:

(CD1) The Helmholtz equation in V_1 and V_2;
(CD2) Radiation conditions in the positive and negative Z-direction;
(CD3) A periodicity condition that: the relation $f(X+d, Y, Z) = e^{\mathrm{i}\alpha d} f(X, Y, Z)$ holds for any component of the diffracted wave, and the phase constant in Y is β;
(CD4) The total tangential component of electric and magnetic field must be continuous across the grating surface $\mathrm{S_G}$.

Dealing with a problem of conical diffraction, we should keep in mind the unique nature of the problem. First, because every field component has a common phase constant β in Y, it is sufficient to match the boundary condition on a cross section between the grating surface and a plane $Y = \mathrm{const}$. The conically-mounted gratings, hence, belong to the class of quasi-3-D structures. Second, because the TE- and TM-wave are not independent, we always need both TE and TM vector modal functions in constructing approximate solutions.

We define the modal functions satisfying **(CD1)**–**(CD3)** by

$$\begin{cases} \boldsymbol{\varphi}_{\ell m}^{\mathrm{TE}}(\mathbf{r}) = \mathbf{e}_{\ell m}^{\mathrm{TE}}\exp(-\mathrm{i}\mathbf{k}_{\ell m}\cdot\mathbf{r}), \quad \boldsymbol{\varphi}_{\ell m}^{\mathrm{TM}}(\mathbf{r}) = \mathbf{e}_{\ell m}^{\mathrm{TM}}\exp(-\mathrm{i}\mathbf{k}_{\ell m}\cdot\mathbf{r}) \\ \quad (\ell = 1, 2;\ m = 0, \pm1, \pm2, \ldots). \end{cases} \tag{8.115}$$

Here,

$$\mathbf{e}_{\ell m}^{\mathrm{TE}} = \frac{\mathbf{k}_{\ell m}\times\mathbf{u}_Z}{|\mathbf{k}_{\ell m}\times\mathbf{u}_Z|}, \quad \mathbf{e}_{pm}^{\mathrm{TM}} = \frac{\mathbf{e}_{\ell m}^{\mathrm{TE}}\times\mathbf{k}_{\ell m}}{|\mathbf{e}_{\ell m}^{\mathrm{TE}}\times\mathbf{k}_{\ell m}|} \quad (\ell = 1, 2), \tag{8.116}$$

$$\mathbf{k}_{1m} = (\alpha_m, \beta, \gamma_{1m}), \quad \mathbf{k}_{2m} = (\alpha_m, \beta, -\gamma_{2m}), \tag{8.117}$$

and

$$\begin{cases} \alpha_m = \alpha + \dfrac{2m\pi}{d}, \quad \gamma_{\ell m} = \sqrt{(n_\ell k)^2 - (\alpha_m^2 + \beta^2)} \\ (\mathrm{Re}\,\gamma_{\ell m} \geq 0,\ \mathrm{Im}\,\gamma_{\ell m} \leq 0). \end{cases} \tag{8.118}$$

[35] $\mathbf{e}^{\mathrm{TE}}$ is perpendicular to the plane of incidence; the fact that the magnetic field accompanying $\mathbf{e}^{\mathrm{TM}}$ is orthogonal to the plane can be seen by manipulation.

Note that the functions in (8.115) are for constructing electric fields. For the magnetic fields we get

$$\boldsymbol{\psi}^{\mathrm{q}}_{\ell m}(\mathbf{r}) = \frac{1}{\omega\mu_0}\mathbf{k}_{\ell m} \times \boldsymbol{\varphi}^{\mathrm{q}}_{\ell m}(\mathbf{r}) \quad (\ell = 1, 2;\ \mathrm{q} = \mathrm{TE}, \mathrm{TM}) \tag{8.119}$$

through Maxwell's equations. Finite linear combinations of the modal functions define approximate solutions:

$$\begin{bmatrix} \mathbf{E}_{\ell N} \\ \mathbf{H}_{\ell N} \end{bmatrix}(\mathbf{r}) = \sum_{m=-N}^{N} A^{\mathrm{TE}}_{\ell m} \begin{bmatrix} \boldsymbol{\varphi}^{\mathrm{TE}}_{\ell m} \\ \boldsymbol{\psi}^{\mathrm{TE}}_{\ell m} \end{bmatrix}(\mathbf{r}) + \sum_{m=-N}^{N} A^{\mathrm{TM}}_{\ell m} \begin{bmatrix} \boldsymbol{\varphi}^{\mathrm{TM}}_{\ell m} \\ \boldsymbol{\psi}^{\mathrm{TM}}_{\ell m} \end{bmatrix}(\mathbf{r}) \quad (\ell = 1, 2) \tag{8.120}$$

Here, the number of modal functions M is neglected.

The unknown coefficients in (8.120) should be determined in order that the solutions satisfy the boundary condition (**CD4**) approximately in the mean-squares sense. For this purpose we first consider the cross section C between the grating surface $\mathrm{S_G}$ and a plane $Y = 0$. This is the same thing as the periodic curve C in Sect. 8.2.5. In a similar way to one in Sect. 8.2.5, we define the primary period S_1, whose boundary C_1 ($\subset$ C), the function space **H** consisting of all the square integrable functions on C_1, and the norm $\|f\|$ of a function $f(s)$. Then, we can state the least-squares problem that determines the unknown coefficients:

LSP 4: conical, dielectric grating. Find the coefficients $A^{\mathrm{TM}}_{\ell m}$ and $A^{\mathrm{TM}}_{\ell m}$ ($\ell = 1, 2;\ m = 0, \pm 1, \ldots, \pm N$) that minimize the mean-square error

$$E_N = \left\|\boldsymbol{\nu} \times \left(\tilde{\mathbf{E}}_{1N} + \tilde{\mathbf{E}}^{\mathrm{i}} - \tilde{\mathbf{E}}_{2N}\right)\right\|^2 + Z_0^2 \left\|\boldsymbol{\nu} \times \left(\tilde{\mathbf{H}}_{1N} + \tilde{\mathbf{H}}^{\mathrm{i}} - \tilde{\mathbf{H}}_{2N}\right)\right\|^2. \tag{8.121}$$

Here, Z_0 denotes the intrinsic impedance of vacuum and $\tilde{\mathbf{E}}^{\mathrm{i}}$ etc. mean periodic functions with respect to x defined in the same way as one in (8.69)–(8.71). The method of discretization and the solution method are found in Sect. 8.2.4.

Appendix 4: Comparison of Modal Functions and Algorithm of the SP

Here we show some results of effectiveness comparison between three kinds of modal functions in solving a sample problem[36]: E-wave scattering from a PC cylinder whose

[36]We can use monopole fields also in the grating problems discussed in Sect. 8.2.5. A countably infinite set of monopoles located periodically in x—i.e., the location is given by $(x_1 + \ell d, y_1)$ $(0 < x_1 \le d;\ y_1 < \eta(x_1);\ \ell = 0, \pm 1, \pm 2, \ldots)$—radiates a plane wave [4, 36] satisfying (**GD1**) and (**GD2**). If we let the monopoles be accompanied by phase factors $\exp(i\ell kd \sin\theta)$, the plane wave meets the periodicity (**GD3**). Increasing the number of monopoles in the first strip region to M, i.e., $(x, y) = (x_1, y_1), (x_2, y_2), (x_3, y_3), \ldots, (x_M, y_M)$, and repeating the same procedure, we have a set of M plane waves, which is the desired set of modal functions [28, 37].

cross section is given by[37]

$$C : r_s = a(1 + 0.2 \cos 3\theta_s). \tag{8.122}$$

Let us normalize every quantity having dimension of length by the total length of C. And, we assume the incident wave comes along the x-axis from the negative x-direction (i.e., $\iota = 0$).

The modal functions considered here are: (**a**) the separated solutions, which we defined by (8.7) in Sect. 8.2.2; (**b**) monopole fields defined by (8.9); (**c**) monopole fields whose poles are located densely near the convex part of C. Because the separated solutions are known widely, we explain the monopole fields below:

(**b**) Equally spaced poles. Let L be a similar curve to C with the ratio of similitude d $(0 < d < 1)$.[38] We arrange M poles on L at regular intervals. Then, the distance between two poles is L/M where L is the length of L.

(**c**) Concentration of poles near the convex parts of L. (**i**) First, we draw the similar curve L. (**ii**) Next,we calculate the curvature $\kappa(t)$ of L as a function of t $(\in \text{L})$, and add some positive bias c in order that the biased curvature (BC) be no less than 0: $\tilde{\kappa}(t) = \kappa(t) + c$ (≥ 0). (**iii**) Thirdly, we define a probability density function by normalizing the BC.[39]

$$f(t) = \frac{\int_0^t \tilde{\kappa}(t')\,dt'}{\int_0^L \tilde{\kappa}(t')\,dt'}. \tag{8.123}$$

Thus we get the number of poles between t_1 and t_2 by

$$n(t_1, t_2) = M \int_{t_1}^{t_2} f(t)\,dt. \tag{8.124}$$

We have solved the problem using the method explained in Sects. 8.2.2 and 8.2.3. We used three kinds of modal functions (**a**), (**b**), and (**c**); and tried at two frequencies: $ka = 10$ and 30. The parameter d was set to be 0.87. To see the accuracy of a solution we calculated two kinds of errors: the normalized mean-square error $E_M(\mathrm{m})$ and

---

[37] Although the employment of polyphase wave functions is effective because of the periodicity, we do not use them for simplicity.

[38] According to the result of numerical computation, an optimum d was in the rage $[0.85, 0.90]$ when the total number of poles was between 40 and 120. If we increased (or decreased) the number of poles, the optimum d approached 0.90 (or 0.85). Note, however, that the trends were observed in solving a particular problem with specific computational parameters and are no more than reference data.

[39] The result of sample calculation has shown that the use of $|\text{BC}|^\alpha$ $(\alpha > 1)$ instead of BC (i.e., further emphasis of the convex part in locating poles) gives better solutions.

the error on energy balance (or on the optical theorem) $e_M(\mathrm{m})$. The former is the same thing as one defined in (8.22) and (8.30)[40] except that the subscript shows the total number of modal functions. The latter shows the deviation from a proportional relation between the forward scattering amplitude and total cross section.[41] The argument m shows the type of modal functions: m = sep, esp, and pcc, which mean (**a**) separated solutions, (**b**) equally-spaced poles, and (**c**) poles concentrated near the convex parts.

Results at $ka = 10$. Because the obstacle size is handy, the E_M errors fall off rapidly: $E_{45}(\mathrm{sep})$, $E_{35}(\mathrm{esp})$, and $E_{31}(\mathrm{pcc})$ are below 1%. As for the e_M errors of the solutions, the situation is different. The solutions with esp or pcc modal functions converge rapidly as $e_M(\mathrm{esp})$ and $e_M(\mathrm{pcc})$ are below 1% at $M \simeq 30$. On the other hand, $e_{31}(\mathrm{sep})$ is about 10%. Increasing M to 70, we have: $e_{70}(\mathrm{esp}) = 9 \times 10^{-5}\%$; $e_{70}(\mathrm{pcc}) = 1 \times 10^{-5}\%$; and $e_{71}(\mathrm{sep}) = 4\%$.

Results at $ka = 30$. The advantage of the monopole fields is clear in this range of frequency. Setting $M \simeq 100$, we have $E_{101}(\mathrm{sep}) = 4\%$, $E_{100}(\mathrm{esp}) = 2 \times 10^{-1}\%$, $E_{100}(\mathrm{pcc}) = 2 \times 10^{-3}\%$, $e_{101}(\mathrm{sep}) = 7\%$, $e_{100}(\mathrm{esp}) = 2 \times 10^{-1}\%$, and $e_{100}(\mathrm{pcc}) = 5 \times 10^{-3}\%$. The pcc modal functions seems to be the best choice in solving the problem. In fact, we can find an accurate solution with a $10^{-5}\%$ e_M error by setting m = pcc and $M = 120$.

These results mean that the potential of a combination of separated solutions is not so strong in describing scattered fields from obstacles deformed strongly from a circle. We have two ways to cope with this issue: (**i**) employment of a set of modal functions other than the separated solutions[42]; and (**ii**) employment of the SP.

The Algorithm of the SP

Here we include a guidance how to apply the SP in the boundary-matching process based on Yasuura's method of modal expansion for convenience. We start from **DLSP 1**, i.e., minimization of the numerator of (8.30), $\|\Phi\mathbf{A} - \mathbf{f}\|^2$. Instead of minimizing it directly, we force a constraint

$$(\mathbf{1}, \Phi\mathbf{A} - \mathbf{f}) = 0 \tag{8.125}$$

on the M-dimensional solution vector $\mathbf{A}$, where the parentheses mean an inner product and $\mathbf{1} = [1\ 1\ \cdots\ 1]^{\mathrm{T}}$ is a J-dimensional constant vector.

An operator of the smoothing procedure (in a discretized form) is a $J \times J$ matrix given by

$$\mathrm{K}^{(p)} = \left[K_{j\ell}^{(p)}\right], \tag{8.126}$$

---

[40] We have applied the rule $J = 2M$ and have omitted J.

[41] The relation is referred to as the optical theorem, which implies energy conservation.

[42] We have employed the monopole fields and have seen their effectiveness [30]. It is worth noting that inclusion of a few dipoles located near the convex part of L in addition to the monopoles improves the efficiency greatly. This might be related to Cadilhac-Petit's opinion [7] in locating the poles near an internal focus.

where p means the order of the SP. The explicit forms of the matrix elements for $p = 1, 2$, and 3 are[43]:

$$\begin{cases} K_{j\ell}^{(1)} = u(j-\ell) - \dfrac{j-\ell}{J} - \dfrac{1}{2}, \\ K_{j\ell}^{(2)} = -\dfrac{1}{2}\left[\dfrac{(j-\ell)^2}{J^2} - \dfrac{|j-\ell|}{J} + \dfrac{1}{6}\right], \\ K_{j\ell}^{(3)} = \dfrac{1}{6}\left[\dfrac{|j-\ell|^3}{J^3} - \dfrac{3(j-\ell)^2}{2J^2} + \dfrac{|j-\ell|}{2J} + \dfrac{1}{6}\right]. \end{cases} \tag{8.127}$$

Thus we can state a method of solution with the SP as follows:

DLSP 3: E-wave, PC, SP. Find the solution vector $\mathbf{A}$ that minimizes the discretized mean-square error

$$E_{MJ} = \frac{\|\mathrm{K}^{(p)}(\Phi\mathbf{A} - \mathbf{f})\|^2}{\|\mathrm{K}^{(p)}\mathbf{f}\,\|^2} \tag{8.128}$$

under the constraint (8.125).

Two ways are possible to solve this conditioned least-squares problem: (**i**) employment of Lagrange's multiplier; and (**ii**) elimination of a modal coefficient by using the constraint. Although (**i**) is a standard way in handling a constraint, we take (**ii**) because our constraint is simple and can eliminate one of the M unknowns to deduce a least-squares problem with $M-1$ unknowns.

References

1. M. Bass (ed.), *Handbook of Optics; Volume II — Devices, Measurements, and Properties*, 2nd edn. (McGraw-Hill, 1995)
2. R.H.T. Bates, Analytic constraints on electromagnetic field computations. IEEE Trans. Microw. Theory Tech. **MTT-23**(8), 605–623 (1975)
3. R.H.T. Bates, J.R. James, I.N.L. Gallett, R.F. Millar, An overview of point matching. Radio Electron. Eng. **43**(3), 193–200 (1973)
4. A. Boag, Y. Leviatan, A. Boag, Analysis of two-dimensional electromagnetic scattering from a periodic grating of cylinders using a hybrid current method. Radio Sci. **23**(4), 612–624 (1988)
5. G.P. Bryan-Brown, J.R. Sambles, M.C. Hutley, Polarization conversion through the excitation of surface plasmons on a metallic grating. J. Modern Opt. **37**(7), 1227–1232 (1990)
6. A.P. Calderón, The multipole expansion of radiation fields. J. Ration. Mech. Anal. (J. Math. Mech.) **3**, 523–537 (1954)
7. M. Cadilhac, R. Petit, On the diffraction problem in electromagnetic theory: a discussion based on concepts of functional analysis including an example of practical application, in *Huygens'*

[43] We got the elements under the assumption that the length of C is 1. This is convenient in mathematical analysis and does not affect applications to obstacles made of a lossless material including PC. In dealing with a lossy material, in particular a metal in light frequency, the normalization should be accompanied by a law of similitude in time-dependent EM field [19] and, hence, the use of actual length might be appropriate.

Principle 1690–1990: Theory and Applications, Studies in Mathematical Physics, ed. by H. Blok, et al. (Elsevier, Amsterdam, 1992)
8. G. Hass, L. Hardley, Optical properties of metal, in *American Institute of Physics Handbook*, ed. by D.E. Gray, 2nd ed. (McGraw-Hill, 1963), pp. 6–107
9. J.P. Hugonin, R. Petit, M. Cadilhac, Plane-wave expansions used to describe the field diffracted by a grating. J. Opt. Soc. Am. **71**(5), 593–598 (1981)
10. H. Ikuno, K. Yasuura, Numerical calculation of the scattered field from a periodic deformed cylinder using the smoothing process on the mode-matching method. Radio Sci. **13**(6), 937–946 (1978)
11. H. Ikuno, M. Gondoh, M. Nishimoto, Numerical analysis of electromagnetic wave scattering from an indented body of revolution. Trans. IEICE Electron. **E74-C**(9), 2855–2863 (1991)
12. T. Inagaki, J.P. Goudonnet, J.W. Little, E.T. Arakawa, Photoacoustic study of plasmon-resonance absorption in a bigrating. J. Opt. Soc. Am. B **2**(3), 433–439 (1985)
13. M. Kawano, H. Ikuno, M. Nishimoto, Numerical analysis of 3-D scattering problems using the Yasuura method. Trans. IEICE Electron. **E79-C**(10), 1358–1363 (1996)
14. A.N. Kolmogorov, S.V. Fomin, *Elements of the Theory of Functions and Functional Analysis* (Dover, New York, 1999)
15. C.L. Lawson, R.J. Hanson, *Solving Least Squares Problems* (Prentice-Hall, New Jersey, 1974)
16. T. Matsuda, D. Zhou, Y. Okuno, Numerical analysis of plasmon-resonance absorption in a bisinusoidal metal grating. J. Opt. Soc. Am. A **19**(4), 695–701 (2002)
17. A. Matsushima, Y. Momoka, M. Ohtsu, Y. Okuno, Efficient numerical approach to electromagnetic scattering from three-dimensional periodic array of dielectric spheres using sequential accumulation. Progr. Electromagn. Res. **69**, 305–322 (2007)
18. R.F. Millar, Rayleigh hypothesis and a related least-squares solution to scattering problems for periodic surfaces and other scatterers. Radio Sci. **8**(8–9), 785–796 (1973)
19. H. Nakano, Frequency-independent antennas: spirals and log-periodics, in *Modern Antenna Handbook*, ed. by C.A. Balanis (Wiley, New Jersey, 2008), pp. 263–323
20. Y. Nakata, M. Koshiba, M. Suzuki, Finite-element analysis of plane wave diffraction from dielectric gratings. Trans. IEICE Jpn. **J69-C**(12), 1503–1511 (1986)
21. M. Neviér, The homogeneous problems, in *Electromagnetic Theory of Gratings*, ed. by R. Petit (Springer, Berlin, 1980), pp. 123–157
22. M. Ohtsu, Y. Okuno, A. Matsushima, T. Suyama, A Combination of up- and down-going Floquet modal functions used to describe the field inside grooves of a deep grating. Progr. Electromagn. Res. **64**, 293–316 (2006)
23. Y. Okuno, A numerical method for solving edge-type scattering problems. Radio Sci. **22**(6), 941–946 (1987)
24. Y. Okuno, The mode-matching method, in *Analysis Methods in Electromagnetic Wave Problems*, ed. by E. Yamashita (Artech House, 1990), pp. 107–138
25. Y. Okuno, An introduction to the Yasuura method, in *Analytical and Numerical Methods in Electromagnetic Wave Theory*, ed. by M. Hashimoto, M. Idemen, O.A. Tretyakov (Science House, 1993), pp. 515–565
26. Y. Okuno, H. Ikuno, Completeness of the boundary values of equivalent sources. Mem. Fac. Eng. Kumamoto Univ. **38**(1), 1–8 (1993)
27. Y. Okuno, H. Ikuno, Yasuura's method, its relation to the fictitious source methods, and its advancements in the solution of 2D problems, in *Generalized Multipole Techniques for Electromagnetic and Light Scattering*, ed. T. Wriedt (Elsevier, Amsterdam, 1999)
28. Y. Okuno, T. Matsuda, T. Kuroki, Diffraction efficiency of a grating with deep grooves, in *Proceedings of the 1995 Sino-Japanese Joint Meeting on Optical Fiber Science and Electromagnetic Theory (OFSET'95)*, vol. 1 (Tianjin, China, 1995), pp. 106–111
29. Y. Okuno, T. Suyama, R. Hu, S. He, T. Matsuda, Excitation of surface plasmons on a metal grating and its application to an index sensor. Trans. IEICE Electron. **E90-C**(7), 1507–1514 (2007)
30. Y. Okuno, H. Yamaguchi, The idea of equivalent sources in the Yasuura method, in *Proceedings 1992 International Symposium on Antennas Propagat (ISAP'92)*, vol. 1E3-2 (Sapporo, Japan, 1992)

31. Y. Okuno, K. Yasuura, Numerical algorithm based on the mode-matching method with a singular-smoothing procedure for analysing edge-type scattering problems. IEEE Trans. Antennas Propagat. **30**(4), 580–587 (1982)
32. R. Petit (ed.), *Electromagnetic Theory of Gratings* (Springer, Berlin, 1980)
33. R. Petit, M. Cadilhac, Electromagnetic theory of gratings: some advances and some comments on the use of the operator formalism. J. Opt. Soc. Am. A **7**(9), 1666–1674 (1990)
34. H. Raether, Surafce plasmon and roughness, in *Surface Polaritons — Electromagnetic Waves at Surfaces and Interfaces*, ed. by V.M. Agranovich, D.L. Mills (North Holland, 1982), pp. 331–403
35. M. Tomita, K. Yasuura, The Rayleigh expansion theorem for the boundary value problem in two media. Kyushu Univ. Tech. Rep. **52**(2), 142–154 (1979)
36. J.R. Wait, Reflection from a wire grid parallel to a conducting plane. Can. J. Phys. **32**, 571–579 (1954)
37. X. Xu, B.W. Chen, R. Gong, M. Zheng, Use of auxiliary source fields in Yasuura's method, in *Proceedings of the 2017 IEEE International Conference on Computational Electromagnetics (ICCEM2017)*, vol. 2C1.2 (Kumamoto, Japan, 2017)
38. K. Yasuura, T. Itakura, Approximation method for wave functions (I). Kyushu Univ. Tech. Rep. **38**(1), 72–77 (1965)
39. K. Yasuura, T. Itakura, Complete set of wave functions – approximation method for wave functions (II). Kyushu Univ. Tech. Rep. **38**(4), 378–385 (1966)
40. K. Yasuura, T. Itakura, Approximation algorithm by complete set of wave functions – approximation method for wave functions (III). Kyushu Univ. Tech. Rep. **39**(1), 51–56 (1966)
41. K. Yasuura, H. Ikuno, Smoothing process on the mode-matching method for solving two-dimensional scattering problems. Mem. Fac. Eng. Kyushu Univ. **37**(4), 175–192 (1977)
42. K. Yasuura, Y. Okuno, Singular-smoothing procedure on Fourier analysis. Mem. Fac. Eng. Kyushu Univ. **41**(2), 123–141 (1981)
43. K. Yasuura, M. Tomita, Convergency of approximate wave functions on the boundary – the case of inner domain. Kyushu Univ. Tech. Rep. **52**(1), 79–86 (1979)
44. K. Yasuura, M. Tomita, Convergency of approximate wave functions on the boundary – the case of outer domain. Kyushu Univ. Tech. Rep. **52**(1), 87–93 (1979)
45. K. Yasuura, M. Tomita, Numerical analysis of plane wave scattering from dielectric cylinders. Trans. IECE Jpn. **62-B**(2), 132–139 (1979)
46. K.A. Zaki, A.R. Neureuther, Scattering from a perfectly conducting surface with a sinusoidal height profile: TE polarization. IEEE Trans. Antennas Propagat. **AP-19**(2), 208–214 (1971)

Chapter 9
Pole Location in GMT

James E. Richie

Abstract This chapter begins with an overview of the variety of methods and techniques used to choose discrete pole locations in the family of Generalized Multipole Techniques (GMT). The heuristic rules and guidelines that are described are often quite successful. In addition, studies of the performance of GMT methods for canonical problems are reviewed. It has been shown that there are at least two sources of error when using GMT: analytically-based error and numerically-based error. The effective spatial bandwidth (EBW) of fields along the boundary of scatterers is described and used to show the conditions necessary to obtain stable solutions from GMT techniques. The EBW for two-dimensional circular boundaries is applied to some examples. In addition, an extension of EBW for non-circular boundaries is described and applied to elliptically shaped boundaries.

9.1 Introduction: GMT and Its Variations

There are many techniques to compute the electromagnetic scattering from objects. The methods considered here are based on the use of discrete sources to model the scattered field and (in the case of non-perfectly conducting scatterers) the internal field. The name used here for these methods is the Generalized Multipole Technique (GMT) [1]. Authors refer to particular implementations of these methods as the Multiple Multipole Program (MMP), the Method of Auxiliary Sources (MAS), the Filamentary Current Method (FCM), and the Method of Fictitious Sources (MFS).

In Fig. 9.1 a scattering object is shown. The object has some permittivity ε, permeability μ, and conductivity σ. There is an incident field denoted as $\mathbf{E}^i$ and $\mathbf{H}^i$. The goal is to determine the scattered ($\mathbf{E}^s$, $\mathbf{H}^s$) and internal fields ($\mathbf{E}^d$, $\mathbf{H}^d$).

One model for determining the unknown fields is the Rayleigh hypothesis. A multipole expansion with all poles at a single origin is used. A large number of poles may be needed at the origin depending on the size and complexity of the object.

J. E. Richie (✉)
Marquette University, Milwaukee, WI, USA
e-mail: james.richie@marquette.edu

T. Wriedt and Y. Eremin (eds.), *The Generalized Multipole Technique for Light Scattering*, Springer Series on Atomic, Optical, and Plasma Physics 99,
https://doi.org/10.1007/978-3-319-74890-0_9

Fig. 9.1 The general scattering problem

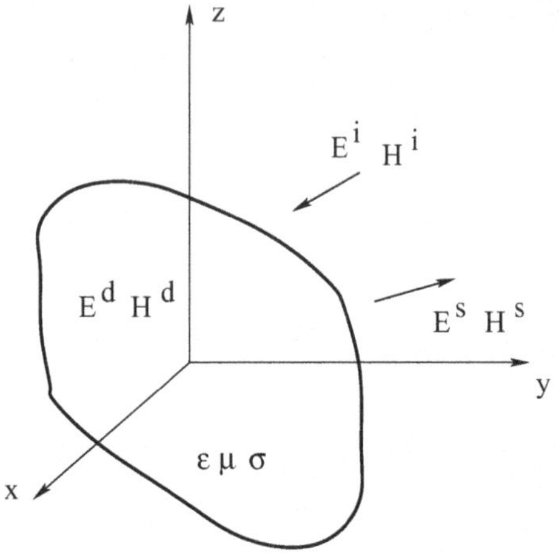

Computing the coefficients may be difficult and the series solution may converge slowly.

A second approach is to use the moment method [2]. In this approach, an integral equation (IE) is formulated that expresses the boundary conditions of the problem assuming an unknown current on the object surface. The current is modeled using basis functions with unknown coefficients. Then, testing functions are applied to the resulting modified integral equation to compute the coefficients. Because the current is on the boundary and the IE is evaluated on the boundary, special care must be taken near the singularity in the kernel.

In the moment method, enough computational complexity must be applied to the basis and/or testing functions to obtain suitable accuracy [3]. To ensure sufficient accuracy, sophisticated basis functions with simple testing functions can be used or vice versa. Similar results are obtained by using a lesser amount of sophistication applied equally to the basis and testing functions.

In the GMT methods, multipoles are used. The sources are not at a single origin as in Rayleigh's hypothesis, and the sources are not on the boundary as in the moment method. Typically, scattered field poles are inside the object and internal field poles are outside the object.

Each multipole has unknown strength. To determine the pole coefficients, the boundary conditions are satisfied at a discrete set of locations on the surface of the object, i.e., collocation or point matching is used. Using a number of boundary condition equations equal to or greater than the number of unknowns, a linear system of equations (LSE) can be solved to obtain the unknown coefficients.

Sufficient accuracy can be obtained using point matching due to the sophisticated arrangement of the multipoles. The pole locations are adjusted according to the complexity of the problem. Borrowing terms from the moment method, sophisticated "basis" functions are used and therefore point matching can be applied with little loss in accuracy.

A number of researchers have applied discrete source methods to electromagnetic scattering. GMT is discussed in [4], where two-dimensional and three-dimensional problems have been solved using the Multiple Multipole Program (MMP) [5]. The MMP implements multipoles at multiple origins and uses an overdetermined LSE to find the pole strengths. The work in [6] is a similar implementation of GMT for a conducting body of revolution.

In [7–9], the Filamentary Current Method (FCM) is described for two/dimensional problems. In FCM, filaments of current (monopoles) are used to model the scattered and internal fields. The same research group also reported some three-dimensional work in [10].

The Method of Auxiliary Sources (MAS) is fundamentally formulated from [11]. The MAS is organized slightly differently than the previous techniques [12]. In the MAS, auxiliary sources are placed on an auxiliary surface (AS). The auxiliary sources are usually the lowest order poles, and the auxiliary surface is typically conformal to the boundary. The sources are homogeneously placed on the AS.

There are a number of advantages to GMT methods. By choosing multiple origins that are away from the surface, the singularity inherent in the moment method is avoided. Another advantage is that the effect of the multipoles becomes approximately localized. Each multipole expansion has a strong effect along the portion of the boundary that is closest to the pole origin. This localization improves convergence of the scattered field model.

In addition, the integration along the bounding surface that is often necessary in moment methods is avoided by using collocation. This results in a computational algorithm that is more efficient than the moment method.

A final advantage of GMT is that the boundary conditions can easily be verified by computing the fields along the surface of the scatterer. This feature of GMT solutions is highly significant because the accuracy of the solution can be inferred from the boundary condition error.

Although GMT and its variations have several advantages, there are some disadvantages as well. One disadvantage is that objects with edges and corners require special care to solve for the fields. A second disadvantage of the GMT methods is that it may be unclear where to place the multipoles. The number, location, and order of the poles is not obvious when solving a specific scattering problem.

9.2 Placement Rules Developed and Utilized

Some researchers have reported "rules" for the placement of poles for solving scattering problems in GMT. Typically, the rules are cast within the specific variation of GMT, i.e., MAS rules, MFS rules, etc. In this section, the variety of heuristic rules proposed for GMT shall be discussed.

A few early guidelines for the use of GMT appear in [6, 8, 9]. In [6], the scatterer is a cone-sphere, which is a body of revolution. The multipoles are placed along the axis of rotation. This is well-known to be a beneficial choice for bodies of revolution. Also,

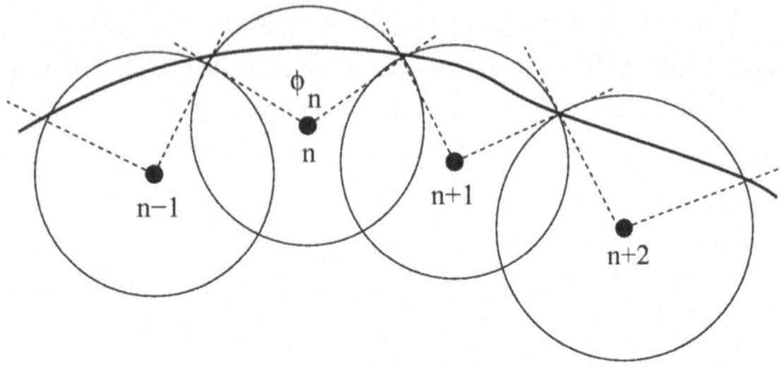

Fig. 9.2 Illustration of the region of influence. The pole origins are the dots, the boundary is solid, and the regions of influence are the circles

it is shown that the order of the poles should increase as the pole origin approaches the tip of the cone. In [8, 9], the scattering of a plane wave by a dielectric cylinder of radius a is modeled using filaments of current. The poles used to model the scattered field are found to be ideally situated between $0.5a$ and $0.8a$; for the internal field, $2a$ is shown to be an adequate pole distance.

Hafner published some heuristic guidelines in [4]:

- pole origins must be outside the domain of the modeled field;
- no pole should be within the region of influence of another pole;
- the minimum distance from the pole origin to the boundary should be on the order of one wavelength or less.

A key concept in this list of rules is the region of influence. Figure 9.2 illustrates the concept of domain of greatest influence for a pole. The boundary is divided into subdomains where each subdomain is linked to the origin of a multipole expansion. A circle is drawn, centered at the origin of the pole and has a radius somewhat larger (roughly 1.2–1.4 times) the minimum distance from the pole origin to the boundary. The region of influence admits an angle, ϕ_n for origin n, as shown in Fig. 9.2.

Hafner provided additional guidelines regarding the poles and match points. Each origin consists of an expansion of poles that begins at the lowest order (0 for two dimensions, 1 for three dimensions) and includes all orders up to the maximum. For smooth boundary regions, the angles ϕ_n should be roughly equivalent. For more complex boundary regions, as the angle ϕ_n increases, the highest pole order required for origin n should increase.

Additional guidance regarding the nature of the pole distance to the boundary is reported in [13]. As the radius of curvature of the boundary decreases, the poles should come closer to the boundary and the poles should come closer together. This was demonstrated using cylinders of elliptical cross section and the packing number concept.

Pole location in the MAS is described differently. In MAS [12], an auxiliary surface that is conformal to the boundary is defined. The lowest order poles for the problem are then placed in a uniform fashion along the AS. As the AS approaches the boundary, the density of the poles increases. The lowest order poles in the two-dimensional case are monopoles (or current filaments). In the three-dimensional case, pairs of perpendicular dipoles, both tangent to the boundary, are used.

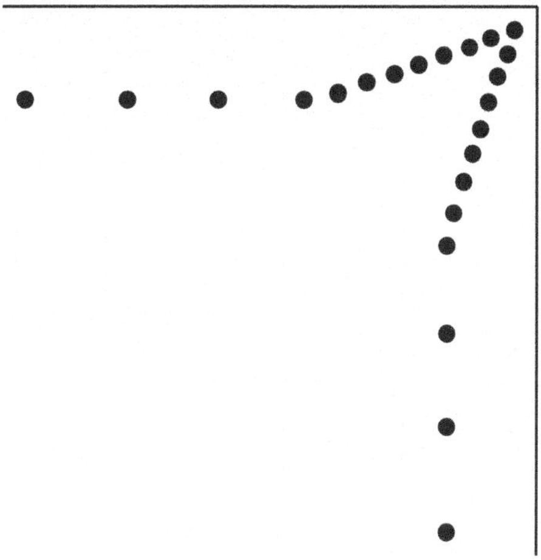

Fig. 9.3 Pole locations near a corner or edge

One disadvantage of GMT techniques is modeling the scattered field for scatterers with edges and/corners. There are two approaches to resolve this problem. The first is to slightly round off the corner of the scatterer itself. In the cone-sphere object case [6], it is reported that a radius of 0.01 λ results in a radar cross section error less than 0.5 dB. A second technique for edges is described in [14] and is shown in Fig. 9.3. In this approach, poles are placed closer to the boundary as the expansion origins approach the edge. As the origins approach the edge, the pole density increases along with the matching point density. Although both approaches can provide acceptable results, [12] suggests that the second approach provides more accurate results than rounding the edge.

9.2.1 Automatic and Semi-automatic Pole Placement Approaches

A very early attempt to automatically place the poles in GMT is reported in [15]. In this method, a network of origins is defined for possible pole locations. To place a pole, all possible origins are tested to determine which location will minimize the residual boundary condition error. This (rather slow) procedure results in a monotonically decreasing boundary condition error as poles are added to the field model.

A detailed discussion of rules for the nearly automatic placement of multipoles appears in [16] and represents a refinement of Hafner's guidelines discussed earlier. The guidelines are based on the geometry of the boundary along with the wavelengths of fields in each region.

The pole placement procedure is based on three ideas:

- the complexity of the fields is correlated with the complexity of the geometry;
- the density of poles should correspond to the complexity of the geometry;
- the pole locations should avoid numerical dependencies that result in ill/conditioned matrices.

Based on these principles, a set of heuristic rules are developed:

- the distance from the pole location to the boundary is chosen according to the local radius of curvature of the boundary;
- every section of the boundary must be within the region of influence of some expansion origin;
- the distance between the multipoles must be larger than the radius of each region of influence and be somewhat larger than the distance of each origin from the boundary;
- a multipole origin must be outside the domain of validity of the field.

These rules ensure that pole origins are neither too close to each other or to the boundary. Reference [16] provides a number of examples that illustrate these rules and some common exceptions for pole placement.

The rules are implemented by constructing an auxiliary surface that the multipole expansions lie on. The algorithm used to define the auxiliary surface depends only on the local properties of the boundary. The exceptions in [16] show that non-local effects can cause pole locations to severely violate the spirit of the rules.

Other automated procedures related to pole location have been reported. In [17] a scheme is described for detecting (and removing) pole locations that cause the matrix to be ill-conditioned. In [18, 19], a genetic algorithm is used to find the optimum pole placement in MAS for perfectly conducting and dielectric cylinders, respectively.

9.2.2 Singularities of the Scattered Field

In principle, the scattered field induced by some incident field on an object can be analytically continued into the object. However, this analytic continuation is limited by the location of the scattered field singularities (SFSs). The SFSs must exist, or the scattered field would be constant everywhere.

The pole locations in GMT methods are the singular points in the scattered field model. The GMT pole locations must enclose the SFSs of the true scattered field to obtain an accurate model for the scattered field. A fundamental analysis of the scattered field singularities and their relationship to the scatterer boundary appears in Chap. 5 of [1].

To illustrate the importance of the SFSs, consider the scattering of a monopole line source in the vicinity of a perfectly conducting cylinder of radius a, as shown in Fig. 9.4. The line source is at a radius ρ_o from the center of the cylinder. It is well known that the analytic continuation of the scattered field for this canonical problem

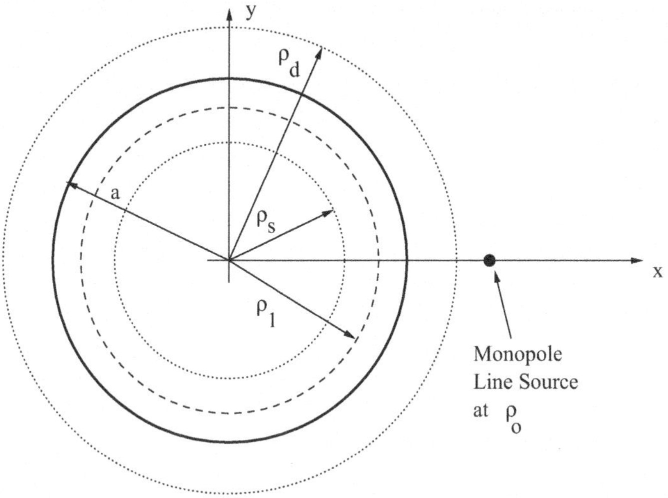

Fig. 9.4 Geometry of scatterer (radius a) with a monopole line source at ρ_o. Pole locations shown at ρ_1 outside the radius of the singularity (ρ_s)

exists within the cylinder for radius ρ between a and a^2/ρ_o (indicated as $\rho_s = a^2/\rho_o$ in the figure).

An early investigation of the performance of GMT on the problem for Fig. 9.4 (perfectly conducting cylinder case) appears in [7]. This study shows that pole locations that do not enclose the singularity provide results that are not stable. Specifically, as the number of poles increases, the solution tends to vary considerably with little improvement in the boundary condition error. However, for pole locations that do enclose the singularity (such as ρ_1 in the figure), the results are very stable and the boundary condition error falls rapidly as the number of poles increases.

One way to understand the role of the SFS's is to consider a point source a distance d above a perfectly conducting ground plane. The SFS is at a distance d below the ground plane. In [20, 21], the MAS is used to investigate pole locations for this canonical problem. An auxiliary surface is placed parallel and under the ground plane at a distance $b < d$. For b near zero (close to the ground plane), a large number of poles are needed to accurately model the scattered field. As b increases, fewer poles are required for similar accuracy. When $b = d$, only one pole is needed.

This result along with some additional canonical examples are used to argue that MAS can be extremely efficient, particularly for large scatterers. This is accomplished by using the SFS locations (or caustic surfaces) as the location of the auxiliary sources.

For general scatterers, a scheme is developed and applied in [21] that computes the locations of the caustics for a more general scatterer. The techniques are also applied to the inverse scattering problem.

9.2.3 Summary Comments

The semi-heruistic rules described earlier along with the examples provided in [16] provide a suitable framework for choosing pole locations that are likely to result in a reasonable solution to a scattering problem. Coupling these rules with knowledge of the scattered field singularities and the special considerations for particular geometric details (such as edges or corners), GMT and its variations are quite successful.

Despite the success of the rules and guidelines described in this section, the use of GMT and its variations continues to rely on the experience and intuition of the person applying the method, particularly when choosing pole locations. Additionally, situations have been observed where the method fails for no apparent reason. These difficulties have led to studies regarding convergence and error analysis of the MAS. In the next section, an error analysis of MAS and discussion of convergence issues for these techniques will be described.

9.3 Convergence and Error Analysis in MAS

The creation of guidelines for pole placement in the GMT methods is one step toward understanding the best implementation of discrete source methods. Another step is to investigate the nature of error and convergence in the GMT methods for canonical problems. In this section, convergence and error analysis of the MAS will be described for the case of a circular cylindrical scatterer.

The Method of Auxiliary Sources is based on the use of auxiliary sources on an auxiliary surface. The fundamental theory considers a continuous source on the auxiliary surface. In practice, an approximation to the theory is used where a discrete collection of N monopoles are placed on the auxiliary surface. Convergence is concerned with the nature of the solution as the number of sources increases. Therefore, as $N \to \infty$, the solution for the fields should converge to the exact solution. However, it is possible for the solution to not converge as N increases. Error analysis is concerned with the characteristics of the error in the implementation of the MAS. This error can have multiple sources. For example, errors can be due to the numerical computations, and errors can be due to the approximation of the continuous source as a set of N poles.

9.3.1 MAS Convergence Analysis: Monopole Line Source Incident Field

One area of recent interest is the nature of the convergence of MAS methods as N approaches infinity. In [22], the convergence properties of the MAS for a line source scattering from a perfectly conducting cylinder (see Fig. 9.4) are reported. Cases where the auxiliary surface does enclose the SFS's and where it does not enclose the SFS's are compared.

When the AS encloses the singularities, the MAS pole strengths are well-behaved and converge as N increases. In addition, the pole strengths in this case also replicate the analytic continuation of the scattered field within the object between the AS and the scatterer boundary.

When the AS does not enclose the singularities, the pole strengths oscillate and do not converge as N approaches infinity. Fields between the AS and the scatterer boundary can not replicate the analytic continuation of the scattered field within the object. The pole strength oscillations are not due to matrix ill-conditioning. Even though the coefficients do not converge, the fields do converge to a close approximation of the actual scattered fields on and outside the object.

In [23], the results of [22] are extended to dielectric circular cylinders with a line source incident field. For the dielectric case, an auxiliary surface (of radius ρ_d) outside the cylinder is used to model the internal field as shown in Fig. 9.4. If $\rho_d < \rho_o$ (as shown in the figure), the pole strengths are stable and convergent. If $\rho_d > \rho_o$, then the pole coefficients are not stable. In both cases, the fields within the cylinder converge toward the solution.

Thus, as the research into the nature of convergence for the MAS shows, there are two distinct types of convergence. The first is concerned with the convergence of the scattered and internal field. The second is concerned with the convergence of the pole strengths. In the preceding analysis, it was demonstrated that the pole coefficients converge if the singularities are enclosed by the AS. The field model can converge to the solution even when the pole coefficients do not converge.

9.3.2 MAS Accuracy Analysis: Plane Wave Incident Field

A second area of recent interest is the analysis of the accuracy of MAS. Consider the problem of a TM^z incident plane wave on a perfectly conducting circular cylinder. Plane wave illumination can be considered as monopole illumination where the distance between the monopole and the cylinder is very large. As the monopole moves away from the cylinder, the singularities approach the origin. Therefore, the auxiliary surface for a cylinder illuminated by a plane wave can range in size from nearly zero to nearly the radius of the cylinder.

In [24], the MAS solution to the plane wave incidence problem is analyzed. The MAS matrix is analytically derived and investigated in a fashion similar to [25]. The condition number and invertability of the matrix are studied. It is shown in [24] that the computational error can be separated into an analytical and a numerical component. The error that dominates the total error in the solution is found to depend on the radius of the auxiliary surface.

The following is a brief synopsis of the results reported in [24]. First, consider the effect of the auxiliary surface (AS) radius. For very small AS radius, the numerical error dominates. This is due to the numerical precision of the computer used to perform the calculations. As the AS radius increases, the numerical error falls. At some point, the increasing radius reaches a region where the analytical error dominates the cumulative error.

As the AS radius increases further, the analytical error increases. For an AS radius nearly equal to the cylinder radius, the numerical error again begins to increase faster than the analytical error. This increase in the numerical error is attributed to the very large entries along the diagonal of the matrix since the poles are very close to the matching points on the cylinder.

For an AS radius where the analytical error dominates the cumulative error, increasing the number of auxiliary sources results in lower boundary condition error. This is the result expected from the fundamental theory that MAS is based upon.

Finally, a set of spikes in the numerical error are observed at very specific and consistent ratios of AS radius to cylinder radius. Using the analytically derived solution, it is shown that these errors are due to zeros of the Bessel function.

In conclusion, it is suggested in [24] that the auxiliary surface should be placed as far from the boundary as possible and yet close enough to the boundary to avoid numerical errors. One convenient method of estimating the optimum location is to monitor the condition number of the matrix, since the condition number is a way to approximate the expected level of numerical error.

The analysis in [24] for TM^z plane wave scattering on a perfectly conducting circular cylinder is repeated in [26] for the TE^z plane wave PEC case, in [27] for the dielectric cylinder, and in [28] for the case of a cylinder with material properties that allow the use of the surface impedance boundary condition (SIBC).

In each case, the results are very similar. When the auxiliary surface has a very small radius, the numerical error dominates because of the numerical precision of the machine. When the auxiliary surface radius approaches the cylinder radius, the large diagonal entries cause numerical errors. Isolated spikes in the numerical error are attributed to zeros of the Bessel functions.

For the dielectric case, the radius of the auxiliary surface to model the internal field is also investigated. In this case, no numerical difficulties are encountered.

It is instructive to correlate the results of the MAS accuracy studies in [24, 26–28] with the heuristic rules discussed earlier [16]. In particular, the large numerical errors are avoided by appropriate application of the rules: poles should neither be too close together nor too close to the surface. In fact, as the poles approach the surface, they ideally should proportionately approach each other to maintain suitable regions of influence.

9.3.3 Summary Comments

The studies described in this section focus on the numerical implementation of MAS and details of the resulting solutions for some well-known canonical problems. However, there are other ways to describe and quantify the characteristics of the scattered field model. In two-dimensional MAS, the fundamental object in the scattered field model is a monopole displaced from the origin. In the next section, the characteristics of the displaced monopole will be compared to the characteristics of the incident field to obtain additional insight regarding pole placement.

9.4 Effective Spatial Bandwidth

In this section, the effective spatial bandwidth (EBW) of fields on a boundary is introduced, described, and computed to ascertain the range of radii that will provide MAS solutions that avoid excessive numerical error (as in [24]) and avoid large oscillations in the MAS source strengths (as in [22]). Two examples will be used to demonstrate that EBW is an effective, relatively practical engineering tool to evaluate the suitability of pole locations.

9.4.1 Introduction

An incident field scatters off of a circular cylinder with radius a, as shown in Fig. 9.5. The cylinder is infinitely long, and the incident field is polarized such that the electric field is in the z direction, parallel to the cylinder axis. There is no field variation in the z direction. The incident field can be a plane wave or a line source in the vicinity

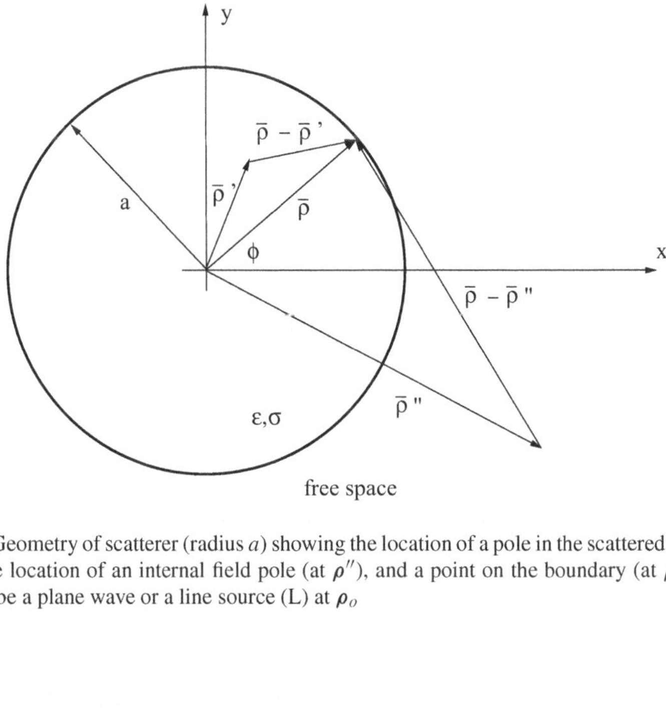

Fig. 9.5 Geometry of scatterer (radius a) showing the location of a pole in the scattered field model (at $\boldsymbol{\rho}'$), the location of an internal field pole (at $\boldsymbol{\rho}''$), and a point on the boundary (at $\boldsymbol{\rho}$). Incident field may be a plane wave or a line source (L) at $\boldsymbol{\rho}_o$

of the cylinder (both are shown in the figure). The line source may be a monopole or a multipole of some order with even or odd symmetry in the angle ϕ. The scatterer in general has conductivity σ and permittivity ε. An $\exp\{j\omega t\}$ time dependence is assumed throughout.

Using Rayleigh's hypothesis, the scattered field is

$$E^s_{z,analytic} = \sum_{n=-\infty}^{\infty} c_n H_n^{(2)}(k\rho)\, \mathrm{e}^{jn\phi} \tag{9.1}$$

where k is the wavenumber outside the cylinder, and $H_n^{(2)}(\cdot)$ is the Hankel function of the second kind of order n representing outward traveling waves. If the incident field is a unit strength plane wave, $E^i_z = \exp\{-jkx\}$, then

$$c_n = -j^{-n}\frac{J_n(ka)}{H_n^{(2)}(ka)} \tag{9.2}$$

where a is the cylinder radius.

In GMT, the scattered field is modeled as a collection of monopoles:

$$E^s_z = \sum_{n=1}^{N_p} a_n H_o^{(2)}(k\,|\,\boldsymbol{\rho} - \boldsymbol{\rho}'_n\,|) \tag{9.3}$$

where there are N_p poles with amplitudes a_n at locations $\boldsymbol{\rho}'_n$, and $|\,\boldsymbol{\rho}'_n\,| < a$. We have assumed only monopoles in the scattered field model. The internal field is also modeled using a set of monopoles at origins $|\,\boldsymbol{\rho}''_n\,| > a$ as

$$E^d_z = \sum_{n=1}^{N_p} b_n H_o^{(2)}(k_d\,|\,\boldsymbol{\rho} - \boldsymbol{\rho}''_n\,|) \tag{9.4}$$

where k_d is the wavenumber inside the cylinder. In both (9.3) and (9.4), the poles are equally spaced in ϕ.

Suppose the incident field is a plane wave such that the solution for E^s_z is given by (9.1) with c_n in (9.2). The scattered field model consists of monopoles displaced from the global origin of the scatterer. A displaced monopole can be written as a multipole expansion at the global origin using the addition theorem (AT):

$$H_o^{(2)}(k\,|\,\boldsymbol{\rho} - \boldsymbol{\rho}'\,|) = \sum_{n=-\infty}^{\infty} J_n(k\rho_<)H_n^{(2)}(k\rho_>)\, \mathrm{e}^{jn(\phi-\phi')} \tag{9.5}$$

where $\rho_<$ ($\rho_>$) is the smaller (larger) of ρ and ρ'.

Consider a single monopole in the scattered field model (9.3). The distance $\rho' < a$, so the coefficients of the AT expansion are governed by the Bessel function $J_n(k\rho')$. If

ρ' is very small, only a few terms in the AT expansion are significant because $J_n(k\rho')$ will quickly go to zero as n increases. However, c_n may have considerable values for many more terms. In this case, the scattered field model may not accurately represent the actual scattered field. This is analogous to the situation described earlier and in [24]. The auxiliary surface (i.e., ρ_n') is too small and the numerical errors dominate the errors in the scattered field computation.

If ρ' is very small, the a_n coefficients are affected as well. If $J_n(k\rho')$ decays more quickly than c_n, then the computed coefficients, a_n in (9.3), will become very large. To match the boundary condition, the sum in (9.3) must remain bounded. Therefore, while the magnitude of a_n diverges, the phase of a_n will provide extensive cancelation to the fields along and outside the boundary, as explained in [22].

On the other hand, consider the case where ρ' is large enough. The AT expansion coefficients will decay at a rate that is similar to c_n. The pole locations admit a scattered field error that is dominated by the analytic error of [24]. The a_n will most likely be stable values at a reasonable magnitude.

9.4.2 Theoretical Development of EBW

In this section, the concept of effective spatial bandwidth of the fields is described. A procedure to quantify the bandwidth is developed and used to aid in pole placement in the GMT.

A plane wave is incident upon the cylinder of the previous section. On the cylinder boundary, the incident field has a constant magnitude and the phase variation is correlated to the position on the boundary. As the cylinder radius increases, the phase of the incident field varies more. If a line source is the incident field, the magnitude of the field along the boundary will also vary according to the distance between the line source and the cylinder boundary.

Suppose a GMT monopole is inside the cylinder. If the monopole is at the origin ($\rho' = 0$), the field along the boundary has a constant magnitude and phase. As ρ' increases, the boundary field variation increases. If $\rho' = a$, then the field of the monopole on the boundary is singular at the monopole.

Thus, if the amount of variation of the field along the boundary can be quantified for both the incident field and the monopoles in the scattered field model, then the GMT pole radius can be chosen so that the amount of variation of the GMT pole matches the variation of the incident field along the boundary. Using GMT poles with similar variation is likely to result in coefficients a_n and b_n that are stable and well-behaved in addition to satisfying the boundary conditions.

The variation of the fields along the boundary of the scatterer will be quantified using the effective spatial bandwidth of the fields. An early study on the bandwidth of scattered fields can be found (for the non-periodic case) in [29]. In the two-dimensional case described here, the scattered field is periodic. The "bandwidth" or Effective spatial BandWidth (EBW) is introduced in [30], and is summarized below.

Consider a field quantity such as E_z^s, denoted $e(\phi)$, around the circular boundary of Fig. 9.5. Define the energy of $e(\phi)$:

$$\mathscr{E}(e) = \int_0^{2\pi} |e(\phi)|^2 \rho(\phi) d\phi \tag{9.6}$$

where $\rho(\phi)$ is the distance from the origin to the point on the scatterer at angle ϕ. The function $e(\phi)$ can have many harmonics. Since the boundary is a circle, the harmonics will be $\exp\{jn\phi\}$ where n is an integer. Our present goal is to find the function $e_N(\phi)$ that has a maximum spectral component $\exp\{jN\phi\}$. One could convert $e(\phi)$ to the spectral domain, eliminate terms with coefficients greater than N, and convert back to the spatial domain to find $e_N(\phi)$. Analytically, it is convenient to use the convolution:

$$e_N(\phi) = \frac{1}{C} \int_0^{2\pi} B_N(\phi, \xi) e(\xi) \rho(\xi) d\xi \tag{9.7}$$

where the subscript N indicates that the function has been bandlimited to a maximum harmonic of order N, C is the circumference of the scatterer, and

$$B_N(\phi, \xi) = \frac{\sin\left[\left(N + \frac{1}{2}\right)(\phi - \xi)\right]}{\sin\left[\frac{1}{2}(\phi - \xi)\right]} \tag{9.8}$$

For a periodic function $e(\phi)$, define $EBW = N$ such that N is the smallest integer with $\Delta_N \leq 0.1\%$, where

$$\Delta_N = \frac{\mathscr{E}(e) - \mathscr{E}(e_N)}{\mathscr{E}(e)} \times 100\% \tag{9.9}$$

In other words, find N such that 99.9% of the energy is within the first N harmonics.

Computation of the EBW for the incident field along the boundary is now possible using the above definition. The EBW for a monopole in the GMT scattered field model can also be found. If the location of the GMT monopole is varied from a very small radius to nearly the radius of the scatterer, there should be a radius where the EBW of the monopole matches the EBW of the incident field.

If the pole location is at a radius where the EBW of the poles matches or is a bit larger than the incident field EBW, then, the GMT solution for the scattered field would be well behaved because the model includes enough field variation along the boundary to match the boundary conditions. The resulting coefficients a_n should also be well-behaved and converge as N_p approaches infinity.

If the pole locations are at a much smaller radius, then the EBW of the GMT poles is significantly smaller than the incident field EBW. In this case, the GMT poles are unable to match the variations in the incident field along the boundary. Therefore, the numerical implementation results in a_n coefficients that have divergent magnitudes and oscillating phases to create the field variation required to match the boundary conditions.

9.4.3 Results

In this section, the EBW of fields along the circular boundary are computed for several different cases.

9.4.3.1 Plane Wave Incident Field

Consider a unit strength plane wave incident on the cylinder, traveling in the $+x$ direction. The electric field is given by $\mathbf{E} = E_z^p \hat{e}_z$, where

$$E_z^p = \mathrm{e}^{-jkx} = \mathrm{e}^{-jk\rho\cos\phi} = \sum_{n=-\infty}^{\infty} j^{-n} J_n(k\rho)\mathrm{e}^{jn\phi} \tag{9.10}$$

where the superscript p indicates plane wave.

The field E_z^p on the boundary of a cylinder with radius a can be limited to a bandwidth of N:

$$E_{z,N}^p(\phi) = \frac{1}{2\pi} \sum_{n=-\infty}^{\infty} [j^{-n} J_n(ka)] \int_{\xi=0}^{2\pi} B_N(\phi,\xi)\mathrm{e}^{jn\xi}\, d\xi \tag{9.11}$$

The integral is $2\pi \exp\{jn\phi\}$ if $|n| < N$ and is zero otherwise. As N increases, the quantity in brackets approaches zero. Once the bracketed quantity becomes small enough so that $\Delta_n < 0.1\%$, the EBW is found.

The EBW for a plane wave incident on the cylinder can also be computed numerically by integrating (9.6) and (9.7). A graph of the EBW of a plane wave as a function of cylinder radius a is shown in Fig. 9.6.

As expected, when a is very small, the plane wave EBW is also small because the phase remains fairly constant around the boundary. As a increases, the phase variation around the boundary increases resulting in a larger EBW.

9.4.3.2 Monopole Line Source Incident Field

The field due to a unit strength monopole line source at ρ_o is written as:

$$E_z^o = H_o^{(2)}(k \mid \boldsymbol{\rho} - \boldsymbol{\rho}_o \mid) \tag{9.12}$$

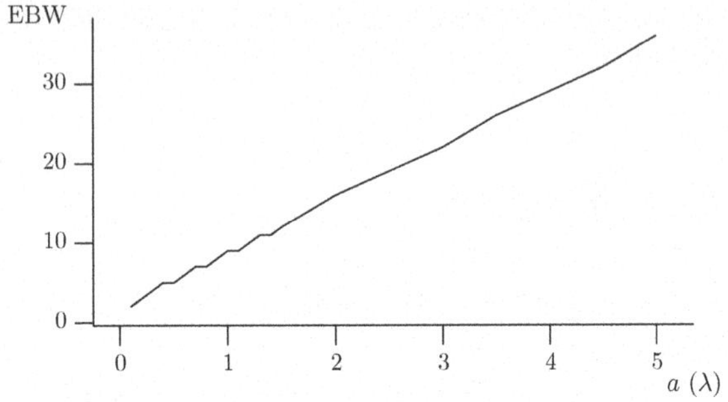

Fig. 9.6 EBW for a plane wave incident on a cylinder of radius a (in wavelengths)

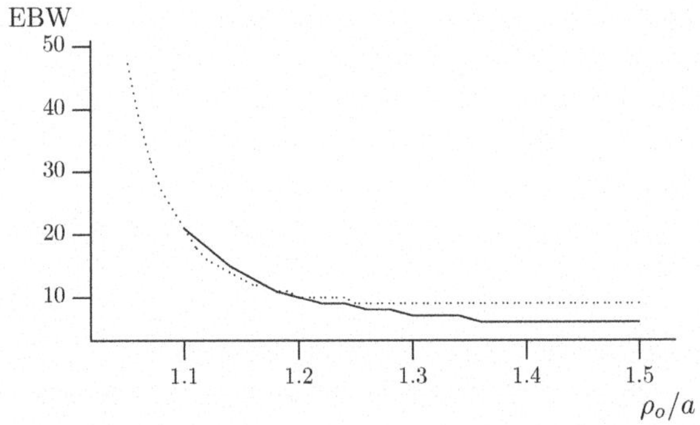

Fig. 9.7 EBW for a monopole line source at $\rho_o > a$ for: $a = 0.5\,\lambda$ (solid); $a = 1.0\,\lambda$ (dotted)

where the superscript o indicates a monopole line source. The EBW for the line source on the cylinder boundary can be computed in the same fashion as the plane wave incident field case. Figure 9.7 shows the results of this calculation for cylinders of radius $0.5\,\lambda$ and $1.0\,\lambda$.

For a line source very close to the scatterer, the EBW is quite large and diverges as $\rho_o/a \rightarrow 1$. As ρ_o increases, the EBW converges to the plane wave value. Convergence to the plane wave result depends on both ρ_o and a. For larger a, a larger ρ_o is needed to reach the plane wave approximation.

9.4.3.3 GMT Monopole Located Inside Cylinder

A monopole inside the scattering cylinder represents one term of the scattered field model. The electric field at $\boldsymbol{\rho}$ for a unit strength monopole at location $\boldsymbol{\rho}'$ is given by:

$$E_z^M = H_o^{(2)}(k\,|\,\boldsymbol{\rho} - \boldsymbol{\rho}'\,|) \tag{9.13}$$

where the superscript M indicates GMT monopole. After applying the AT, the bandlimited field for the monopole can be computed as:

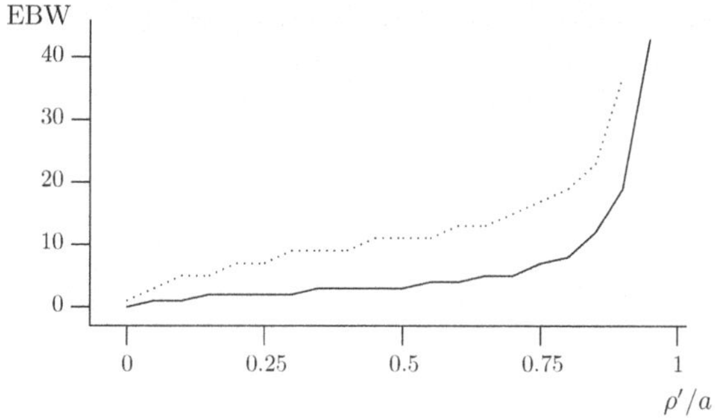

Fig. 9.8 EBW for a monopole inside the cylinder: $a = 0.5\,\lambda$ (solid); $a = 1.0\,\lambda$ (dotted)

$$E^M_{z,N} = \frac{1}{2\pi}\sum_{n=-\infty}^{\infty} J_n(k\rho')H_n^{(2)}(ka)\mathrm{e}^{-jn\phi'}\int_{\xi=0}^{2\pi} B_N(\phi,\xi)\mathrm{e}^{jn\xi}\,d\xi \tag{9.14}$$

where the same integral as found in (9.11) is seen. If $\rho' = 0$, the monopole is at the origin. The field around the cylinder will be constant and the only non-zero term in the series of (9.14) will be the $n = 0$ term, as seen by the $J_n(k\rho')$ factor. As ρ' increases, the number of terms in (9.14) with a significant $J_n(k\rho')$ will increase as well. The number of significant terms in the series is an indication of the EBW for $E^M_{z,N}$. Therefore, as ρ' increases, the EBW increases as well.

The EBW for a GMT monopole can be computed by numerically integrating as before. The EBW for a monopole within cylinders of radius $0.5\,\lambda$ and $1\,\lambda$ are plotted as a function of ρ' in Fig. 9.8.

As seen in Fig. 9.8, the EBW for the GMT monopole increases slowly at first and diverges as $\rho' \to a$. A larger cylinder results in a higher EBW for a similar monopole location.

9.4.3.4 GMT Monopole Located Outside Cylinder

For penetrable scatterers, the internal field is modeled using poles outside the scatterer as in (9.4). The EBW for a GMT monopole outside the cylinder is identical to the monopole line source results shown in Fig. 9.8. If the monopole is placed far from the scatterer, the field is approximately a plane wave. Then, the model for the internal field is approximately a plane wave expansion using a finite set of plane wave propagation directions.

9.4.4 Examples

In this section, some example calculations are used to evaluate the usefulness of EBW when choosing pole locations. The EBW for the incident field, denoted EBW^{inc} will be compared to the EBW for the scattered field (EBW^{scat}). GMT will be used to compute the solution to the scattering from a cylinder under a variety of situations. The GMT model for the scattered field is (9.3) and the internal field is (9.4). Solutions are obtained by choosing a set of N_p matching points, equally spaced in ϕ along the boundary. The ϕ values are identical to the ϕ values for the pole locations. A LSE with square matrix is formulated by forcing the total tangential electric field to zero. The LSE is solved to obtain the unknown coefficients.

To evaluate the accuracy of the solution, define the average boundary condition error (in percent) as:

$$\bar{\varepsilon}\% = \frac{1}{360} \sum_{\phi=1^\circ}^{360^\circ} \frac{|E_z^i(\phi) + E_z^s(\phi)|}{|E_z^i(\phi)|} \times 100\% \tag{9.15}$$

Recall, if the scattered monopole radius is too small, the a_n tend to diverge in magnitude and oscillate in phase. Typically, one would prefer a solution that has stable and converging coefficients. Solutions with stable coefficients shall be called "suitable" or "stable" solutions.

To quantify the stability or suitability of GMT solutions for the general problem, define a measure, V, that depends on the scattered field coefficients:

$$V = \frac{|a_n|_{max}}{\left|\frac{1}{N_p} \sum_{n=1}^{N_p} a_n\right|} \tag{9.16}$$

Note that a similar scheme using $V = |a_n|_{max}$ is implemented in [23].

The measure V takes the maximum coefficient magnitude and divides by the sum of the coefficients. If the coefficients are very large with nearly equal and opposite phases, V will be large and diverge as N_p increases. If the coefficients are well behaved, then V will remain nearly constant as N_p increases.

9.4.4.1 Example 1. Plane Wave Incident on a PEC Circular Cylinder

A perfectly conducting (PEC) cylinder of radius $a = 1.0\,\lambda$ is excited by a plane wave. The spatial bandwidth for the incident field is $EBW^{inc} = 9$. A GMT monopole within the cylinder has an equivalent EBW when ρ' is approximately $0.4\,\lambda$.

Figure 9.9 shows the GMT results for various ρ' using $N_p = 30$ monopoles. The average boundary condition error ($\bar{\varepsilon}\%$) indicates a "best" solution for ρ' near $0.125\,\lambda$. However, the suitability measure V diverges strongly for $\rho' < 0.4\,\lambda$. Recall that the EBW for the monopoles and the incident field match at $\rho' = 0.4\,\lambda$.

Next consider results for increasing N_p. Figure 9.10 shows $\bar{\varepsilon}\%$ and V for ρ' at $0.25\,\lambda$ and $0.45\,\lambda$. For both ρ' values, the average boundary condition error decreases as N_p increases. The suitability does not diverge in either case; however, for

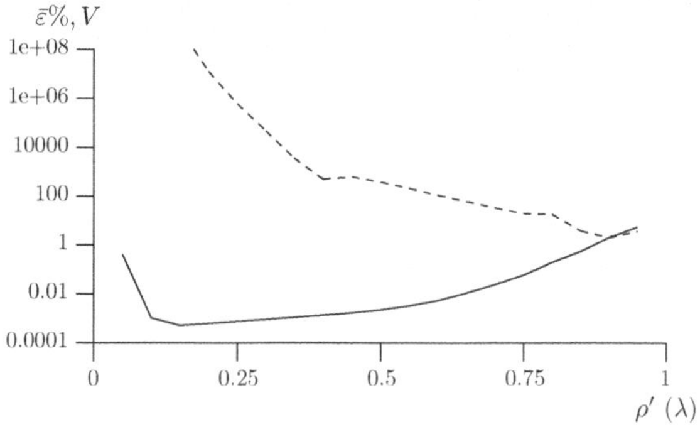

Fig. 9.9 GMT results for Example 1: solid, $\bar{\varepsilon}\%$; dashed, V

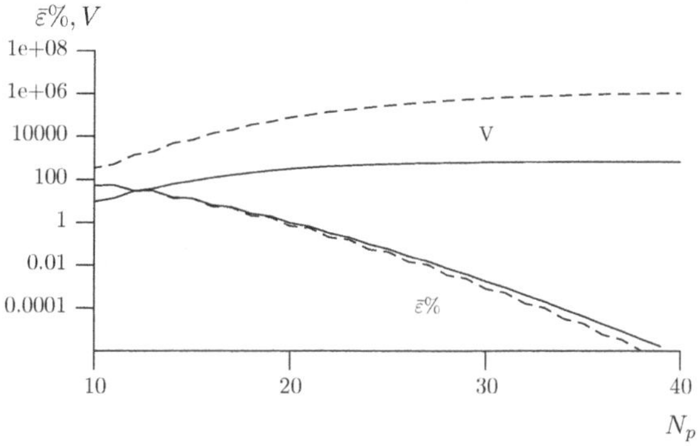

Fig. 9.10 GMT results for Example 1: solid, $\rho' = 0.45\,\lambda$; dashed, $\rho' = 0.25\,\lambda$

$\rho' = 0.25\,\lambda$, $|V|$ is two orders of magnitude larger. In general, the non-divergent behavior of $|V|$ indicates that the poles are enclosed by the auxiliary surface. The suitability degrades as ρ' decreases because of the numerical error described earlier.

In this example, it is clear that using ρ' greater than $0.4\,\lambda$ is useful to avoid numerical errors. This result demonstrates the validity of EBW as a guide to choosing the pole radius.

9.4.4.2 Example 2. Monopole Line Source Incident on a PEC Circular Cylinder

A perfectly conducting cylinder of radius $a = 1.0\,\lambda$ is in the vicinity of a monopole line source. The location of the line source is $\rho_o = 1.12\,\lambda$ at $\phi = 0$. The EBW of the incident field is $EBW^{inc} = 16$. A GMT monopole within the cylinder has an equivalent EBW when ρ' is approximately $0.75\,\lambda$. The singularity for the configuration is at $a^2/\rho_o = 0.893\,\lambda$. Note how the radius to satisfy the EBW requirement is *smaller* than the singularity radius.

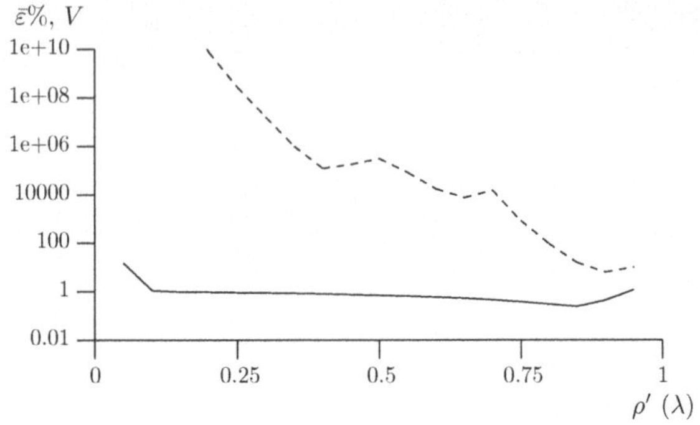

Fig. 9.11 GMT results for Example 2: solid, $\bar{\varepsilon}\%$; dashed, V

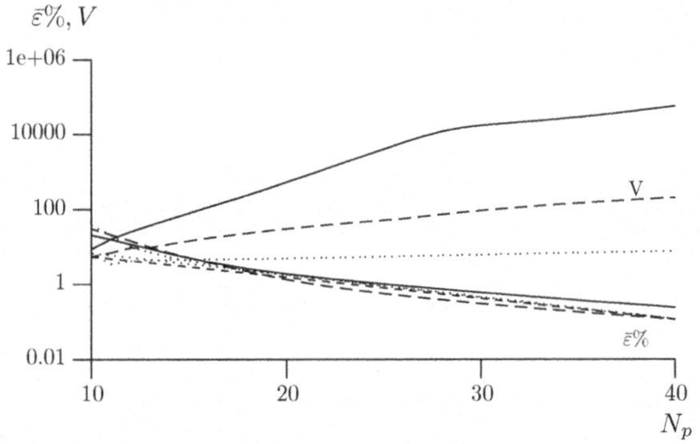

Fig. 9.12 GMT results for Example 2: solid, $\rho' = 0.6\,\lambda$; dashed, $\rho' = 0.8\,\lambda$; dotted, $\rho' = 0.9\,\lambda$

Figure 9.11 shows the average boundary condition error and V as a function of GMT monopole radius for the monopole line source example with $N_p = 30$. The boundary condition error does not fall much below 1% because N_p is small; however, V illustrates divergent behavior as ρ' gets smaller. The value of V is quite small for ρ' near $0.9\,\lambda$, as expected from the singularity radius.

Figure 9.12 shows $\bar{\varepsilon}\%$ and V for Example 2 as N_p increases from 10 to 40. In all cases, the average boundary condition error decreases as N_p increases. When $\rho' = 0.6\,\lambda$, the pole radius is smaller than the radius dictated by both the singularity and EBW. The V increases to very large values as N_p increases. For $\rho' = 0.8\,\lambda$, the pole radius is large enough to satisfy the EBW requirement but not the singularity requirement. In this case, V is smaller. When $\rho' = 0.9\,\lambda$, both conditions are satisfied and V remains quite small and fairly constant over the entire range of N_p.

In general, Example 2 illustrates that enclosing the singularity results in the best situation for the GMT monopoles for a scattering problem. However, in many cases, the location of the singularity may be difficult to ascertain; in these cases, the EBW can be used as a way of estimating the pole locations that come close to enclosing the singularity and hence, obtain a suitably stable solution to the problem.

9.4.5 Comments on the Dielectric Cylinder Case

All of the results in the preceding section pertain to the scattering from a perfectly conducting cylinder. In this brief section, the results obtained for a dielectric cylinder will be discussed [31]. For a dielectric cylinder, the boundary conditions change; there is now a condition of continuity of the tangential magnetic field as well as continuity of the tangential electric field.

In the dielectric case, the internal field is modeled using (9.4). At the boundary, only the ϕ component of the magnetic field is needed:

$$H_\phi = \frac{1}{j\omega\mu}\frac{\partial E_z}{\partial \rho} \tag{9.17}$$

A set of N_p matching points are chosen along the boundary. There are $2N_p$ unknowns and each matching point admits two equations. Thus, a square matrix is obtained and used to find the unknown coefficients.

The effective spatial bandwidth calculations are independent of the material properties of the scatterer. Thus, the EBW results shown previously are still valid. The EBW guidelines for the scattered field monopole locations are the same.

The internal field monopoles at ρ_n'' follow guidelines that can be inferred preceding, depending on the incident field type.

For a monopole line source, ρ_n'' must be between the cylinder radius a and the line source distance ρ_o. The EBW of the GMT monopoles at ρ_n'' follows the same analysis as the monopole line source. Therefore, if ρ_n'' is larger than ρ_o, then the internal field model will have a smaller EBW than the incident field.

For a plane wave incident field, [31] reports that ρ_n'' may be quite large without encountering large boundary condition errors and still provide suitably stable solutions. As the monopole moves far from the cylinder, the field in the vicinity of the cylinder will become approximately a plane wave. Thus, having a collection of monopoles far from the cylinder is equivalent to a discrete plane wave expansion for the internal field.

9.5 Effective Spatial Bandwidth for Non-circular Cylinders

In this section, the generalization of the EBW calculation to non-circular two-dimensional geometries is briefly introduced. First, a modified kernel for the bandlimiting operation is presented. Then, results are presented using a perfectly conducting cylinder with elliptical cross section.

9.5.1 Introduction

The effective spatial bandwidth for some field quantity on the scatterer boundary is computed by performing the bandlimiting operation and comparing the energy in the original function and the bandlimited function. For non-circular boundaries, the EBW can be computed by modifying the bandlimiting operation. The key is the kernel in the integration over the boundary, $B_N(\phi, \xi)$. Properties of the kernel include that the kernel is constant for $N = 0$ so the integration returns the average value of $e(\phi)$. Also, the kernel must be periodic over the arc length of the boundary. Since the kernel must be periodic, EBW will again be integer values.

The appropriate modified kernel can be written:

$$K_N(s, \sigma) = \frac{\sin\left[\left(N + \frac{1}{2}\right)\frac{2\pi}{C}(s - \sigma)\right]}{\sin\left[\frac{1}{2}\frac{2\pi}{C}(s - \sigma)\right]} \tag{9.18}$$

where the boundary is parameterized by s and C is the circumference of the boundary. Both s and σ are distances between 0 and C, relative to a point on the boundary identified as $s = 0$. The integrations (9.6) and (9.7) are also modified to integrate s or σ from 0 to C.

The EBW of a plane wave over a non-circular scatterer boundary, for example, has some interesting features. The variation in the incident field will be a function of the angle of incidence. Thus, EBW will be a function of the angle of incidence of the plane wave. Similarly, the EBW for a GMT monopole within the scatterer will be a more complex function of pole location.

9.5.2 Scattering from a Perfectly Conducting Elliptical Cylinder

As an example, scattering of a plane wave by an elliptical cylinder as shown in Fig. 9.13 will be discussed. The cross section is an ellipse with major axis $2a$ and minor axis $2b$. The axes are aligned with the x and y axes. An ellipse can be parameterized as points (x, y):

$$x = a\cos t \qquad y = b\sin t \tag{9.19}$$

where t is from 0 to 2π. An ellipse is also characterised by its eccentricity, e:

$$e = \sqrt{\frac{a^2 - b^2}{a^2}} \tag{9.20}$$

where $e = 0$ indicates a circle ($a = b$) and $e = 1$ corresponds to a flat plate from $x = -ae$ to $x = ae$. Denote ellipses that have the same focal points as a family of ellipses. Then, ellipses with equal ae products are in the same family.

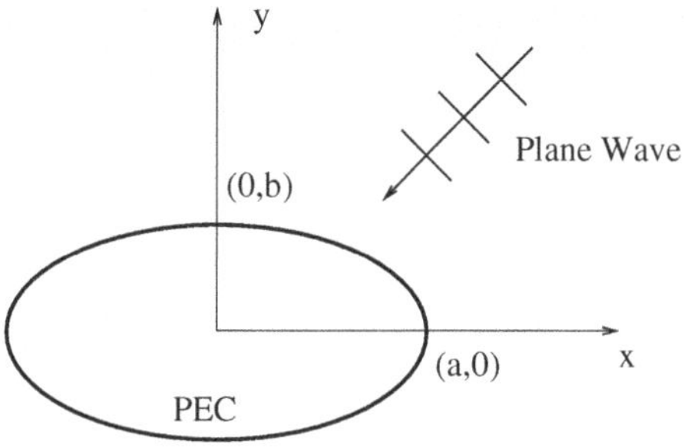

Fig. 9.13 Geometry for the elliptical cylinder scatterer

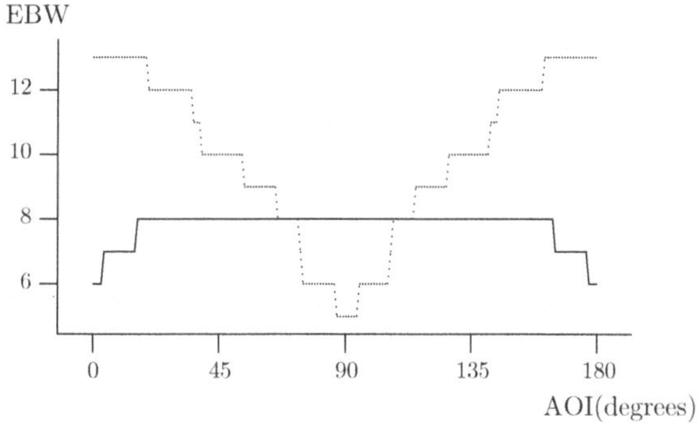

Fig. 9.14 EBW for a plane wave incident on an elliptical cylinder, $a = 1\,\lambda$: solid: $b = 0.5\,\lambda$; dotted: $b = 0.1\,\lambda$

The incident field is a plane wave arriving at some angle of incidence (AOI). For example, a plane wave traveling in the $+x$ direction has AOI $= 0°$ and in Fig. 9.13, AOI $= 225°$. The electric field is in the z direction.

The focal points $(ae, 0)$ and $(-ae, 0)$ are also the singularities for the scatterer geometry. Therefore, the scattered field model should have an auxiliary surface that encloses the line segment between the focal points.

The EBW for a plane wave incident on an elliptical cylinder with $a = 1\,\lambda$ for $b = 0.1\,\lambda$ and $b = 0.5\,\lambda$ is shown in Fig. 9.14. The figure shows the EBW results as a function of AOI. For $b = 0.5\,\lambda$, there is a small amount of change in EBW with AOI. Near 0° and 180°, the EBW falls slightly from the 90° result of $EBW = 8$. However, the $b = 0.1\,\lambda$ case is more interesting. The range of EBW values is from 2 to 13. Near AOI $= 0°$, the plane wave propagates parallel to the narrow dimension of the ellipse. Large phase variations of the incident field occur along the x direction and result in a large EBW. Near AOI $= 90°$, the phase variations along x are much smaller and a lower EBW is observed.

The EBW for a GMT monopole inside the cylinder positioned at $(x', 0)$ is shown in Fig. 9.15. The scatterer again has $a = 1\,\lambda$, and data for $b = 0.1\,\lambda$ and $0.5\,\lambda$ is shown. In both cases, the EBW begins smaller and rises sharply as x' approaches a. In addition, the EBW is higher when $b = 0.1\,\lambda$ for all x'. One interesting item to note in Fig. 9.15 is that the EBW of the interior monopole does not monotonically increase as x' increases. For example, for the $b = 0.5\,\lambda$ case the EBW falls for x'

Fig. 9.15 EBW for a GMT monopole at $(x', 0)$ within an elliptical cylinder, $a = 1\,\lambda$: solid: $b = 0.5\,\lambda$; dotted: $b = 0.1\,\lambda$

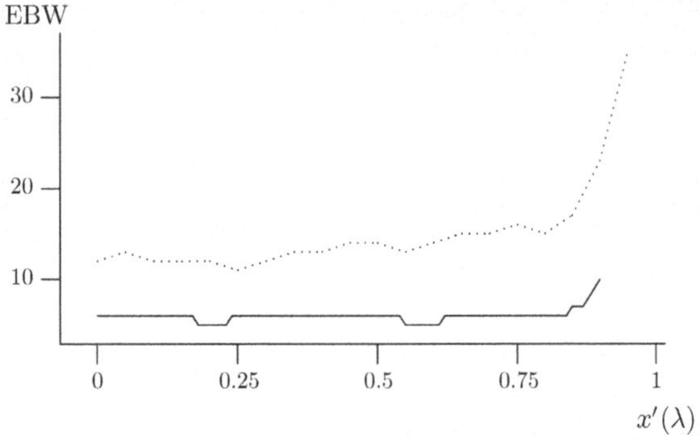

Fig. 9.16 EBW for a GMT monopole at $(x', 0)$ within an elliptical cylinder, $a = 1\,\lambda$: solid: $b = 0.5\,\lambda$; dotted: $b = 0.1\,\lambda$

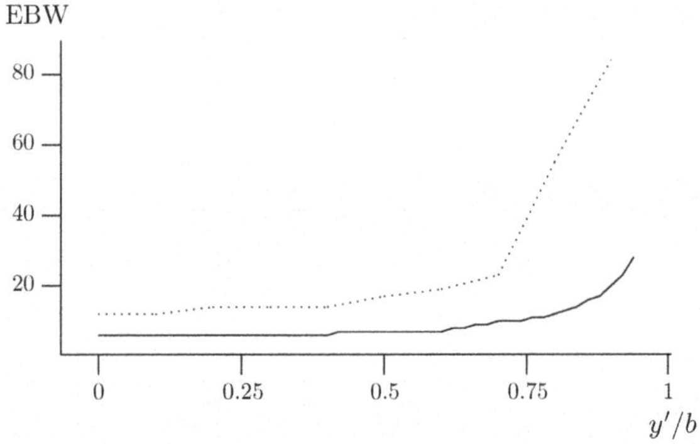

near $0.2\,\lambda$ and $0.6\,\lambda$. The nature of this drop is unclear. Figure 9.16 shows results for a GMT monopole at $(0, y')$. Similar trends are observed.

To complete the elliptical cylinder example, the EBW results will be applied to the GMT solution for plane wave scattering by an ellipse with $a = 1\,\lambda$ and $b = 0.5\,\lambda$. The AOI shall be 90° which admits an $\text{EBW}^{inc} = 8$. The GMT monopole results indicate that $\text{EBW}^{GMT} = 8$ when $x' \doteq 0.88\,\lambda$. This result is consistent with the notion that the singularities must be enclosed by the auxiliary surface. Using $a = 1\,\lambda$, the focal distance:

$$ae = \sqrt{3}/2 \doteq 0.866 < 0.88 \tag{9.21}$$

The monopole results also indicate that $\text{EBW}^{GMT} = 8$ when $y' \doteq 0.32\,\lambda$.

The GMT scattered field model consists of 50 monopoles placed on an ellipse with $a' = 0.88\,\lambda$ and a variable b'. It has been observed that poles and observation points equally spaced in t (see (1.19)) provides better results when compared to poles equally spaced in arc length. In all simulations, the match points and pole locations are equally spaced in t along each ellipse.

The boundary condition error and suitability measure V are shown in Fig. 9.17 for simulations where b' sweeps from nearly zero to nearly $0.5\,\lambda$. The boundary

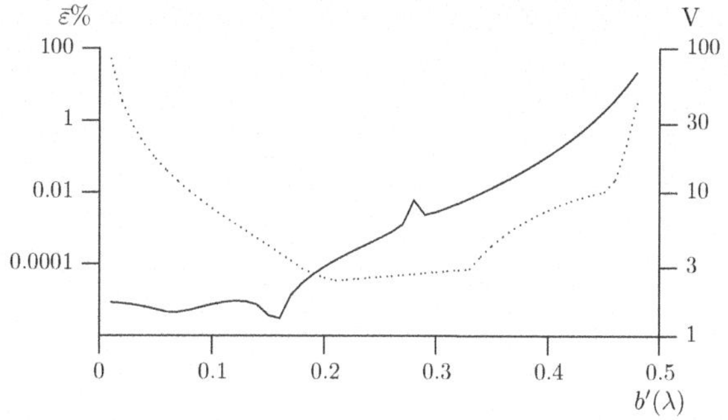

Fig. 9.17 GMT results for an elliptical cylinder, $a = 1\,\lambda$, $b = 0.5\,\lambda$, and $a' = 0.88\,\lambda$: solid: $\bar{\varepsilon}$%; dotted: V

condition error is quite small for small b' and increases for b' larger than $0.15\,\lambda$. The cause of the spike near $b' = 0.275\,\lambda$ is not known.

An unusual dip in the boundary condition error occurs near $b' = 0.15\,\lambda$. In general, it has been observed that auxiliary surfaces that are in the same family as the scatterer boundary provide the lowest boundary condition error. Using

$$ae = a'e' \quad \rightarrow \quad e' = \frac{ae}{a'} \doteq 0.984 \quad \rightarrow \quad b' \doteq 0.156 \tag{9.22}$$

it is seen that the dip in boundary condition error near $b' = 0.15\,\lambda$ is when the auxiliary surface is in the same family as the boundary.

The suitability measure V shown in Fig. 9.17 is larger for small and large b'. Strictly, the minimum is near $b' = 0.2\,\lambda$; however, V does not increase significantly until $b' > 0.32\,\lambda$. This is consistent with the EBW calculations, which showed that the EBW for a y-displaced monopole matches the plane wave EBW at $y' = 0.32\,\lambda$.

In this section, EBW was implemented for scatterers of elliptical cross section. The EBW calculations result in an auxiliary surface that encloses the singularities of the scattered field. However, EBW calculations result in global information. In more complex situations, local characteristics of the scatterer boundary and the incident field may dominate decisions regarding pole placement.

9.6 Conclusions

The family of GMT methods to solve scattering problems has many advantages; however, one major disadvantage is a lack of knowledge or understanding regarding the placement of the poles. Heuristic rules and guidelines have been developed and are quite successful.

In addition, analysis of the performance of the technique (particularly MAS) has been accomplished by studying the canonical problem of scattering by a circular cylinder. Accuracy analyses and investigation into the different types of convergence are described along with the relationship between pole placement and the scattered field singularities.

Finally, the effective spatial bandwidth for tangential fields along the boundary of scatterers has been described. The location of the GMT poles is chosen so that the EBW of the poles matches (or slightly exceeds) the EBW of the incident field. The EBW concept was extended to non-circular boundaries and the elliptical cylinder was used as an example to illustrate the usefulness of the effective spatial bandwidth.

References

1. T. Wriedt (ed.), *Generalized Multipole Techniques for Electromagnetic and Light Scattering, Computational Methods in Mechanics*, vol. 4 (Elsevier Science B. V, New York, 1999)
2. R.F. Harrington, *Field Computation by Moment Methods* (R. E. Krieger, Malabar, 1968)
3. A.F. Peterson, S.L. Ray, R. Mittra, *Computational Methods for Electromagnetics* (IEEE Press, New York, 1998)
4. C. Hafner, *The Generalized Multipole Technique for Computational Electromagnetics* (Artech House, Boston, 1990)
5. P. Leuchtmann, F. Bomholt, IEEE Trans. Electromagn. Compat. **35**(2), 170 (1993)
6. A.C. Ludwig, in *IEEE AP-S International Symposium Digest*, Dallas, TX (1990), pp. 48–51
7. Y. Leviatan, IEEE Trans. Antennas Propag. **38**(8), 1259 (1990)
8. Y. Leviatan, A. Boag, IEEE Trans. Antennas Propag. **35**, 1119 (1987)
9. Y. Leviatan, A. Boag, A. Boag, IEEE Trans. Antennas Propag. **36**, 1026 (1988)
10. Y. Leviatan, A. Boag, A. Boag, IEEE Trans. Antennas Propag. **36**(12), 1722 (1988)
11. I.N. Vekua, *New Methods for Solving Elliptic Equations* (Wiley, New York, 1967)
12. D.I. Kaklamani, H.T. Anastassiu, IEEE Antennas Propag. Mag. **44**(3), 48 (2002)
13. K. Beshir, J.E. Richie, IEEE Trans. Electromagn. Compat. **38**(2), 177 (1996)
14. S. Eisler, Y. Leviatan, IEE Proc. Pt. H **136**(6), 431 (1989)
15. P. Leuchtmann, IEEE Trans. Mag. **19**(6), 2371 (1983)
16. E. Moreno, D. Erni, C. Hafner, R. Vahldieck, J. Opt. Soc Am. A **19**(1), 101 (2002)
17. A.K. Bandyopadhyay, C. Tomassoni, A.S. Omar, in *IEEE MTT-S International Symposium Digest* (2004), pp. 1381–1384
18. I.I. Heretakis, P.J. Papakanellos, C.N. Capsalis, J. Electromagn. Waves Appl. **16**(11), 1555 (2002)
19. I.I. Heretakis, P.J. Papakanellos, C.N. Capsalis, IEEE Trans. Antennas Propag. **53**(3), 938 (2005)
20. R.S. Zaridze, R. Jobava, G. Bit-Banik, D. Karkasbadze, J. Electromagn. Waves Appl. **12**, 1491 (1998)
21. R.S. Zaridze, G. Bit-Babik, K. Tavzarashvili, D.P. Economou, K.K. Uzunoglu, IEEE Trans. Antennas Propag. **50**(1), 50 (2002)
22. G. Fikioris, IEEE Trans. Antennas Propag. **54**(7), 2022 (2006)
23. C.A. Valagiannopoulos, N.L. Tsitsas, G. Fikioris, J. Opt. Soc. Am. A **29**(1), 1 (2012)
24. H.T. Anastassiu, D.G. Lymperopoulos, D.I. Kaklamani, IEEE Trans. Antennas Propag. **52**(6), 1541 (2004)
25. K.F. Warnick, W.C. Chew, IEEE Trans. Microw. Theory Tech. **48**, 1652 (2000)
26. H.T. Anastassiu, D.I. Kaklamani, J. Electromagn. Waves Appl. **18**(10), 1283 (2004)
27. H.T. Anastassiu, D.I. Kaklamani, Radio Sci. **39**(5), RS5015 (2004). https://doi.org/10.1029/2004RS003028
28. H.T. Anastassiu, Prog. Electromagn. Res. PIER **52**, 109 (2005)
29. O.M. Bucci, G. Franceschetti, IEEE Trans. Antennas Propag. **35**(12), 1445 (1987)
30. J.E. Richie, IEEE Trans. Antennas Propag. **58**(11), 3610 (2010)
31. J.E. Richie, IEEE Trans. Antennas Propag. **59**(12), 4861 (2011)

Index

T. Wriedt and Y. Eremin (eds.), *The Generalized Multipole Technique for Light Scattering*, Springer Series on Atomic, Optical, and Plasma Physics 99,
https://doi.org/10.1007/978-3-319-74890-0

The manufacturer's authorised representative in the EU is Springer
Nature Customer Service Centre GmbH, Europaplatz 3, 69115 Heidelberg,
Germany. If you have any concerns regarding our products, please
contact ProductSafety@springernature.com

Printed and bound by CPI Group (UK) Ltd, Croydon, CR0 4YY
30/09/2026
02225654-0010